FORDING A RIVER ON THE ROAD TO THE KLONDIKE

OVERLAND ROUTE TO KLONDIKE

ALASKA

AND THE

KLONDIKE GOLD FIELDS

CONTAINING

A FULL ACCOUNT OF THE DISCOVERY OF GOLD; ENORMOUS
DEPOSITS OF THE PRECIOUS METAL; ROUTES
TRAVERSED BY MINERS; HOW TO
FIND GOLD; CAMP LIFE
AT KLONDIKE

Practical Instructions for Fortune Seekers, Etc., Etc.

INCLUDING A

GRAPHIC DESCRIPTION OF THE GOLD REGIONS; LAND OF
WONDERS; IMMENSE MOUNTAINS, RIVERS AND
PLAINS; NATIVE INHABITANTS, ETC.

By A. C. HARRIS

THE WELL-KNOWN AUTHOR AND TRAVELER

INCLUDING

Mrs. Eli Gage's Experiences of a Year among the Yukon Mining
Camps; Mrs. Schwatka's Recollections of her husband as
the Alaskan Pathfinder; Prosaic Side of Gold
Hunting, as seen by Joaquin Miller,
the Poet of the Sierras

EMBELLISHED WITH MANY ENGRAVINGS REPRESENTING
MINING AND OTHER SCENES IN ALASKA

PREFACE.

KLONDIKE is the magic word that is thrilling the whole country. It stands for millions of gold and great fortunes for hundreds of miners, who have risen from poverty to affluence in the brief period of a few months. Thousands are reading of fortunes made in the Klondike Gold Fields, and thousands of others are turning their longing eyes toward the new El Dorado.

The old Spanish dreams of a wonderful realm somewhere in the Western Continent, made of gold and precious stones, seem almost on the point of being realized. Not since 1849, when the marvellous discoveries of gold were made in California, has there been such excitement among all classes of people.

Everybody wants to know the real facts concerning the new discoveries. On every hand there is an eagerness for the most reliable information, which is furnished by this new and comprehensive work, containing a full description of Alaska and the Gold Regions. The author writes from personal experience and observation, as he has been an eye-witness of the scenes, incidents and facts which he describes and narrates.

The work gives a complete account of the rise of the gold fever, the excitement produced by the news of unlimited deposits of the precious metal; the rush of miners seeking fortunes at Klondike; hasty preparations for the long and perilous journey; and the formation of companies eager to take possession of the region abounding in untold wealth. The thousands of prospectors hurrying to the Gold Fields give us a picture of the rush to California when the discoveries of gold were made in that State in 1849.

How to get there is a question fully answered in this volume. The different routes are described, together with the best modes of transportation. This work tells you what is required for the trip; the clothing, food and implements that are needed; the hardships and dangers to be encountered; the difficulties arising from extreme cold in winter, and all the trying experiences awaiting the gold-seekers.

Alaska is a land of wonders. It is a vast region and one of the least known, yet one of the most remarkable countries in the whole world. Its history is fully related; its purchase by our Government from Russia; its slow development and its peculiar characteristics. It has vast tracts of primeval forests; mountains of awful sublimity; rivers that rival the largest in other parts of the world; Arctic snows and summer foliage and flowers; deep cañons and grand water-falls; solitudes peopled only by polar bears and other fur-bearing animals; and weird scenes that startle the beholder and fill him with awe.

These are all vividly described, together with the towns and settlements; the appearance, habits and customs of the native inhabitants; the climate in different parts of the country, and the progress of civilization up to the present time The mineral resources and wealth of Alaska are fully treated, showing it to be a country rich in natural products. Its important fisheries and possibilities for agriculture are all set forth, together with its industries, including its famous traffic in seals.

How to mine for gold is a subject on which the information is most complete and valuable. The reader follows the miners to their camps; learns the process by which they extract the precious metal from the recesses where it is stored; how it is separated from the ore; what machinery is employed, and what are the most successful methods for obtaining the coveted prize.

CONTENTS.

CHAPTER IV.

HOW TO GET THERE.

CHAPTER V.

A LAND OF WONDERS.

CHAPTER VI.

WOMEN AT THE MINES.

CHAPTER VII.

POET OF THE SIERRAS' VISION.

CHAPTER VIII.

HISTORY AND PURCHASE OF ALASKA.

CHAPTER IX.

TOPOGRAPHY.

CHAPTER X.

FLORA, FAUNA AND CLIMATE.

CHAPTER XI.

INDUSTRIES AND INDUSTRIAL DEVELOPMENT.

CHAPTER XII.

RESOURCES AND WEALTH.

CHAPTER XIII.

GOLD MINING IN ALASKA.

CONTENTS.

CHAPTER XIV.

RESUME OF MINING LAWS.

CHAPTER XV.

GOLD CRAZES OF OTHER DAYS.

CHAPTER XVI.

SIDE-LIGHTS.

CHAPTER XVII.

CAMP LIFE AND MORALS.

CHAPTER XVIII.

DOMESTIC LIFE IN THE WILDS.

CHAPTER XIX.

ETHNOGRAPHY.

CHAPTER XX.

NATIVE RELIGION AND TRAITS.

CHAPTER XXI.

SPREAD OF THE CHRISTIAN FAITH.

CHAPTER XXII.

BRITISH COLUMBIA AND NORTHWEST TERRITORY.

CHAPTER XXIII.

ADVENT OF WINTER.

CLAIM No. 3 ON MILLER CREEK, OWNED BY JOSEPH BEAUDREAU, FROM WHICH GOLD WORTH OVER $100,000 WAS TAKEN

MINER'S TENTS AND LOG HOUSES

OFFICIAL MAP OF THE KLONDILE AND YUKON REGION.   (United States Survey)

CHAPTER I.

Land of the Argonauts.

A Country Frozen by the Lapse of Time—Discovery of Gold Not New—News
is Flashed Over the World and Creates a Furore—Old Diggings are Soon
Abandoned—Effect of the Find on the People of the United States and
on the Money Centres of the World—Region which may Properly be
called the Land of Gold once Thought so Worthless the Russians Offered
to Give it Away for Nothing—Testimony as to the Richness of the
Deposits—The Popular Demand for Information as to the Country, its
Inhabitants, Scenery, Resources and the Like—Camp Life and Experiences.

ALASKA is the land of the Nineteenth Century Argonauts ;
and the Golden Fleece hidden away among its snow-
capped and glacier-clad mountains is not the pretty creation
of mythological fame, but yellow nuggets which may be transformed into the coin of the realm. The vast territory into which
these hardy soldiers of fortune penetrate is no less replete with
wonders than the fabled land into which Jason is said to have
led his band of adventurers.

There is this difference, however, between the frozen land of
of the North and the fabled land of mythology. There is
nothing conjectural about Alaska or its golden treasure. Jason
led his band into an unknown country without the certain knowledge that the treasure he was seeking was there. The men and
women who brave the perils of the wilderness to seek their
fortunes in Alaska, go with a certainty that the treasure *is* there.
It is a mere matter of finding it when once they have reached
the fields.

What is more the Land of Gold, as we may properly term
Alaska, has proved and will prove to tourist and prospector
as rich in delights and marvels as the land which has come

2   17

down to us in legend. It seems to be a spot chosen by nature as a field of adventure. The person, therefore, who goes from the South to the Yukon Valley will be sure to find, even though disappointed in the quest for which primarily he went, enough of the beautiful and marvelous to pay him for his trip.

Frozen by Lapse of Time.

And first a word about this land of bleakness and grandeur. Captain Butler, an English officer who crossed the great country some little time ago, writes in the most enthusiastic terms of its scenery, and one cannot do better than quote his picturesque words. Says he:

" Nature has here graven her image in such colossal characters that man seems to move slowly amid an ocean frozen rigid by the lapse of time—frozen into those things we call mountains, rivers and forests.

"Rivers whose single length roll twice 2,000 miles of shore line! Prairies over which a traveler can steer for weeks without resting his gaze on aught save the dim verge of the ever-shifting horizon! Mountains rent by rivers, ice-topped, glacier seared, impassable! Forests whose sombre pines darken a region half as large as Europe!

"In summer a land of sound; a land echoed with the voices of birds; the ripple of running water; the mournful music of the waving pine branch! In winter a land of silence; its great rivers glimmering in the moonlight, wrapped in their shrouds of ice; its still forests rising weird and spectral against the auroral lighted horizon; its nights so still that the moving streamers across the northern skies seem to carry to the ear a sense of sound."

The land thus strikingly described has been deemed since early in 1887 the Eldorado where nature has apparently strewn her golden gifts most lavishly.   It is to this land that thousands have wended their way in the hopes of wresting from their hidden beds enough of these treasures to lift them to opulence.

Not a New Discovery.

The knowledge of these gold fields in the North is not new. From early in the days of the Russian occupation it has been known that there were vast deposits of the precious metal in Alaska, practically under the Arctic Circle.

Year by year the gold fields have attracted adventurous fortune seekers, who have gone thither in ever-increasing numbers. Following the discovery of the rich deposits in the Klondike region, however, there has been an influx of people into these frozen wilds, such as has never been known before.

The first chance discovery was for a long time virtually held in secret, not intentionally, but because the lack of transit facilities made it difficult to get the news to civilized communities. When at length, however, the story of the find was brought south, and with the story was brought specimens of nuggets and gold dust which had been found, the news was put upon the wires and flashed through the length and breadth of the land, and the excitement caused gave every promise of a repetition of the memorable scenes which made Cariboo and Cassiar famous a generation ago.

In New York, in Chicago, in London, in Paris, throughout the world, the attention alike of rich and poor, was directed to the marvelously rich, but almost wholly unknown wilds of Alaska. People talked of the days of '49 and devised a new slogan, "The days of '97." The rich immediately began to organize new companies and map out new enterprises, such as made fortunes for thousands in days of other gold excitements ; and multitudes of the poor, dissatisfied with their opportunities

in districts longer settled and better improved, made haste to
provide their outfits and take passage to the Yukon.

In former days it was "Pike's Peak or Bust." Now the
watch-word became "On to the Klondike."

In the gold mining regions of Alaska there were, in 1893, not

MAP OF
·ALASKA·
·AND ITS·
GOLD FIELDS

more than about 300 miners all told. This number was doubled
practically the following year. Owing to the glowing reports of
successful operators, the number of miners attracted by 1895 was
3000. Probably twice that number of miners and prospectors
invaded the country in 1896.

In 1897 came this furor that caused the Klondike district to
rank with the great historical gold fields of the world. This

year witnessed the greatest influx of people into the territory on record, and there was every prospect that the year following would see the number quadrupled, possible many times over.

Old Diggings Abandoned.

And in the excess of enthusiasm and the wild hurrah raised when the new fields on the Klondike were discovered the old diggings were virtually abandoned. For ten years, at least, men worked placers in the Yukon district. Leaving Juneau early in the spring, they went out over the Chilkoot Pass and down the little chain of lakes on the other side, making long portages, it is true, and enduring some hardships, to the Yukon River. They returned to Juneau in the fall, year after year, bringing with them from $2000 to $3500 each in gold dust, the product of the summer's work.

But they were improvident, these men who won gold from the beds of rivers, and when the spring came they were stranded financially, many of them without a grub-stake, but they " won out " some way and got back again to return—unless they had crossed the divide forever—and repeated the same old story of excess and extravagance.

They never grew money wise, these grizzled veterans of the rocker, the gold pan, the pick and the shovel, but after all they are of God's people.

Quartz lodes were worked in ten or more districts, some of which are large and contain many district claims. The ten districts referred to are as follows : Sheep Creek region, which yields ore containing silver, gold and other metals; Salmon Creek, near Juneau, silver and gold; Silver Bow Basin, mainly gold ; Douglas Island, mainly gold ; Fuhter Bay, on Admiralty Island, mainly gold ; the Silver Bay mining district, near Sitka, gold and silver; Besner's Bay, in Lynn Canal, mainly gold ;

Fish River mining district, on Norton Sound ; Unga district and Lemon Creek.

But the furor over Klondike brought revolution. A change came over the spirit of the miners' dreams.

This country has been seized with the gold fever many times in the last half century, but never since yellow deposits were discovered in the Sacramento Valley was there such universal interest as was displayed over the discovery of gold on the Yukon and the Klondike. In many districts men and women talked of nothing else than of the new find. They were enthusiastic beyond bounds.

Experienced miners who had spent years in Alaska came to the front with words of caution and advice to let these enthusiasts know that the road to wealth in the Alaskan gold fields was even more beset with hardships in the way of cold, hunger and toil than the fields to which they were accustomed, and with which they had become dissatisfied. The friendly counsel, however, was disregarded. The one cry was "On to the Klondike," and one and all were apparently seized with the mad fever to leave civilization and seek wealth in the wilds.

Made His Blood Boil.

"What makes my blood run faster in my veins is to think that I have walked all over that gold and that now others are digging it. It prevents me from sleeping at night.

The speaker was Francois Mercier, a resident of Montreal, who can claim the honor of having been one of the first band of hardy pioneers who raised the American flag over the now celebrated gold fields of Alaska, and who spent seventeen winters in that desolate country.

Thousands besides Mercier found it difficult to sleep, and Alaska suddenly arose from an obscure district, which had often been called the "back dooryard of the United States," into the most talked of region of America. People then began to learn something of the history, the resources, the climate and the future of the country.

They were surprised to find that this vast territory, which was purchased in 1867 by Secretary Seward for half a cent an acre, had already paid $103,000,000. This was the returns of thirty years on an investment of $7,200,000. This enormous sum they then learned had been derived from furs, herring, salmon, cod, ivory, whalebone and gold. Gold, of course, was the most interesting item.

They found at the time of the last census the United States had taken out $76,000,000 in the precious metal. They found that since then the mines of the country had enriched the world's gold supply by about $27,000,000.

Came Like a Whirlwind.

It is no wonder, therefore, that the discovery of gold in the Yukon region should have come like a whirlwind among the people and that there should have been such an exodus from the southern States to the frozen regions of the North. The figures that came to light then about the Alaskan territory were giant figures, but they were the exact truth.

From the days when the Czar of Russia, in his zeal for discovery, sent his minions to find the fabled land of Vasco da Gama to the time of the discovery, the regions lying under the Arctic Circle had wooed but few, and those few were those who had drifted thither from adjacent territory. The real settlement of Alaska may, in a sense, be called the influx of people that resulted from the excitement incident to the discovery of gold on the Klondike.

It was an easy matter to compute what had come to the

United States from Alaska up to that time, but it was then said throughout the land, and in thousands of organs, that the sum which would be added to the world's wealth within a few years by this territory passed all surmise. Thus hope fanned conjecture and desire. The wealth to be expected was thought to be a pile of money as mountainous and as sublime as the country itself.

It is of interest to note in this connection that this territory of Alaska which was not then declared to be the world's storehouse of gold, was once offered to the United States by the Emperor Nicholas, of Russia, for nothing, if our government would merely pay for the transfer papers and agree by thus accepting the gift from Russia to bar England from coast territory on the Pacific. It is also of interest to note that almost similar propositions were repeatedly made, for the simple reason that no one suspected that enormous wealth lay hidden under the snows of this Arctic region.

Precaution of the Russians.

More properly speaking, some did suspect the existence of the boundless treasure. But those who did, discretely kept it to themselves, so that the news did not reach the people who might have profited by it.

It is a singular fact that the existence of gold in quantities along the tributaries of the Yukon was known to a few men a century and a half ago. The truth has been held back by the fur trading companies. They were not after minerals, and they feared the ruin of their industry, which was in itself a gold mine. Trappers, explorers, and men who lived with the Indians were forbidden to tell what they knew on pain of death.

The Russia Fur Company did summarily shoot one man who grew excited with drink and blabbed. That death is still remembered in Alaska, having been passed from mouth to mouth, as is the manner of unlettered peoples. Other fur companies have done nothing to develop the country and have kept their lips sealed. They foresaw the effect of a torrent of immigration. Such things cannot be hidden, however. The secret is out at last.

No, such things cannot be kept hidden. They came out, and the world had the secret as soon as the first ship from the North reached Seattle with the men who had "struck it rich," and brought back with them evidence of their good luck in the shape of gold dust and nuggets.

Then a state of affairs resulted comparable with the days of '49. It was said that the world's richest deposit of gold had been discovered. To the average man in the coast States, who had been nurtured virtually on stories of vast fortunes easily made in California, this news was not more acceptable than exciting.

It was true that the Yukon region was 2000 miles away, across a trackless desert, over snow-bound mountains, and through passes beset with dangers. But the fabulous tales of wealth that were brought south made the distance and the danger practically sink into insignificance and stimulated all with a desire to brave the unknown and investigate for themselves the great mineral belt in the Klondike region.

Evidence of Authorities.

This popular excitement was backed up by the testimony of men competent to speak of the country and its resources. They declared unqualifiedly that the gold districts on the Yukon and Klondike were but a speck in the gold territory of Alaska. They said that the placer mining which had resulted in such wealth thus far, was but an indication of the larger wealth to be acquired by a different process of mining.

When the miners find it no longer profitable to wash out the gravel they can attack the conglomerate, where they will be able to accomplish something by hand labor. Finally, there is the original source of gold, the veins in the hills. These must be of enormous value. They must lie untouched until the proper machinery for obtaining the gold is erected. A clear, scientific, and authoritative explanation of the geological conditions of the Klondike and neighboring gold-bearing rocks is furnished by Professor S. F. Emmons, of the United States Geolological Survey. Professor Emmons said:

"The real mass of golden wealth in Alaska remains as yet untouched. It lies in the virgin rocks, from which the particles found in the river gravels, now being washed by the Klondike miners have been torn by the erosion of streams. These particles, being heavy, have been deposited by the streams, which carried the lighter matter onward to the ocean, thus forming, by gradual accumulation, a sort of auriferous concentrate.

Richness of the Soil.

"Many of the bits, especially in certain localities, are big enough to be called nuggets. In spots the gravels are so rich that, as we have all heard, many ounces of the yellow metal are obtained from the washing of a single panful. That is what is making the people so wild—the prospect of picking money out of the dirt by the handful literally."

Hardly had the news of the great find been flashed over the world when Director of the Mint Preston was asked for his views as to the Alaskan gold fields and their influence. His words but added fuel to the flames that were then consuming the masses. Said he:

"That gold exists in large quantities in the newly discovered Klondike district is sufficiently proven by the large amount recently brought out by the steamship companies and miners returning to the States who went up into the district within the last eight months.

" So far $1,500,000 in gold from the Klondike district has been deposited at the mints and assay offices of the United States, and from information now at hand there are substantial reasons for believing from $3,000,000 to $4,000,000 additional will be brought out by the steamers and returning miners sailing from St. Michael's the last of September cr early October next.

" One of the steamship companies states that it expects to bring out about $2,000,000 on its steamer sailing from St. Michael's on September 30th, and has asked the government to have a revenue cutter to act as a convoy through the Behring Sea. In view of the facts above stated I am justified in estimating that the Klondike district will augment the world's gold supply in 1897 nearly $6,000,000."

Demand for Information.

As might be expected, the prominence given to Alaska by the discovery of the gold fields, resulted in a demand for a detailed statement of information as to the country in all its relations. So little was the country known, however, and so meager were the reports that had been brought to civilized communities concerning it, that the multitude found it difficult to obtain the information desired.

How were they to get there? What was there of interest or of importance connected with the history and purchase of the country? What could be learned of the various industries of the territory? What of the fauna and flora? What of the mineral wealth. Under what conditions and amenable to what laws would the prospectors have to work? What outfits were required for safety, comfort and convenience? What conditions of domestic life would those who left their homes in the south have to face in the unknown regions to which they contemplated going? What of the topography of the country they would have to traverse?

These and a thousand of other things became matters of prime importance, and it is to place such information in the hands of the public that this volume is issued.

A Land of Wonders.

Literally the land of Alaska is a Land of Wonders, a land differing markedly in its natural features from the districts of the south and bound to excite the admiration and awe of visitors by its natural features. These are so unlike the natural phenomena to be beheld in other parts of United States territory that the person who ventures into the region of the gold fields will find himself practically in a new world.

As will be seen in the following chapters, it is a country of almost boundless extent where the rivers, the mountains, the plains, the glaciers, everything, is in keeping with the distances that have to be traversed by the tourist or the prospector. It is a land of strange sights and stranger experiences, where much that is never dreamed of in the south will be found to be the commonplaces of an unknown people. As will be seen in the following pages, it is the land of sunless days and moonless nights; where Nature apparently has transposed the natural order of things, as is observed in southern latitudes, and inaugurated a new regime for visitors to wonder and marvel at.

Everything is mapped out on a gigantic scale and is clothed in such a way with its covering of ice and snow, and its strange forestation, and is overarched with such peculiar skies, that the voyager will not marvel less at what he sees than, to revert again to the opening passage from mythology, Jason and his band of adventurers marveled at what they are supposed to have seen in the fabled land of the Golden Fleece.

The Lesson of History.

The story of the history and purchase is not without its touch of romance and its lesson of wisdom. There is certainly food for thought in the narrative of a region so boundless in extent that was once thought so valueless as to be offered as a gift, owing to the ignorance of the people owning it as to its actual wealth. Secretary Seward always maintained that it was his crowning glory to have purchased the Alaskan territory. He and his staunch supporter, Senator Charles Sumner, always declared that the country had a future which would make it a profitable investment for the United States to purchase it at a far higher figure than had to be given.

The wisdom of their decision in the matter was shown within a few years after the transfer was made from Russia to the United States, and, as will be set forth in a chapter to follow, long before ever gold was discovered in the Klondike region the purchase money of the United States was returned over and over again, and the wisdom of Seward and his friends was established beyond a doubt.

Incident to the purchase and transfer of the territory, grave international questions arose which are well worthy of the attention of any one interested in the history of the country and the development of its latest possession. These are all carefully set forth in the following pages and will be deemed an acceptable contribution of information by those who, influenced by the excitement incident to the recent discovery of gold, may wish to invade the northern regions.

The fauna and flora of the territory, too, are of deep interest, especially from the fact that for many years one of the chief sources of wealth in the country was the furs. The Russians, who first owned the country, were not slow to recognize the value of the fur-bearing animals and to develop the industry of hunting them for their pelts. Following the initial steps taken by the Russians, John Jacob Astor sent his army of hunters and trappers into the northwest and carried the business far beyond the limits ever dreamed of by the Russians who began it.

Of late years, however, trapping in Alaska has, in a measure, fallen into abeyance, and in those regions where the miners have begun their work the difficulty of securing fresh meat has caused them to drive away all game from the districts invaded. Still it is of importance to those likely to go to the gold fields to know that there is still ample field for the hunter, and that fortunes are even yet to be made in trapping the animals for their furs.

Touching on furs Mr. Olgivie writes :

" The principal furs procured in the district are the silver-gray and black fox, the number of which bears a greater ratio to the number of red foxes than in any other part of the country. The red fox is very common, and a species called the blue is very abundant near the coast. Marten, or sabie, are also numerous, as are lynx ; but otter are scarce, and beaver almost unknown.

Value of the Fox Skins.

" It is probable that the value of gray and black fox skins taken out of the country more than equals in value all the other furs. I could get no statistics concerning this trade for obvious reasons.

" Game is not now as abundant as before mining began, and it is difficult, in fact impossible, to get any close to the river.

"A boom in mining would soon exterminate the game in the district along the river."

Directly connected with the discovery of gold and of vast

importance to prospective miners, there is much to be learned relative to the necessities of those visiting the territory. Prime among these items of interest is the matter of getting to the diggings. Many have been deterred from making the trip by the reported inaccessibility of the gold-bearing region, and the interminable stretches of the country that have to be traversed by all who seek fortunes in the wilds.

Route after route has been mapped out until there is scarcely a way by which it would be possible to go from Sitka to the Yukon, that has not been laid down as more or less practicable. It is safe to say that many of the routes outlined for the benefit of the public are thoroughly impracticable. The mere enumeration and explanation of the many courses prospective miners may follow, is not less an item of interest than of importance.

Features of the Journey.

To reach the distant fields, it will be necessary for any one to take an ocean voyage on landlocked arms of the sea, traverse trackless prairies, skirt mountain ranges, thread rivers lined with falls and rapids, that are a constant menace to life, and even, in a region for a large share of the year covered with an unbroken blanket of ice and snow, go in sledges or on snow-shoes in a way that adds to the fatigues and dangers of the journey.

Many are the wild schemes that have been devised by so-called "tenderfeet," of getting from civilization to the camps, and those who have had their interest awakened to the extent of wishing to undertake the journey to Alaska, will welcome a careful statement of the most desirable ways of getting there, and an outline of the principal courses which may be followed in the undertaking.

Another matter of importance, and one that is replete with interest and romance, is the domestic life of the mining region.

The camps of the North, thus far at least, have been unique in
the great mining enterprises of the world. It is probable that
the days of '97 will be attended by no such forms of life and
forms of depravity as marred the days of '49. Many women,
and these in a large measure women of culture and education,
have gone to the north to grace the camp life with their presence. They have gone, however, with a legitimate and honorable purpose in view, and the inaccessibility of the region, and
the dangers and hardships that are reported to attend the journey to the diggings have had the result of keeping away the lawless
classes.

Camp Life Comparatively Pure.

As a consequence, camp life is pure and better in every way
than it was in the days of the gold excitement in California, and
those who read the following pages will be pleased with the
remarkable contrast that is pointed out.

Immediately on the discovery of gold and its announcement
to the world, grave questions arose as to the international
boundary between the United States and the British territory, and
it became a matter of importance to miners and prospectors to
study the mining laws of two countries, partly to provide against
personal annoyance and partly to protect their individual interests. On the opposite sides of the boundary line different sets of
laws and regulations were in force, and miners were expected to
observe the laws obtaining in the respective districts. That these
laws were often disregarded, goes without saying.

Canada, in a grasping spirit of gain, proceeded without delay
to modify her mining laws for her own benefit and to the detriment of Americans who went to the Klondike district. The
old dispute as to boundary and territorial jurisdiction arose, and
for a time there was the prospect of a grave international dispute. Not content to live and let live, Canada undertook to

TENTS OF GOVERNMENT SURVEYORS IN ALASKA.

impose a tax on all Americans crossing the real or alleged boundary line, and this measure was bitterly opposed by the miners.

Would Keep the Gold.

Further than this, the Dominion Cabinet devised a scheme to limit the flow of gold to the United States from the diggings, and this too caused a protest in the entire region, from the fact that a large percentage of the miners were Americans who had gone thither on the mere chance of winning fortunes, and who naturally objected to being taxed for their enterprise and to being placed in leading strings as to the disposition of whatever they might acquire. In the following pages a digest of the mining laws of both countries, together with the history of the contention that arose and its development to the time of publication, is given :

In the wild rush for the diggings incident upon the news coming to the more settled States, thousands of people with no experience whatever in mining life set out immediately to tempt fortune in the territory. Many of the outfits they provided for themselves were very curious, and it became necessary for those furthering the enterprise of the fortune seekers in a commercial way, to make a schedule of the necessary outfits they should provide for themselves.

For the most part these specially devised outfits received publication in the daily press, and then from lack of novelty were allowed to fall into abeyance and practically be forgotten. As a result, many of those who took their traps and started for the overland journey from Juneau and St. Michael's, found themselves, when on the way, practically destitute of the things which experience showed to be necessary for effective work.

The fortune seekers were likewise equally without knowledge of the methods of working claims, should they secure them

Very few of the thousands who took their way to the Klondike region, knew the first thing of how to mine gold. They were obliged to trust to fortune and pick up from those already in the field the rudiments of the new calling to which they proposed to devote themselves. Many, to their sorrow, deplored the fact that ignorance or oversight had led them to overlook this important preparation for their work.

"If I had had but a manual telling me what to provide and how to do the work on arriving at the diggings, I should have deemed myself a fortunate person." This was a saying of almost daily occurence wherever the work of mining was undertaken by "tenderfeet" from the south. Naturally they worked at a disadvantage as compared with the men of experience, who flocked to the new fields from Weare, Circle City and other camps where mining had been followed for a length of time. In the following pages all this information, which those who early flocked to the diggings lacked, has been gathered together for the instruction and convenience of those who may propose to make the journey in the future.

Food Question Paramount.

Food is the great problem of life in this district. Cold does not cause much worry, for men can wrap themselves warmly enough to guard against loss of life from exposure, but few things grow in that northern clime and there is a lack of animal food which can be sacrificed to support the life of man. Hence enormous prices are charged for provisions.

Reports sent back by the miners in the Klondike region show that potatoes are twenty-five cents a pound and bacon forty cents. These are the cheapest articles of diet, and others sell at proportionate prices based upon the cost of their transportation to the gold fields as well as upon their power to sustain life.

Starvation is the real danger that confronts the miner who goes there in search of gold. Although ten dollars a day is paid for labor, no man is given work unless he brings some provisions with him, this being due to the fact that the claim owner cannot afford to supply his workman with food nor even sell him any from his own scanty store.

The rapid growth in the population of Alaska has made this problem seem of sufficient importance to Congress to appropriate $5000 to pay for an investigation of the food resources, and in addition, under the present law, the experiment stations which will be established will be entitled to $15,000 per annum for their support.

Field for Enterprises.

Apart from all consideration of the discovery of gold and the excitement incident to it, the Territory of Alaska has a deep interest for Americans in many lines of commercial enterprise. The remoteness of the country and its inaccessibility, owing to poor methods of transit, has thus far had the effect of shrouding the region in a certain mystery, which lack of interest, apparently, has not cleared away. The rise of the Klondike fever has opened up to the public the fact that the gold fields are only one of a number of interests that claim attention. This is shown by such reports as the following, which was made by one who spent many years in the interior of the country.

" It is a prevalent idea that the Alaskan Territory produces only gold and things of the sea, but this is wrong. Even in Klondike, which is far removed from the mollifying influences of the Japanese current, hardy vegetables grow in profusion, although cauliflower and asparagus will not ripen. Hay is as high as a man's head. When the country comes to be better known it will be found capable of making many things for humanity now unthought of.

TRADERS BARTERING WITH NATIVE INDIANS

CHILKOOT PASS SHOWING SNOW-CAPPED MOUNTAINS

" Although for some undiscoverable reason, reports have gone abroad that there is no game, the fact remains that there is plenty of it. Moose, elk and cariboo, or the American reindeer, abound. Every river is stocked with fish. No man should starve who has a hook and a flint-lock musket. When we were school children we used to read of the musk-oxen of Alaska, but none are there. The musk-ox is not found in America anywhere west of the great continental divide, or Rocky Mountains."

Another Fine Possibility.

Professor Allen thinks Alaska has before it a great future as a stock-raising country, and declares that stock can be raised there as successfully as in Montana or Wyoming. At present, however, there are practically no domestic animals in the country, the chief being reindeer. Explorers will experiment and learn what domestic animals are best adapted to the climate.

Sheep, pigs and goats can live there with proper treatment, and it is thought that, in the islands of the coast, they will flourish all winter on the wild grasses, even if left to their own devices. Farther north and in the interior it is probable they would have to be sheltered during the two or three months of the severest weather. Poultry can probably be raised to advantage.

It has been the aim in the following pages to gather together, from every possible source, such information relative to the mineral wealth, the fisheries, the agricultural development, the ethnology of the country and all similar lines of interest such as would naturally be sought by a public whose interest had been aroused by the recent developments in the Territory, and to give as fully as possible the story of the rise of the Klondike fever, with all the wealth of romantic experiences and fortunate discoveries that has been made public since Alaska stepped so prominently into notice.

The narrative, in a sense, will of necessity read like a chapter of fiction, for the camp life of the Klondike, like the camp life of similar regions, has its light and shade, its amenities and hardships, its peculiarities and its streaks of fortune, that will ever be of interest to those who have a love of the unusual and the unexpected. Miners' experiences, in a district so remote, must ever have the element of oddity, and this, coupled with the peculiar characteristics of life in a region which is little less than a new world, makes the story of the Yukon, as the following pages will show, one virtually of romance.

What Gold Seekers Will Find.

The Argonauts of 1898 will see that their contemplated journey is as likely to be one of good luck as of disappointment ; that the journey is as likely to be one of delight as of hardship; and that, while they are leaving home with all its comforts and conveniences, and society with its pleasures, for a country devoid for the most part of the experiences of ordinary life, they are going to a wilderness, nevertheless, in which they will find, disguised it may be, cut short it may be, a fair quota of what they have been used to.

Further, the Argonauts of 1898 will not be content with the answers to their questions that literature will give them. They will want and long to read the great unwritten book of Alaska on the plains and glaciers, along the rivers and passes of the vast territory. Their desire will simply be whetted by printed stories and their longing will be that of Joaquin Miller. Says the Poet of the Sierras :

" You want to ask questions. You wonder why the other islands of black-white mountains, a thousand of them on either hand, so stupendous, so steep, so sublimely majestic, mysterious, solemn and silent, are so voiceless, so utterly empty and still.

"You want to ask questions of Alaska, but Alaska is the sphinx with a forehead of gold. We have now steamed up the straits and out and away from under the mantle of fire and gold that hung above Juneau and Douglas City—a mantle woven in some sort from the smoke and chemicals of the great gold mine —and the morning is crisp, blue, white, clear as a bell.

"If one cared to look on the gray side of the situation, he might easily write of the location and all the land about "the abomination of desolation." But, on the contrary, the scene is grand, grand, sublimely grand, and the air is sweet, healthful and invigorating as wine. The heavens' breath smells wooingly here. You never saw snow so white anywhere as here.

"White as snow; whiter than any miller can whiten. This is because this is a land of granite; no dust in the air as in California or Colorado; no tall trees to scatter bits of bark and leaves and litter through the air and over the snow. One constantly thinks of the transfiguration all along this land of whiteness and blue; white clouds, white snow, blue seas and blue skies. Heavens! Had I but years to live here and lay my hand upon this color, this fearful and wonderful garment of the most high God!"

CHAPTER II.

Spread of the Klondike Fever.

Arrival of the Portland with More than a Ton of Gold on Board—Miners
Tell of their Marvelous Strikes—Gold and the Aborigines—First Great
Gold Craze—Prospecting in Early Days—Rich Gold Discovery on
Bonanza Creek—Argonauts Flock to the Steamers—Scenes at the
Wharves—Companies Formed in Response to the Rush—Millions of
Money and Thousands of Men—Craze in Wall Street—Royalty Affected
—Money in Grub-stakes—Joaquin Miller Under Way—"Lucky"
Baldwin After Mother Lode—Bright and Dark Sides of Story.

WHEN the steamer Portland reached Seattle from St.
Michael's, Alaska, on July 17, 1897, bringing not
only the verified news of the great gold discoveries in
the upper Yukon region, but nearly a million and three quarters
in gold "dust" as freight, beside a cabin full of bronzed miners
to bear witness to the Golconda-like find, not only the Pacific
coast, but the whole northern country as well, whether British
or American, began to go stark, staring mad over the well-nigh
incredible reports from the new diggings. Some of the miners
had with them $75,000 and even twice that sum, and not a man
had less than $3000, every ounce taken from the placers of the
Klondike within the year.

Over a Ton of Gold.

More than a ton of gold was on board the steamer as it
came up the sound. In the captain's cabin were three big
chests full of the yellow "dust," and the large safe had no
room for more of the precious nuggets which had been taken
out of the ground in less than three months of last winter.
In size the nuggets ranged from that of a pea to a guinea
hen's egg.

40

Surely, it was enough to set the land wild with excitement. And yet, it was no news there was gold in and near Alaska, and in fabulously paying quantities. The marvelous tales of wealth sent out by the California pioneers were no less wonderful than those brought back by men who had braved the last cold season in the frigid mineral belt. The great Klondike strike was made in the early winter of 1896-97, but nothing was known of it in the United States until June 15, 1897, when the Excelsior arrived in San Francisco laden with Klondike miners who were in turn laden with gold. Then came the Portland and the " craze."

" Chechockoes " Make Their Piles.

In speaking of the miners who came out on the Portland, Captain Kidston was enthusiastic.

"These men," said Captain Kidston, "are every one what the Yukoners call 'Chechockoes' or newcomers, and up to last winter they had nothing. To-day you see them wealthy and happy. Why, on the fifteen days' trip from St. Michael's I never spent a pleasanter time in my life. These fortunate people felt so happy that anything would suffice for them, and I could not help contrasting them with the crowd of gold hunters I took with me on the last trip up. They were grumblers, without a cent in the world, and nothing on the boat was good enough for them. Some of these successful miners do not even own claims. They have been working for other men for $15 a day, and thus have accumulated small fortunes. Their average on this boat is not less than $10,000 to the man, and the very smallest sack is $3000. It is held y C. A. Branan, of Seattle, a happy young fellow just eighteen years old. There is no country on earth like the Yukon."

Gold has been a familiar metal to the Alaskan aborigines for a time that is old even in their legends, but, lacking civilization, they lacked also the knowledge of the highest use of the precious metal, and the yellow nuggets which they gathered from the beds of their Arctic streams played no other part than that of savage ornaments until the land passed under the dominion of the white man.

The earliest white voyagers to the Aleutian coasts had their cupidity kindled, like the soldiers of Cortez and Pizarro, by the bits of gold shining here and there among the barbarous trappings of the natives who came, half-menacing, to the iron-girt coasts to barter with them for the rare treasures of sharp knives and gaudy fabrics, but, beyond the trivial ounces secured in shorewise trade, it was years after white sails had become familiar sights, winging their intricate way among the devious channels of the island-dotted coast, that civilized men began to think it worth the peril to brave the dangers of the iron land in quest of the golden stores Nature had so lavishly treasured in the strongholds of her cliffs and torrents.

Behring Found Gold.

When Behring, after whom the great Northwestern sea beyond the Aleutian Island is named, discovered and explored the Alaskan coast in 1741, he found gold, but he found, as befitted the climate and people, more furs and, with auriferous supplies nearer home in the convict-worked mines of the Czar's domain, the country was granted for fur-gathering purposes alone by the Emperor Paul to the Russo-American Fur Company, and with it remained until the Seward purchase in 1867 transferred it to the United States for a consideration (long since repaid in full) of $7,200,000.

Mineral riches were hinted at, however, by the early explorers. In 1885 the director of the mint credited Alaska with $300,000 in gold and $2000 in silver, most of the precious

metal coming from Douglas Island.   In 1896 the total output
of lode and placer mines in Alaska was put at $4,670,000 and
in 1897 the gold output, it is estimated, will reach $10,000,000,
or nearly twice that of Colorado in 1892.

The first great gold craze in the extreme Northwest came in
1858.   The Kootenai region was famous a few years ago, per-
petuating the fame of the Frazer River mines.   The Cariboo
region on the fifty-third parallel, proved a steady and constant
producer.   Placers were also worked on the Peace river.   In
the 60's there was a period when the annual production of the
northwest province exceeded $2,000,000, the highest figure
being $3,735,850.   Through the exhaustion of the known
deposits, however, the product fell off until, in 1890, it was less
than half a million.

Prospecting in 1883.

Charles McConky, Ben Beach, George Marx and Richard
Poplin set out from Juneau in the spring of 1883 to prospect
the interior for gold.   The rich deposits which were making the
Treadwell mine famous had stimulated inquiry among practical
miners, and science had answered that the mother lode lay
somewhere waiting to be tapped in the fastnesses of the giant
Rockies.   The quartette meant to find it.   Crossing the divide
in the early spring, they reached the lakes which constitute
the head waters of the Yukon River, while they were yet frozen,
and remained there building their boats preparatory to going
down the river as soon as the opportunity availed.   The boats
built and the ice having disappeared, they continued their
journey on the unknown waters of the Yukon.

Upon arriving at the mouth of Stewart River and being favorably impressed that their fortunes lay in that direction, they
proceeded to stem this stream in the hopes of finding things

more favorable, as they had seen nothing that they had considered diggings up to that time.   They had traveled about four miles up this river when they came to a bar that carried gold of a fine order, and then continued up the river, finding many bars which were afterwards worked to the satisfaction of the owners.

Dr. C. F. Dickenson, of Kadiak Island, which lies just at the

A PROSPECTOR'S TENT.

mouth of Cook's Inlet, says : " When I left Kodiak, two weeks ago, the people were leaving all that section of country and flocking in the direction of Klondike.   In a way, the situation is appalling, for many of the industries are left practically without the means of operation.

" Mines that were paying handsomely at Cook's Inlet have been deserted.

OPENING UP A RICH CLAIM

GROUP OF FORTUNE SEEKERS IN THE GOLD FIELDS

" In my opinion there are just as good placer diggings to be found at Cook's Inlet as in the Klondike region.

" There is not a foot of ground in all that country that does not contain gold in more or less appreciable quantities.

" There is room there for thousands of men, and there is certainly no better place in the world for a poor man."

There is good reason for believing from the reports of men well acquainted with the whole region that there is gold to be found anywhere in Alaska. The streams flowing into the great salt channel which bounds the coast below Sitka bear many auriferous evidences, and several of them, as for example in the neighborhood of Fort Wrangel, have been worked successfully heretofore. Some, indeed, have been literally " washed " out.

J. W. McCormick's Strike.

The richest gold placers in the upper Yukon were discovered by a white man in August, 1896. The find was due to the reports of Indians. J. W. McCormick, a Scotchman, who had been in the employ of William Ogilvie, Dominion Land Surveyor, for seven years in the same region, was the lucky prospector. He located a claim on the branch of the Klondike, which has since become known to fame as Bonanza Creek. McCormick located late in August, 1896, but had to cut some logs for the mill to get a few pounds of provisions to enable him to begin work on his claim. The fishing of Klondike having totally failed him, he returned with a few weeks' provisions for himself, his wife and brother-in-law (Indians) and another Indian in the last days of August, and immediately set about working his claim. As he was very short of appliances he could only put together a rather defective apparatus to wash the gravel with. The gravel itself he had to carry in a box on his back from thirty to one hundred feet. Notwithstanding this the three men, working

PLACER MINING ON THE KLONDIKE RIVER.

very irregularly, washed out $1200 in eight days, and McCormick asserts with reason that had he had proper facilities it could have been done in two days, besides having several hundred dollars more gold which was lost in the tailings through defective apparatus.

On the same creek two men rocked out $75 in about four hours, and it is asserted that two men in the same creek took out $4008 in two days with only two lengths of sluice boxes.

A branch of Bonanza named Eldorado has prospected magnificently, and another branch named Tilly Creek has prospected well; in all there are some four or five branches to Bonanza which have given good prospects. There were about one hundred and seventy claims staked on the main creek in the summer of '97, and the branches are good for about as many more, aggregating say three hundred and fifty claims, which will require over one thousand men to work properly.

Spread of Klondike Fever.

The Klondike fever spread wherever telegraph wires and newspapers disseminated the wonderful news of the marvelous diggings.

The Londoner, educated to gold fevers by the Rand and Barney Barnato, began besieging the trans-Atlantic transportation companies for intelligence about Alaska and the gold region of his own Northwest Territory. Experienced gold miners from South Africa thought they saw a bigger strike than the one which had lured them to the Cape of Good Hope. The new Canadian Trans-Atlantic line began work at once on a fleet of new boats.

In America, capitalists and poor men, Argonauts and "tenderfeet" went well-nigh crazy—literally daft with the mania for gold. In the cities of the Pacific coast employes in all industries threw down their tools and abandoned their pursuits to go to

Alaska and dig in the river bed for the shining nuggets. In Tacoma and Seattle telegrams were received from New York and London inquiring how many hundred men could be equipped on short notice for a journey to the gold fields. The street car employes of Tacoma, at a mass meeting, selected nine men to go to the Klondike for the benefit of the rest to prospect and locate claims, and raised a sufficient sum to equip and maintain them.

Hardly had the news of the Klondike strike got fairly started in its meteor-like circuit of the country than Seattle and Tacoma began to fill with men and women hurrying to the diggings. In a week beds could not be had at the hotels, and still the throngs of gold-seekers poured in from all directions except the West, and struggled and schemed and, in a bloodless way, fought' for fabulous priced chances to sail for the Yukon mines. First cabin, steerage, 'tween-decks or "on deck"—it was all one to these feverish Argonauts so long as they found themselves under way to Eldorado.

Scene on " Steamer Day."

Here is a sample description of a Tacoma scene on "steamer day," August 7th, when the Willamette cast off for Alaska:

" The most excited and largest crowd of people that has ever gathered on the ocean docks in this city, on any occasion, gathered to-day to see the steamer Willamette off for Alaska. Four hundred people boarded the vessel here, and their friends and relatives and thousands of sight-seers gathered to see the start. The passengers came from all parts of the State and a sprinkling from all over the United States. The baggage was carried mostly on horseback, only a few mules being used. The pack trains marched through the city in droves, and Grand Army men said it reminded them of war times.

"All sorts of outfits for making money were taken aboard, from a bakery to gambling tables. Nearly every person aboard has a list of from six to three dozen persons who had been promised letters. Fathers parted from families and young men from their sweethearts at the docks. Not a few of the men have pledged their families and friends that they will not return from the Eldorado of the North, until they have amassed a fortune, if it takes ten years to accomplish it.

"Aboard this vessel, Tacoma sent forward its first installment of physicians and surgeons to the Klondike. The doctors will dig for nuggets, if they cannot get patients."

Here is another scene on "steamer day," described by an eye-witness:

"The Alki started for Alaska this afternoon with 125 passengers, 800 sheep and 50 horses. Crazed with the gold fever and the hope of reaching Klondike quickly, the passengers bade good-bye to thousands on shore, who were crazed because they could not go. Food, comfort, sleep were ignored in the fierce desire to get to the gold fields. Those

OFF FOR THE MINES.

4

who could not go to Alaska stayed on the dock all day, shaking hands with those who were going, and gazing with eyes of chagrin and envy on the lucky ones as the steamer started for the North.

"There was grim pathos in the scene on the dock while the goldhunters were waiting for permission to go on board. Some were taking passage who would surely never leave Alaska alive. They had heard stories of the returned miners, that health was an absolute requisite in the terrible climate of the Klondike district. They smiled and knew better.

The Ruling Passion.

"One man said he was suffering from lung trouble, but that he might as well die making a fortune as to remain on the shores of Puget Sound and die in poverty.

"Not an inch of room was left on the Alki. It was tested to its utmost capacity. Excited men, drunk with visions of fortunes, were huddled among the sheep, horses and baggage. Space was valuable, and a cattle pen had been constructed on the main deck, which had hitherto been reserved for passengers. The sheep were put on board only after the crowd had been driven back from the steamer. On the main deck the horses and sheep will stay until the journey by water is ended. When port is reached the pen will be reduced to its original state and the lumber put to new use."

The same day the Willamette steamed out of Tacoma the Queen sailed from Seattle with 400 passengers for Dyea. And over twenty steamers were then due to sail before September 1st and passage on any one was already at a premium. New charters were being made daily and three schooners and even two scows were pressed into service in Seattle the day the Queen sailed. It is estimated Seattle has supplied already 3500 prospectors and Tacoma 1600.

Chicago became a centre for Klondike news and outfitting at the start of the craze. Over five hundred men had either left the Windy City, or were practically ready to leave, for the Klondike, at the end of the first week in August, and the fever had only been in the air three weeks. All sorts and descriptions of men were in the ranks of prospectors—lawyers, doctors, merchants, bankers, farmers and city men, stalwart giants and men whose physique gave promise rather of a grave beside the trail than of lasting long enough to " wash " a fortune out of the frozen Alaskan gravel. And there were women, too, in plenty, considering the hardships to be encountered, who were just as anxious to get into the wilderness to locate claims as any man who wore boots in the crowd.

Deny Women and Weaklings.

In fact, so great did the rush of women and of men of seemingly weak physique become, that many transportation agents at last refused to book any but those evidently the most robust, lest they should die enroute to Dawson. This order was later revoked as to women.

Among those who went from Chicago in early August were William H. Hubbard, in the party of Mrs. Eli Gage and her brother, W. W. Weare, going to Dawson to take the management of the banking system to be established by the North American Transportation and Trading Company in every mining camp in Alaska ; Dr. G. E. Meryman, Gustave Peterson and his two sons, Daniel Wright, Joseph Roman, F. J. Richardson, Mortimer Stevens, Dr. C. W. Chamberlain and wife, F. M. Sessoies and wife, F. H. Searle, E. H. Craig and Miss Alice Ross. Miss Minnie Goddard, the well-known organist and pianiste of Aurora, Ill. ; Miss Grace Allaire, daughter of the late Dr. Allaire, of the same city, and Mrs. Ira W. Lewis, of

Dixon, Ill., were three refined and dainty who left with a party of Chicago to cast in their lot with the masculine argonauts in the land of frozen gravel and marvelous " pans."

Montreal sent out three parties the first and second weeks in August, numbering altogether some fifty men. They were in charge respectively of Ernest Genest, representing the Canadian-Yukon Company; C. J. McQuaig, for the Montreal-London Gold and Silver Development Company, limited ; and W. H. Scroggie, the St. Catherine Street dry-goods merchant, whose companoins were principally his employes.

Ex-Governor John H. McGraw and General E. M. Carr left Seattle for Alaska on the first steamer out after the Portland arrived with its golden cargo—as luck would have it, the steamer was the treasure boat, the Portland itself. They went as the representatives of the Yukon, Caribou, British Columbia Gold Mining Development Company, limited, capital $1,000,000. J. Edward Addicks, of Delaware, is the head of the company and Senator John L. Wilson is interested in it.

Craze in Wall Street.

On July 31st, so early had the Klondike fever reached the great money centres of the land, the following report from Wall Street was sent over the country :

" Wall Street has been seized by a genuine ' '49 ' gold fever as a result of the discoveries in the Klondike. Men who have mined and made money ; men who have mined and lost money ; men who have always thought they might speculate a little in mining, and men who have had a complete abhorrence of mining —all seem to be affected the same way. More than half a dozen banking concerns, and as many individuals in Wall Street, whose standing in the financial world is the very best, have actually turned away from $5000 to $125,000 each which clients and

MINERS' CABINS NEAR DAWSON CITY

MUIR GLACIER, GLACIER BAY, ALASKA

customers wished to invest, under their guidance and supervision in the great gold fields of Alaska. Ladenburg, Thalman & Co., H. L. Horton & Co., Kean, Van Cortlandt & Co., R. P. Loundsberry & Co., and Charles Head & Co., are some of these firms who have more money offered them for investment in the Klondike than they have desired. The prejudice against mining is waning. Only recently bankers who dabbled in mines were looked upon with about as much suspicion by their customers and the money world as a bank clerk or cashier who regularly played faro, roulette and the races. But that is wearing off and the best concerns are beginning to mine in one way or another. Among these various down-town banking and business houses who are either interested in the Klondike, who have sent a representative there for themselves or customers, or who have made up their minds to do so, are R. P. Loundsberry & Co., N. Guggenheim Sons, Kean, Van Cortlandt & Co., Nicholas Chemical Company, H. B. Hollins & Co., H. L. Horton & Co., Charles Head & Co., and Seligman & Co.

Heard from Grub-stakers.

Seven men living near Trenton, N. J., "grub-staked" by business men of Trenton and merchants of Philadelphia, started in April for the Alaska gold fields. W. J. Hibbert headed the expedition. He writes that they have laid claim to eighty miles of dredger land, and have received a grant of twenty-one placer claims, which will be added to the dredger lands. He says that the ground is rich, and within a mile and a half of their claim a man by the name of Lereno, after working five days, found, on clearing up, that he was worth $40,000 in gold. Another story told by Hibbert in his letter is that another miner, after two months' work, was $150,000 to the good.

Daniel Guggenheim, of the firm of M. Guggenheim &

Sons, who has large smelting interests, when seen at his Long Branch cottage, confirmed the reported discoveries in the Yukon country, and prophesied that the new fields would yield far in excess of even present roseate indications. He said:

"For some time my firm has had expert mining engineers at work in Alaska, and their reports leave no doubt that the Yukon gold fields will prove the richest in the world. My opinion is that as soon as the country has been opened up and shipping facilities furnished the output of gold will be simply enormous. As the production of gold increases silver will be enhanced in value. This I regard as certain."

English Royalty Affected.

English royalty fell before the golden idol of the Klondike. No less a personage than the Duke of Fife, son-in-law of the Prince of Wales, subscribed to an incorporation formed in London for the purpose of exploring the Klondike region and purchasing such mines as its accredited representatives may decide are worth the investment.

The enterprise will be known as the Klondike Exploration Company, limited. It is stated that the company in which the Duke of Fife is interested will operate along lines similar to the British South Africa Company.

But great as was the number, considering the time available for catching a good hard case of the Klondike fever, who had succeeded in getting away for the diggings in person before the marvelous news from the Northwest was yet a month old; they were but a fraction of the total, who had fallen ready victims to the "placer malady."

Many hundreds of men and many more hundreds of women, who were crazy to own some kind of an interest in the wonderful gold fields, but who were prevented by other business, by family cares, by sickness of a strictly pathological kind, by poverty, or by other insuperable reasons, from taking personal part with the adventurers going into the Klondike, had syndicated their money with their friends and arranged to send "grub-stakers" into the new Galconda, hoping thus vicariously, at least, to partake of the profits, if they could not share in the hardships and the hazards of gold seeking.

It is estimated that at least five times as many people put up their money on "grub-stakes" as attempted to become adventurers in person, and it would require a much larger figure to express the probable ratio of the money applied to outfitting representative prospectors and the cash spent in personal equipment by intending argonauts.

Besides this, in estimating the prevalence of the gold craze in terms of dollars and cents, account must be taken of the mushroom-like appearance of "Mining Co-operations" and "Placer Syndicates" and "Poor Men's Chances," to say nothing of the host of legitimate incorporated mining or prospecting or development concerns, which by presenting shares at low figures, draw tens of thousands of dollars from thousands of pockets into their coffers and which quite as emphatically represented the virulence of the Klondike fever as did the steamer lists, or the names of those who meant to brave the Chilkoot Pass with the slogan of "Klondike or Bust."

Table of New Companies.

No better illustration of the extent and vigor of the Klondike craze can be given than is exhibited in the following table of companies organized or in process of formation for the development of the gold fields in the upper Yukon region. The total capitalization of the different syndicates foots up $164,512,500. After allowing for the regular syndicate grain of salt, the remaining total is still vast enough to indicate that no small portion of the American temperate zone has gone daft over the reported strikes in the Arctic mountains.

The stream of humanity, setting toward the north pole, is a veritable exodus toward a new Land of Promise. Up to August 8th, over 8000 men are officially reported to have started for the Klondike, or made arrangements to do so.

Statistics of Millions.

Here are the naked figures :

Companies.	Town.	Capital-ization.	No. who have left for gold fields.
Bohemian Klondike Syndicate	Baltimore . . .	Not decided	120
Three Syndicates	Boston	$50,000	150
Cudahy-Healy-Yukon Klondike Mining Company	Chicago. . . .	25,000,000	500
Alaska Transportation and Development Company	Chicago. . . .	5,000,000	. . .
Transportation and mining company in process of organization, not yet named .	Chicago. . . .	100,000,000	. . .
Wilkins Syndicate	Cleveland . . .	4,000	. . .
Unnamed syndicate	Cleveland . . .	400	. . .
Two companies	Cripple Creek .	300,000	30
Alaska-Klondike Gold Mining and Development Company	Col. Springs. .	1,000,000	. . .
Council Bluffs Mining and Exploration Company	Council Bluffs .	100,000	8
Six companies	Denver	2,825,000	35
Indiana Mining Company	Indianapolis .	200,000	. . .
General Mining and Developing Co. .	Kansas City . .	Not anncd.	10
Herald Employees	Lexington . .	1,000	12
Lincoln Gold and Improvement Co. .	Lincoln . .	50,000	11
Acme Development Company . .	New York . .	150,000	100
Yukon-Caribou British Columbia Gold Mining Development Company . . .	New York . .	5,000,000	. .
Northwest Mining and Trading Company.	New York . .	5,000,000	. . .
Exploration Syndicate	New York . .	100,000	. . .

The Gold Syndicate	New York	5,000,000	. .
The New York and Alaska Gold Explo-			
ration and Trading Company	New York . .	1,000,000	
Norse-American Gold Company (Ltd.) .	New York .	750,000	. . .
The Philadelphia and Alaska Gold Mining			
Syndicate	Philadelphia .	500,000	52
Alaska Gold Company	Pittsburg . .	1,000,000	. . .
Pittsburg-Alaskan Company	Pitisburg .	25,000	. . .
Four transportation companies	Portland, Ore.		
Two trading companies	Portland, Ore.	500,000	520
Six mining companies	Portland. Ore.		
Register employees	Richmond, Ky.	1,200	. . .
McDonald Syndicate	St. Louis . . .	50,000	. . .
Minnesota-Ontario Gold Mining Co. . .	St. Paul . .	1,000,000	. . .
Klondike Mining Company, St. Paul . .	St. Paul . . .	900	. . .
Yukon-Klondike Mining and Investment			
Company	St. Paul .	5,000,000	. . .
Eight companies	San Francisco .	800,000	1,400
Unnamed syndicate	San Francisco .	1,000,000	. . .
Klondike Commercial and Transportation			
Company	Seattle	1,000,000	3,500
Seattle and Yukon Commercial Company.	Seattle	1,000,000	.
Alaska Transportation Company	Seattle . . .	100,000	. . .
Dodwell and Corlill Steamship Company.	Tacoma	250,000	1,600
Twenty-one syndicates	Tacoma . . .	755,000	. . .

Old Miners Catch the Fever.

Old miners on the Pacific slope supplied some of the earliest victims of the fever and some of the first recruits in the rapidly-swelling army of the gold seekers. The rush to the Klondike seriously affected the mine owners on the mother lode in the vicinity of Senora, Jackson and Sutter Creek, California, and threatened to cause the closing down of the mines in Calaveras, Amador and Tualumne counties. A large party of skilled miners from this region sailed from San Francisco for Alaska on August 7th, and another party was then forming which expected to go in by way of Dyea before the winter grasp of September was upon the passes. The Oneida and Kennedy mines, near

SCENE IN THE SOUTHERN PART OF ALASKA.

Jackson, had lost the majority of their men before the news by the Portland was ten days old.

Joaquin Miller Among the First.

Nor was the rush to the new diggings confined to the wage-earning miners. One of the first of the '49ers to respond was Joaquin Miller, "the Poet of the Sierras." The steamer Portland made port from St. Michael's with its wonderful cargo of yellow dust and nuggets on July 17th, and on the 26th of the same month the venerable and veteran miner of the earliest California and Nevada and Idaho gold fields had forsaken his cozy home nestled among the foothills of Oakland, and was steaming out of the harbor of Victoria, B. C., on the good ship City of Mexico, bound with pick, pan and pack like any other lover of roughing it, on the long road to Dyea and over the Chilkoot Pass to the Klondike.

Some of his impressions enroute will be found elsewhere in this volume, and their bright, buoyant wording shows the Klondike fever could set the blood throbbing as fiercely in senile veins as in the arteries of the most recklessly sanguine lad of a " tenderfoot " that ever went to the mines to learn that all is not gold that glitters. One of the aged poet's fancies was to pack his own outfit in and earn his living by day's work, and to make his election sure he carried a ridiculously small sum of money with him, though he had a buckskin bag all ready for the " dust " he expected certainly to find even more lavishly distributed in the Yukon valley than in California in the golden days when the bed of every stream held a yellow fortune.

E. J. Baldwin, of San Francisco, better known as " Lucky " Baldwin, millionaire hotel man, miner, landowner, turfman and orange grower, himself a California argonaut of the days of '49, who had had hard attacks in his time of the Washoe and

Frazer River gold fevers, was another of the first "big" men on the coast to catch the Alaska fever.

The millionaire announced his intention to go to the Klondike, not to seek the great nuggets and coarse grains of gold found in the creek beds, but to find, if possible, the ledge, the mother lode from which all this treasure comes. He will not go in until spring, however.

" I will not stop at Klondike," said he, "but will push right into the mountains, where I am sure there must be rich quartz ledges. Ample machinery will be shipped to Dawson or elsewhere, if I succeed in locating a paying claim. I think the big fortunes will be made in the quartz districts and not in the placers, which will be sure to give out if so many thousands of people will persist in rushing into the country.

"I am going next spring," continued Mr. Baldwin, "and expect to take twenty-five or thirty husky young men with me who can work and endure the hardships. I am seventy-one years old, but still feel strong enough to do a little prospecting. It is also my intention to take a lot of machinery along for lode mining. My notion of the situation there is that the placer mining they are carrying on is an indication that there is gold in large quantities back in the mountains. I shall hunt out these deposits, and, equipped with modern machinery, will do a regular mining business. I am convinced the gold is there; consequently, I will be taking no long-risk chances."

"Lucky's" Idea of an Outfit.

Mr. Baldwin also gave his ideas of the provisions a man starting to the Klondike should provide himself with. He excluded coffee and ham from the supplies, would fill a box with articles of this sort, giving the amount for one month's use:

Chocolate, 7 ½ pounds, or tea, 3 ¾ pounds; rolled oats, 7 ½

INDIAN WOMEN AND PAPOOSES, ALASKA

INDIAN WOMEN SELLING THEIR WARES, JUNEAU, ALASKA

pounds; navy beans, 22½ pounds, or bacon, 37½ pounds; flour, 30 pounds; salt, 3¾ pounds; pickles, 60 cents' worth; cayenne pepper, ¼ pound for eighteen months, four cakes dry yeast.

Wonderful Letter of G. H. Cole.

Some of the stories told about the marvelous golden wealth of the Klondike would be ample excuse for the worst recorded cases of the fever. Here is one written from Dawson City by G. H. Cole to his wife in Seattle, which speaks for itself. Mr. Cole says:

"This is a wonderful country. There is enough gold here to load a steamboat. Lots of men have made all they want since last fall, and gone out. There is hardly a day but there is from one to half a dozen come from the mines with all the gold they can carry. One man had so much he had to get several men to help him carry it out. He gave the mine to a friend to do what he wanted with it. He was a Seattle man.

"Some of the men who have been out to the mines say there is more gold here than they ever saw in their lives, and some of the old miners, who have been in most all the mining countries in the world, say it beats anything they ever saw. Around some of the camps they have it piled up like farmers have their wheat, and in other camps they have all their cooking utensils full of gold and standing in corners as if it were dirt. Some are taking out $100,000 a day. Old miners say there has been enough gold located to dig up for the next twenty years."

Many and queer are the schemes that have grown out of the Klondike craze, and the more and the queerer they are the more virulent is the attack. The very air is full of schemes; some alluring, some preposterous, more merely audacious. The gold fever marked the heyday of the dreamer and the enthusiast, not to say the crank.

But some attention is worth paying to these projects of vision-aries if for no other reason than to show how far-reaching and insidious is the Klondike mania—for dreamers have little merit unless there are enough of people who believe in dreams.

"If I were to give you the details of some of the schemes that have been submitted to me recently for making money in the Klondike," said one Chicago capitalist, "you would think some insane asylum had been thrown open, and the inmates turned loose. Some of the ideas are not bad in themselves, but are impracticable owing to the conditions of the country. Others are simply the rankest form of lunacy, while others yet are downright swindles. People who would not even think of sug-gesting a fraud in connection with ordinary business have no hesitation in boosting up a fraud in a mining boom. As a rule, however, the irresponsible schemers are merely wild-eyed cranks, who have an honest confidence in their own plans."

Traps for Ready Money.

Inventors, speculators, promoters, and prospectors are going about like modern genii with propositions for making everybody immensely rich. Acquiring great wealth depends solely upon immediate use of a little ready money. Shares in the Consoli-dated Trans-Alaskan Gopher Company, offered at one dollar each, will return dividends of ten dollars a minute as soon as the com-pany gets to work. The idea is to take contracts for tunneling claims with trained gophers. Nothing is impossible, nothing chimerical.

Men with seedy garments and faces bearing all too plainly the marks of hunger and want, rub elbows with portly, well-fed individuals and talk glibly about millions to be had in various ways. Newspapers are full of advertisements calling for finan-.cial aid in developing Alaskan projects, offices of transportation lines are besieged by hundreds of impecunious beings who seek to make their wits pay the price of passage to the Eldorado, and on every street corner people are encountered with Klondike schemes in varying forms of development.   Women have the craze as badly as men ; and some of their hobbies are, if any-thing, even more outlandish.

But while the schemes and yarns of visionaries, charlatans and cranks are worth laughing at for their absurdity or avoiding for their concealed rascality, there is another side to the story which appeals to earnest men with almost irresistible force.   That is the record of the men who have " struck it rich " in the placers of this very Klondike—of the men who have gone in poor and come out in a few short months, or even weeks, rich for life ; of the men who took stock in the tales of the fabulous wealth wait-ing in that frozen Yukon valley gravel to be " washed " out, and who, with wise forethought, prepared themselves for a fierce battle with the Arctic elements and then braved the hardships and privations of the wilderness to emerge in time laden with their golden fruits of victory.

From Alaska Mining Record.

Elsewhere in this volume will be found a more detailed account of those who " struck it rich " on the Klondike ; to show that there is a bright side to the picture, the following from the *Alaska Mining Record*, of Juneau, of June 30th, is sufficient.   It relates to the arrival of Jack Hayes, the mail carrier from the Yukon. :

" Much excitement prevails all through the Yukon district over the Klondike discoveries, and all kinds of stories of the riches there are told, many of which Mr. Hayes says are true.   It is true that two tenderfeet, railroad men from Los Angeles, Cal.— Frank Summers and Charles Clemens—have struck it rich.

They went in a year ago and located on the Klondike last fall. Clemens sold his interest for $35,000 cash, and his partner, Summers, held on two weeks later and got $50,000. The money to pay the men was taken out of the dump which had been lifted from the shaft on the claim during the winter. These two men had each panned out $2500 on their claim while prospecting it. The man that bought Clemens' interest bound the bargain with a $232 nugget which had been taken from the Klondike. Neither man had had any experience in mining.

"Alec McDonald took one pan from his claim which tipped the scales to the tune of $800, and offered a wager of $1000 that he could pick his dirt and in twenty minutes get a pan that would go over 100 ounces ($1600). No one cared to cover the wager.

"Dick Lowe is panning for a living, and is taking out the modest sum of $100 a day.

"Two 'tenderfeet' from Chicago, named Wier and Beecher, leased a piece of ground for sixty days, paid a royalty of $10,000, and divided $20,000. The miners have only advanced up the Klondike nine miles, and at that distance there are several claims that will produce $1,000,000 apiece.

Assays Enormously Rich.

The latest reports from this cold gold clime consist of specimens which were sent to California for assay tests, and they show enormous returns of gold.

The gold find, however, in this Alaskan Territory is not new, although the facts are just beginning to be appreciated by the public. The unanimous verdict of investigators in this northern country has always been that gold abounded in great quantities, but the difficulty has been to get it out and away with any degree of profit. Mining on a small scale has been practically impossible.   The adventurer without money would have no chance to strike it rich, even if he could manage to raise the sum necessary to take him to the country.   The rigors of the winter preclude any work in that season, and the absence of any commercial facilities in the new mining districts prevents any digging that is not connected with some large organized plan. But for the company or individuals with capital and enterprise the prospect seems to be of the best.   The introduction of improved machinery—which has already begun—and the enlargement of the transportation facilities on the long Yukon River will soon bring these golden riches within easy reach of the States.

Natural Exaggerations.

The stories of finds, however, must be taken with usual reservations.   There will be natural exaggerations not only of the richness of the gold but of the character of the hardships that must be endured.   Alaska is no balmy California.   There is no comforting warmth most of the year to sustain the spirits of the wearied seeker after wealth.   The battle for gold there includes a battle with a hostile nature which has guarded her treasure house with icy blasts for all these centuries.   It is no place for the laggard if all reports be true, but for the man of courage and determination it seems to be a land of great promise.

One of the evidences of the Klondike craze is freighted with ill omen to the owners of salmon canneries and of whaling vessels.   Startling rumors have come from the north that parties of fishermen and sailors are coming across country from the mouth of the Mackenzie River into the Klondike, and, should this prove true, many vessels now staunch and trim will be rotting on the Arctic coast when the snows of next winter have cleared away.

At Herschel Island, which is situated in the Arctic Ocean

5

near the mouth of the Mackenzie River, a large number of salmon fishers have made their headquarters. During the summer months, when the Mackenzie River is open, these fishermen, in their myriad of small craft, go up the river in quest of salmon. There are a number of canneries on the Mackenzie. Over 100 deep-sea vessels are annually needed to bring the seasons pack down from the Arctic. It is believed the fishermen and crews which went north to bring back the pack have heard of the wonderful gold strikes and, taking the provisions with which their vessels were stored have deserted and struck out for the gold fields.

Owners of whaling vessels which winter at Herschel Island are as much alarmed as are the canning companies. There are at least 300 men belonging to the whaling fleet, and it is probable that they and the fishermen are now delving into the Klondike soil for gold.

Days of '49 and '97.

In many ways the " days of '49 " in California and the " days of '97 " in the Klondike are alike. To the average man the treasures of the coast State were seemingly as inaccessible as those of the Yukon and its tributaries. The one lay beyond 2000 miles of trackless desert and snow-clad mountains beset with savage hordes whose bloody welcome to the gold seeker narked the trail from the Missouri to the coast with the whitening bones of "pale-face" prospectors ; the other lies 7000 miles by water, or 4000 miles by land and water, from civilization, beyond mountain passes as hazardous to scale as those of the Swiss Alps and guarded from the greed of man by the icy rigors of the Arctic climate hardly less effectually than were the riches of California by the sanguinary red man.

The tales of fabled wealth which set the world crazy to go to the California mines were not less wonderful than those which returning argonauts bring from the upper Yukon country, and both are confirmed by the yellow nuggets whose mute testimony to the modern Cathay is unimpeachable. And the excitement in America is greater than in the wildest days of the South African or the Australian strikes.

Both in California and in the Klondike, the first mining was in placers, " poor man's mining," because no expensive machinery is required—only a pick, spade and pan, with nature's sluiceway of a nearby stream for water.

And, again, the "tenderfoot" often struck it rich where the old miner had trouble to find enough "dust" to buy his daily food.

It was every man's gold mine. Nature had no favorites.

No wonder people went gold crazy.

Fever Reaches a Climax.

The symptoms of the climax of the first attack of the Klondike fever came relatively soon after the yellow malady became epidemic. The fever began on July 27th, 1897; by August 15th the worst was over, and the tens of thousands of poor men who wanted to be rich in a hurry, and of rich men who wanted to be richer, of adventurers who were always ready for anything exciting, and of level-headed business men who had been crazy for only a few brief days over the marvelous tales of wealth to be had for the washing, had begun to convalesce and reason that if the Klondike was really as fabulously rich as it was reported to be, there would likely be some gold left at the diggings when spring came, and the perils to health and even life on the long journey "in" were somewhat diminished by mild weather.

Would-be argonauts who could not get passage to Dyea or Juneau on the overcrowded steamers began to content themselves perforce to stay at home; and weary and disgusted prospectors, who had been stranded by the stampede at the mouths of the mountain passes, began to pour back to winter amid creature comforts in the homes of civilization, and pack up at leisure for another venture in the spring. People found time to get cool, and they took it.

But what a craze it was while it lasted! Even the days of '49 were fairly eclipsed by the universality of the gold insanity of '97. Every city in the Union contributed to the horde of gold hunters pressing and pushing and scrambling on to the new Eldorado. Even the little hamlets of the land sent their quota, and men swarmed by thousands around the wharves of San Francisco, Portland, Tacoma, and Seattle, and "put up" their last cent for a fighting chance in the mad rush for the Yukon placers. Canada sent its thousands through the States and along its own routes, and across the Atlantic the fever spread 'till even the great house of Rothschild was infected and sent a confidential agent to inspect the wonderful gold fields in its behalf.

London Gets the Craze.

A London correspondent of a New York newspaper wrote in these words on August 1st:

"Were it not so late both in the London and the Yukon season, the fashionable thing for society young men to-day would be to make up a party to dare the dangers of the Chilkoot Pass and explore the Yukon River, even at the risk of gold-laden aristocrats meeting mythical pirates on their homeward journey. The gold fever has spread here far wider than the narrow limits of so-called London society, and there would have been a mad rush to the diggings from England of all the men and boys who could beg, borrow, or steal $200 had not one or two explorers sounded a shriek of alarm, and the Emigration Information Office issued a plain warning to the effect that it would be quite useless

SKINNING SEALS, ST. PAUL'S ISLAND, ALASKA.

STREET VIEW IN SITKA, SHOWING THE OLD GREEK CHURCH

to start hence before next April.   Meanwhile such terrible pictures are being painted, in colors laid on so thickly, and the
deadly perils of White Horse Rapids and Chilkoot are so strongly
emphasized that thoughtful men are not without the keen suspicion that the worthy Canadians are doing their best to scare
away intruders and keep their own treasure at home."

New York and Chicago.

New York and Chicago had the fever hard.   Men who had
mined and made money, men who had mined and lost money,
men who had always thought they would like to speculate in
mining, and men who had abhorred the very word, were stricken.
Bankers, brokers, business men and nonentities, from James R.
Keene to plain John Smith, went wild.   Before July was out,
companies representing an aggregate capitalization of $18,000,000
had been organized in New York City alone to traffic, or dig, or
grub-stake in the Yukon Basin.

Men who were blind on every other subject saw the wonderful
Alaskan rainbow of promise and rushed off to find the pot of
gold at its Klondike end with the infantile assurance of the tot
in the nursery tale.

Perhaps the date of the placer discovery—coming at the
close of a period of general business depression, had something
to do with the virulence of the fever.   Anyway, a fortnight after
the news of the strike steamed into port the country was stark,
staring, raving mad.   "Klondike" was the topic at the lunch
counters, men talked "outfits" on the street cars and " L " trains,
women found themselves abandoning the fashions to read up on
routes and fares to Dawson City, farmers drove to town in the
middle of a " hay day " to hear the latest from "the diggings,"
and technical mining phrases became the cant of the day.

Nothing could head off the enthusiasm of the horde of would-be

miners.   They sailed out of the Pacific coast ports, crowded like
animals in and upon vessels known to every sailor as long unsea-
worthy, and periled their lives over the " Boneyard of the Pacific"
or through the devious, rock-studded, fog-enshrouded channels
of the Sitka route; they trusted to captains who had never been
out of sight of land and to pilots who had never sailed the courses;
they heard, unmoved, warnings of deadly hardships enroute and
of probable starvation at the mines ; they gave up good positions
and spent small fortunes for transportation, and with scuppers
awash sailed away in death traps to the frozen North.

So reckless did the mad stampeders to the Klondike become
at last that the highest public officials were forced to take notice
of the epidemic folly and try to head it off.

Secretary Bliss' Warning.

Secretary of the Interior Bliss, on August 10th found it necessary to issue the following warning, a state paper almost without
a precedent on this continent :

" *To Whom It May Concern :* In view of information received
at this department that 3000 persons with 2000 tons of baggage
and freight are now waiting at the entrance to White Pass, in
Alaska, for an opportunity to cross the mountains to the Yukon
River, and that many more are preparing to join them, I deem it
proper to call the attention of all who contemplate making that
trip to the exposure, privation, suffering, and even danger incident thereto at this advanced period of the season, even if they
should succeed in crossing the mountains.   To reach Dawson
City, when over the pass, 700 miles of difficult navigation on the
Yukon River without adequate means of transportation will still
lie before them, and it is doubtful if the journey can be completed before the river is closed by ice.

" I am moved to draw public attention to these conditions by the gravity of the possible consequences to people detained in the mountainous wilderness during five or six months of an arctic winter, where no relief can reach them, however great the need.

"C. N. BLISS,

"*Secretary of the Interior.*"

The Hon. Clifford Sifton, Canadian Minister of the Interior, had already issued a notice to the public of the Dominion that the government would not be responsible for getting provisions into the Yukon during the coming winter tantamount to warning the gold seekers to stay out till spring.

Mad Rush Goes On.

Yet, in the face of all these official warnings, chronicled and spread broadcast by the same press and in the same columns in which the other Klondike news was daily printed, twenty-one steamers, three sailing vessels and two scows, each laden to the utmost carrying capacity, had put out from Pacific coast ports for Alaska before the warnings were a fortnight old.

The North American Transportation and Trading Company repeatedly issued public warnings of the hazards attending an attempt to get into the mines during the remainder of the season of 1897, and finally raised the fare for the last trip of the steamer Portland to $1000, only guaranteeing to get passengers to Dawson City by way of St. Michael's by June 15, 1898. Yet the passenger list was full of names of men who were willing to spend a winter in the Yukon ice or on the cheerless shores of Norton Sound, even at that price.

And those who could not muster patience to go by that route, with Secretary Bliss' warning ringing in their ears, swarmed at the wharves where other steamers were preparing to start with their herded loads of self-deluded gold-seekers, and paid $500 bonus, where they could find a taker, for the privilege of the voyage to overcrowded Dyea or Juneau. They knew the Canadian mounted police were on guard at the passes over the mountains, turning back all who had not a year's provisions in their outfits, but they bid high for the chance to go, just the same. They knew they stood a chance of having to winter at Juneau or Dyea, and eat up their supplies, but they spent their last cent to get there, just the same. It ceases to be a "play" rush for gold and became the wild exodus of a rabble in which men totally unfitted for the rough work and hardships of the miner's life, and unmindful that failure would be the lot of hundreds, and that many would find graves among the frozen placers or along the desert trails, joined with the enthusiasm of devotees.

Said by P. B. Weare.

" There is barely a chance of any of the gold-seekers getting across the divide so as to reach the Klondike region this year, to say nothing about the perils of the long trip beyond, but still the rush goes on," says P. B. Weare, of the North American Company, early in August. " We advise the people now not to attempt to get to Dawson City this year, but it doesn't seem to be any use talking. We hear from our representatives in Alaska and they say it is no use trying to stop the march—in some cases to certain death."

" They go on the theory that the first there will be first served," said John Cuhahy in speaking of the race for wealth ; " but I believe some of the first to go now will be the first dead."

Still the rush to the harvest of hardship and death went on.

Then the shock of disillusion came, and it brought some people to their senses. Word came back from the North that gold-seekers were making famine on the bleak Alaska mountains as fast as they knew how. Winter storms had begun to obliterate

the trails and bury the passes.   Old timers said again the reckless argonauts could not get through to the Klondike, and that
Arctic tempests would cut off their return and force them to fight
for life all winter in famine-stricken camps—and this time the
warning was heeded.

The object lesson from Dyea which was shown to the world
on the morning of August 10th was too fearful not to be heeded.

Misery at Dyea.

Hal Hoffman, writing from Juneau under date of August 3d,
said of Dyea and Skagway, the ports at the head of Iynn Canal,
these graphic and awful words:

" These are the last salt water ports and the points of debark-
ation for the mountain trails and passes.   The number of Indians
and whites and packers and horses is totally inadequate to move
the vast quantities of freight over the mountains, and a blockade
that is daily assuming more formidable proportions has resulted.

"Tons of supplies are piled high on the beach, and they will
likely remain there for an indefinite length of time.   Every
incoming steamer dumps scores of excited gold seekers and tons
of freight on the beach.   The confusion is indescribable.   Much
of the freight is dumped on a long sand spit at Dyea at low tide,
as there are no wharfs at that place.   Before the supplies can be
sorted, claimed, and removed, the tide has risen and ruined or
carried entirely away large quantities of supplies.

" By far the largest portion of the supplies must be packed
over the passes by their owners if they are packed at all.   Only
about one hundred and fifty Indians, fifty white men, and ten
horses are now packing over the Dyea trail.   It is good to be
an Indian now at Dyea.   He is making at least ten dollars a
day.   He lets the palefaces in search of gold bid against each
other for his services as a packer, and calmly takes up the burden

of the highest bidder. His squaw and his children also carry heavy packs up the steep mountain trail.

"The white man with his ten horses is making $100 per day. It is estimated that there will be fifty additional white packers and forty more horses on the trail in a week or ten days, but on the other hand the rush still keeps up, and the end is not in sight. The end is too far away to see. It is back in New York, Chicago, and San Francisco, and has not started yet. Every man who has set foot in Juneau, Dyea, or Stagua has friends back East who are coming.

"When the rivers freeze overland travel to Dawson must stop, except at the greatest peril, till spring smiles again. The Yukon and Lewis have been known to freeze by the middle of August, but while this is an exception it is more than a possibility. Unless an unexpectedly large number of horses and packers arrive soon many men will camp on the route to the Yukon, and eat the supplies in idleness through the long winter.

"Many men are starting for the Yukon with inadequate supplies and little money. It takes gold to hunt gold. One can hardly make a necessary step on the journey here without it costing $10 for each step.

Timber Runs Short.

"There is a great scramble for timber at Lake Bennett, with which to build boats. A little saw mill there is capable of an output of 800 feet of lumber per day. Ten dollars per hundred was first asked, and now twenty dollars for lumber. The whip-saw of gold-seekers is heard throughout the woods. Owing to the great rush there must be more delay at the lakes.

"Prospectors in the Valley of Yukon have returned here from Dyea, and will wait till spring before attempting to make the Klondike. But not so the tenderfoot. He is swarming for the summit in many instances with an outfit unsuitable in kind and quantity. He is leaving here every day with pretty red, frail two-wheeled carts and wheelbarrows, piled high with much superfluous baggage which he cannot hope to push over the mountain trails.

"His vehicle will smash, and his supplies scatter and break before he is out three hours from Dyca. But you can't make

NATIVES OF ALASKA BUILDING HOUSES.

him believe it. He is so excited he can't or won't listen to reason. His one idea is gold and he is going after it with sacks and carts to bring it back in. As these outfits pass through the streets from wharf to wharf old prospectors laugh.

"It looks as though the Canadian customs officials will have an opportunity to report back to their government that they are unable to collect customs duties without reinforcements.

"All the incoming gold hunters are incensed at the action of the Canadian authorities, at Ottawa, in levying a duty on supplies they are taking into the mines. The rougher element among them is intemperate in its language, and has made threats to ignore the customs officials, peaceably if possible, but forcibly if necessary.

"The general prospect, as viewed from the border of the land of gold at this time, is that the route to the Yukon will be strewn with bones as well as blasted hopes.

Hurts Alaska Industries.

"The Klondike craze is having a disastrous effect on the industries of Alaska. The great salmon cannery at Chilkat has been compelled to close down from lack of fishermen in the middle of a very fine season. Nearly every white man in the cannery deserted and started for Dawson City. Manager Murray tried to get men to take the vacant places, but soon gave up the attempt.

"Men are insulted now when asked to work for a cannery.

"The Klondike fever is at a very high pitch in Alaska, as well as elsewhere. The Chilkat cannery is controlled by the Alaska Packers' Association, which operates nearly all the canneries on the coast. Employes are leaving the canneries for the Klondike. The probability is that work at nearly all of them will be abandoned soon, owing principally to a lack of fishermen.

"At Douglas City, across the channel, about fifty men have given notice to quit work next pay day. They are employed in the big Treadwell Gold Mine and Mills. Others are leaving without notice and heading for Klondike. Every shift one or more men are missed. It is feared that so many desert that the mines and mills cannot be worked.

"The fever has also seized the men in the mines and stamp

OLDEST FUR STORE IN ALASKA, AT SITKA

WHALE TOTEM POLE, FORT WRANGEL. ALASKA

mills at Berner Bay.   A large number have thrown up their jobs there and started for the Klondike."

Could anything better express the utter folly of some of the gold-seekers, who were probably types of a large class, than this, clipped from a letter written from Dyea?

"Such is the innocence of some of the 'tenderfoot' prospectors that they have taken bicycles to Dyea.   They have found the park commissioners neglected to boulevard the trail to Dawson and the bicycles being, even in an extremity, unfit for food, are now very cheap."

One of the possible and much-feared episodes in the Klondike sensation may yet add a bloody page to the history of North Pacific navigation, and cause to be re-enacted in American waters some of the fierce buccaneering scenes of the Straits Settlements on the Spanish Main.

Chinese Pirate Scare.

Word was received early in August by the officials of the North American Transportation and Trading Company that a band of Chinese pirates had been organized for the express purpose of intercepting and looting the steamer Portland on its last trip down from St. Michael's in October, 1897.   It was known that a large number of Klondike miners intended to come out in the Portland, bringing their dust with them, and the last company shipment of gold would also be brought down on the same boat.   Altogether, it had been reported, about $2,000,000 of yellow treasure would be aboard, and the company officials were informed a pirate crew recruited from the Highbinders in the Chinese slums of San Francisco, aided by a few renegade white men, would lie in wait to loot and destroy the treasure ship and murder its crew and passengers at some point between St. Michael's and Dutch Harbor.

P. B. Weare, of the company, communicated his fears to Secretary of the Treasury Gage, and the latter at once ordered Commander Hooper, of the Revenue Service, to send a cutter to convoy the treasure ship safely into the Pacific.

The Portland is a staunch vessel, well armed and carries a good crew, and when aided by the fighting tars of the Bear or Rush, is expected to not only come through safely but to give the Mongolian marauders a hot reception if they venture out.

Craze Is Epidemic.

Another effect of the Klondike fever was to cause a similar malady of strictly local extent to break out in a dozen places which had not had a case of genuine gold fever in years.

California promptly " saw " the Klondike and " went it one better " with some remarkable strikes in the Trinity County placers. The largest nugget reported was said to be worth $42,000, and weight 2400 ounces. Little Rock, Arkansas, went wild over the reputed rediscovery of some old Spanish mines in the neighborhood. Nevada got a latter-day Washoe shock in an old mine in Elko County. The Kootenai and Cariboo districts suddenly discovered that they contained mineral enough to warrant a population of 100,000 in a few years, and hearalded the fact to the world. Colorado got up a boom over some sylvanite quartz at Silver Cliff, an old camp. Rat Portage, Ontario, suffered a depopulating exodus over some reported rich finds in the Rainy Lake and Seine River country. Deadwood put in a claim to notice by announcing a new lead in Ragged Top, which assayed $1048 a ton in gold. Altoona, Pennsylvania, temporarily forgot the coal rumpus while it discovered gold ore going $625 a ton on Tussey Mountains. Elizabethtown, Kentucky, got up a little excitement over a gilt bottomed farm near Summit. Columbia, Missouri, ran across a lot of gold in the banks of Dry Creek. Ashland and Marinette, Wisconsin, came in neck and neck with stories of gold discoveries. Marquette, Michigan, found it was roosting on top of a gold lead forty feet wide and hadn't suspected it before. Peru came to the front with a revival of the famous mines of the Incas. Mexico owned up to having gold in the Yaqui country. Russia declared there were fabulously rich new mines in Okhotsk, just across from Alaska. And China came in late in the game and announced the biggest find of all.

It mattered not that the Missouri gold was pronounced pyrites and some of the other "discoveries" mere stock jobbing schemes—it showed how the fever spread.

About Bogus Stock Companies.

A word to the people who did not catch the stampede craze hard enough to get them out of the country, but who are left behind with the "Alaska Mining and Klondike Development Stock Companies : "

The man who goes in person to the Klondike takes great risks, but his success or failure will depend largely on himself in the long run. At any rate, he knows what he is staking on the issue. But the man who would stay at home and still be a Klondiker has to reckon not only with nature, but with rascals.

There will be stock companies innumerable, organized ostensibly to exploit the Northwest. Some will do it. They will be directed by men who will set honestly about the business of trade and transportation and mining, who will handle honestly the funds intrusted to them, and who, by enterprise and square dealing, will make dividends for the stockholders.

There will be other companies organized to exploit the pockets of the people at home. They will not move a boat, they will not grub-stake a miner, they will not sell a shovel, a pick, or a pan.   Their directors will get money from the unsuspecting and use it for their own purposes.   If the boom holds out and grows to sufficient size they will play the part of the adventurers who turned the city of Panama into a modern Babylon with the money contributed by the people of France.

In short, sending capital into the Klondike will be even more precarious than going yourself, for the risks of nature will be added to the risk of man's rascality.

Yet capital is needed in the Klondike, and those who send it there under the proper sort of management will make legitimate profits, and possibly big ones.

CHAPTER III.

" Strike it Rich" on Klondike.

Gold-seekers who " Made their Pile " in the Placers—Tales Brought Back
by Returning Argonauts—Fabulous Stakes made by Novices—The
"Tenderfoot" Has His Day—Clarence J. Berry, the " Barney Barnato "
of the Diggings—His Wonderful Streak of Luck—Gives the Credit to
His Wife—Captain McGregor's Wonderful Panning Results—Fortune
Favors an Indiana Boy—Some of the Dark Sides, by People who Saw
Them—Miners Go Insane—Death on the Glacier—Hard Work and Lack
of Supplies—Advice of a California Pioneer.

THAT men, even a few, have "struck it rich" and "made
their pile" on the Klondike, or anywhere else on the
Upper Yukon, has put the whole question of gold prospects in Alaska beyond cavil or doubt with the masses, for the
coming close season at least. Much good advice will be given
—and wasted—before the ice moves in 1898 in the upper channels in the Alaskan rivers, but not a word of it, nor all of it
together will be potent to overcome the attraction there is in the
list of those who have washed fortunes out of the frozen Klondike gravel.

That tons and more of new gold, a million and three-quarters
of dust and nuggets, that the Portland brought in July, and the
men who had "struck it" who came with her, and the stories
they told of other lucky ones who were still washing away at
the auriferous soil—these things settled it. Alaska is Eldorado
and the cry is " Klondike or Bust."

It seemed strange as the passengers landed from the Portland
to gaze upon a small satchel tightly grasped in a brown hand,
and realize that it contained probably over $10,000, the reward
of untold hardship. The blanket securely strapped and the
leather gripsack seemed favorite packages for the yellow metal.

6   81

This time of '97, unlike all other times, Fortune played no favorites. 1897 on the Klondike was the "tenderfoot's" year for gold. The inexperienced men have been the lucky ones, individuals in several instances taking out approximately $150,-000 in two months and a half, while the old miners, after spending years and suffering hardships and privations innumerable in the far Northwest, had only a few thousands to show for all their pains and perils.

Clarence Berry's Strike.

Clarence J. Berry, of Fresno, California, was one of the luckiest of the "tenderfeet;" in fact, his strike was a proverb in the entire region, and he is known among the Yukoners as "the luckiest man on the Klondike," and the "Barney Barnato of the Klondike," though he is unlike the South African Crœsus in all but luck. A few years ago, Berry said, he did not have enough to pay house rent, and did not dare ask Ethel Bush, of Fresno, to share his poverty. But he brought back from the Klondike, on the Portland, $130,000 in gold nuggets, and the prettiest wife in the territory and a helpmeet, too, for Mrs. Ethel Berry, nee Bush, didn't begin the honeymoon under the midnight sun by asking her husband for pin-money. Not she. She just took a pan and washed out $10,000 or so on her own account.

Clarence Berry was described by Mrs. Eli Gage, who was a passenger with him and his wife on the Portland, as being "the most modest millionaire," she ever saw. But he was willing to talk Klondike after he had turned his dust and nuggets over to Wells, Fargo & Co., at Seattle, on July 17th.

"Yes, I am a rich man," said he, "but I don't realize it. My wife and little ones will, though. I took out my gold last winter in box lengths twelve by fifteen, and in one length I found the sum of $10,000. The second largest nugget ever found in

Alaska was taken out of my claim. It weighed thirteen ounces and is worth $230. Why, I have known men to take out $1000 from a drift claim, and some have taken out several thousand. This gold was found in pockets, and it is not an ordinary thing to make such marvelous finds.

"Yes, there is plenty more of gold there. I expect to take many more thousands from my claim; others on this boat expect to do the same. Those who have good claims will undoubtedly be millionaires in a few years. The gold will not give out for a long time. There is room for more miners in Alaska, but they must be strong men, must have money, and should know about mining. The hardships are many. Some will fail to make fortunes, where a few are successful. A man may have to prospect for many years before he finds a good claim. That means that he needs money and strength to help him along; but if he sticks to it he will come out all right."

Captain McGregor's Big Pans.

Captain John G. McGregor, of Minneapolis, Minn., a placer miner for thirty years, and one of the pioneers at Confederate Gulch, Montana, has been in the Klondike a year. In August he wrote home that his men were washing gravel that occasionally goes $3000 to the pan, and that $1000 is common. He has several miners working for him, and expects to bring out as his own profits next June not less than $1,200,000.

Frank Phiscator, of Gallen, Indiana, came in on the Portland with $50,000, which he washed out in forty days. He left Indiana a year before for the Pacific Slope to begin life anew, having failed in the fruit business. He had never heard of the Alaska gold mines until he reached Seattle, which place he reached "broke." He was grub-staked by a friend who went through from Michigan with him, and together they started for the new Eldorado. For days after they left Circle City they were lost in a blinding storm, and for three days found refuge in a hole in the hardened snow. They reached the Klondike in the dead of winter, and when the weather moderated they were prepared for business. In forty days they sluiced and washed out $125,000 of gold, of which Frank received as his share $50,000.

William Stalley and C. Worden were Phiscator's companions, and they divided $75,000 between them.

William Sloane, a merchant of Nanaimo, B. C., went North for pleasure one year ago. He had no money. A friend induced him to go to Klondike. He came back with $52,000, the amount he received for his claim. He says he will not return, but advises others who want gold to go.

Dougal M'Arthur's Romance.

Young Dougal M'Arthur came down from Klondike with $25,000 in dust and a story no one could doubt. He said:

"I left the good old country when a mere boy, determined, if possible, to carve out a fortune for myself. Coming to America I drifted from place to place with varying success and finally, six years ago, determined to try my luck in Alaska. It was hard working at first, but I soon got used to it, and I determined to stay there until I struck something that would pay me for my trouble.

"At Forty-mile camp I made some money and then I drifted over to Circle City. There I did not do so well, but I kept pegging away, believing like Micawber, that something would turn up after a bit. Well, last fall came the news of a tremendously rich strike on the Klondike. We—that is, my partner, Neal McArthur and myself—pulled up stakes and started for the new discovery. Neal went ahead and was fortunate in locating a good claim. My part of the work consisted in hauling our pro-

OLDEST HOUSE IN JUNEAU CITY, ALASKA

INDIAN BURYING GROUND NEAR JUNEAU, ALASKA

visions and camping outfit over the snow and ice to the new location. I was compelled to make two trips, and it was the hardest work I ever did in my life.

"I reached Dawson City finally just two days before Christmas. Neal had prospected the claim and found it rich beyond our fondest anticipations. Before we could begin work there

SCENE NEAR DAWSON CITY.

was an offer to buy it and we sold out for $50,000. It was a lucky turn of the wheel of fortune for us. Without practically a stroke we cleaned up $25,000 apiece.

"Now we are going home to see our people. My own folks have not heard from me in a long time, and maybe they think I am dead. It will be a joyful home-coming for all."

Among the first people to come back to civilization were **Mr.** and Mrs. Lipton, who, though they had been at the diggings only since April, 1896, returned with $60,000. Most of the party were "tenderfeet," and had spent but one season at the mines, yet some of them had taken out from $10,000 to $25,000 in a few weeks. In the nine miles advance up the Klondike, it is said, there are several mines that will yield over $1,000,000, one piece of ground on the Eldorado, forty-five feet wide, having yielded $90,000. The Berry claim has produced $145,000 in a few months, and there is a pile of gravel on the dump, ready to be washed as soon as sufficient water can be obtained, which contains as much more.

Sample " Piles " on the Portland.

Among the passengers on the Portland, July 17th, Clarence Berry, Frank Phiscator, and Frank A. Kellar, of Los Angeles, each had from $35,000 to $100,000. Henry Anderson and Jack Morden, of Chicago ; William Stanley, of Seattle ; and R. Mc-Nulty and N. E. Pickett, each had at least $20,000. M. Mercer, J. J. Hillerman, and J. Moran, had each from $12,000 to $15.000. The average pile of dust on board the Portland was probably $12,000, and these people, the captain said, are only a handful.

Michael Hickey, of Great Barrington, Mass., brought down $60,000, which he had taken from Klondike placers in the last eighteen months. Hickey is a widower. He left Great Barrington for Alaska in the spring of 1896. In his letters home he has not complained about the hardships he has met. He spent the winter of 1896-97 in the gold regions.

William Stanley, of Seattle, "struck it" rich. He came down with $90,000. His two sons are in the Klondike, looking after their claims, out of which they hope to make at least $300,000.

Henry Anderson, a native of Sweden, had no money when

he left Seattle two years ago. Now he has $45,000 and states that he received it for a half interest in his claim.

Pack Horne, a pugilist who use to work for variety theatres on Puget Sound for ten dollars per week, displayed $6000, the result of a year's work.

T. J. Kelly and son, of Tacoma, went north in the fall of 1896. The father brought back $10,000 and the son is holding the claim.

Gold Breaks the Gripsack.

John Wilkinson, a passenger on the Portland, had his gold in a leather gripsack, and in carrying it out of the social hall of the steamer, in spite of the fact that he had three straps around the bag, the main handle piece broke, and he had to secure a broader strap before he could carry his treasure ashore.

Henry Anderson, another passenger, refused to talk, hurrying aft to get away, but it was said by his companions that he brought down $65,000, and that he had a claim like a river of gold. He sold out a half interest for $45,000 cash. In six hours' shoveling he secured 1025 ounces from his claim.

Thomas Moran, of Montreal, brought out as the proceeds of five years' work $20,000, and still has interests in several claims. Moran will go back. Victor Lord, an old Olympia logging man, brought out $10,000 after four years on various parts of the Yukon. He owns a half interest in two claims, and will return in the spring. M. N. Murcier, of Shelton, Mason & Co., came out with about $160,000.

Among the passengers via the Portland were Fred. Price, August Galbraith, L. B. Rhoads, Thomas Cook and Alexander Orr. Each one had from $5000 to $12,000. Joseph Ladue, the owner of the townsite of Dawson City, was also aboard. Land is selling there, he reported, at $5000 a lot.

Fred. Price, who brought out a snug fortune, said: " I was

located on the Bonanza with Harry McCullough, my partner. I brought down $5000 in gold dust and made $20,000, which is invested in more ground. There were good stakes on the boat coming down—from $5000 to $40,000 among the boys. I refused $25,000 for my interest before I left. My partner remains, and I shall return in the spring after seeing my family in Seattle. I was in the mines for two years. One can't realize the wealth of that creek. There are four miles of claims on the Eldorado, and the poorest is worth $50,000. The Bonanza claims run for ten miles, and range from $5000 to $90,000."

August Galbraith said: "The development of Alaska has only just begun. If I were not an old man, I would have remained where I was. There is no doubt in my mind that all of the country for hundreds of miles around Dawson is rich in gold. It is the best place that I know of for a poor man to go. If a man has $500 when he starts, well and good, for it may be useful if he should not be fortunate the first season."

Rock Lined With Gold.

L. B. Rhoads said: "I am located on Claim 21, above the discovery on Bonanza Creek. I did exceedingly well up there I was among the fortunate ones, as I cleared about $40,000, but brought only $5000 with me. I was the first man to get to bedrock gravel and to discover that it was lined with gold dust and nuggets. The rock was seamed and cut in V-shaped streaks, caused, it is supposed, by glacial action.

"In those seams I found a clay which was exceedingly rich There was a stratum of pay gravel four feet thick upon the rock which was lined with gold, particularly in these channels o streaks. The rock was about sixteen feet from the surface."

Alexander Orr, who brought out $12,000 in dust, said: "Ii winter the weather is extremely cold at Dawson, and it is neces sary that one be warmly clad. The thermometer often goes sixty or seventy degrees below zero. Ordinary woolen clothes would afford little protection. Furs are used exclusively for clothing. Dawson is not like most of the large mining camps. It is not a " tough " town. Murders are almost unknown. A great deal of gambling is done in the town, but serious quarrels are an exception. Stud poker is the usual game. They play $1 ante and oftentimes $200 or $500 on the third card."

Thomas Cook expressed himself as follows : " It's a good country, but if there is a rush, there is going to be a great deal of suffering. Over 2000 men are there at present, and 2000 more will be in before the snow falls. I advise people to take provisions enough for eight months at least. If they have that, it is all right. The country is not exaggerated at all. The mines at Dawson are more extensive and beyond anything I ever saw."

William Sloan, of Nanaimo, B. C., sold his claim for $52,000 and came home to stay. A man named Wilkenson, of the same place, had $40,000.

The smallest sack of gold among the Yukoners aboard the Portland on July 17th was $3000. It belonged to C. A. Branan, of Seattle, a youth of eighteen years.

Over $100,000 for a Boy.

The richest strike was made by a twenty-one-year-old boy named George Hornblower, of Indianapolis. In the heart of a barren waste known as Boulder Field he found a nugget for which the transportation company gave him $5700. He located his claim at the find and in four months had taken out over $100,000.

Henry Lamprecht wrote from the Klondike to say that there are miles of rich pay dirt all through the region. Men have taken a tub of water into their cabin and with a pan "panned out" $2000 in less than a day. This is said to be equal to about $40,000 a day in the summer with sluice boxes. They get from $10 to $100 a pan average and a choice or picked pan as high as $250, and it takes about thirty minutes to wash a pan of dirt. ·

Three hundred thousand dollars' worth of gold from the Klondike found its way to Minnesota in the possession of Peter Olafson and Charles Erickson, two Scandinavians, who returned to Two Harbors after putting in five years in Alaska.

A little over five years ago the two men, aged twenty-seven and thirty years, respectively, were employed in the blacksmith shops of the Duluth and Iron Range Railroad at Two Harbors. They heard of the gold fields in Alaska and decided to go there and seek a fortune. For three years they labored in vain, but two years ago they discovered a rich placer bed on the Stewart River, and later located claims on the .Klondike. In the two years they say they cleaned up $150,000 each.

A new mint record for one day's receipts at the San Francisco Mint was made August 3d, when $3,775,000 in gold was deposited at the branch mint for coinage. This represented the accumulation of six weeks. Three-quarters of a million of this was owned by the Alaska Commercial Company and was mainly from the Klondike. A large portion of the balance was also from the rich northern placers, and was deposited by various miners and smelting companies to whom it had been sold. This is said to be the largest sum deposited at a mint in a single day.

Allan McLeod's Big Stake.

Allan McLeod, of Perth, Scotland, came back with $92,500. His hands and feet were tied up in bandages, and his clothing was ragged and dirty as a result of a long sojourn in Alaska.

He looked anything but prosperous, yet in his pocket reposed a draft for $92,500, and an attendant took care of a deer hide sack heavy with gold nuggets.

Mr. McLeod is a baker by trade, a restaurant cook and proprietor by circumstance, a gold miner by accident and a rich man by luck. Inflammatory rheumatism, contracted in the gold fields, made a temporary cripple of him and rendered his journey painful, yet he had a light heart as he pictured the surprise he would give his old friends in Scotland when he landed with his treasure.

Sold Out For $5,000.

"I went to Alaska early last summer," said Mr. McLeod, "with a crowd of miners who came up the Sound from San Francisco. I was out of money and work, or I doubt whether I would have accepted the offer they made me to go along as cook. We reached Cook's Inlet June 20th, and things looked so discouraging we went back to Juneau. There we bought supplies and started for Dawson City, 750 miles away. We camped there, and I did the cooking for the boys. They did very well, but the gold fever took them farther east, and I remained to cook for another gang of miners. I made good wages, and finally had enough to start a restaurant. In two weeks I sold the place for $5000, and went placer mining with a half-breed for a partner.

"We had good luck from the start, and I would have remained but for a severe attack of inflammatory rheumatism. It would have killed me but for the nursing of my partner. He carried me most of the way to Juneau, where I got passage on a fishing schooner to 'Frisco. I am satisfied with what I've got in money, and hope to get rid of my rheumatism before long. Great fortunes are being found by many men, and no one knows the extent of the gold fields that are constantly developing."

A San Francisco paper, under date of July 23d, prints the fol
lowing :

"Five French Canadians who were successful on the Klon
dike, and are now bound for Montreal, are at the Commercia
Hotel in this city.   They came from Seattle, having reached tha
city by the steamer Portland.   They could not get the prices fo:
their nuggets that they wanted there, nor will they accept the bic
made by the Selby smelting works in this city.   As the Sar
Francisco mint is closed pending the change of administration
these five miners will carry their bullion to Philadelphia and ex·
change it there for coin of the United States."

J. O. Hestwood Sees Millions.

J. O. Hestwood, of Seattle, is a typical returned Argonaut
He is a small man, weighing not over 140 pounds, and has ligh·
blue eyes, clear skin and a firm square jaw.   He has been ;
preacher, teacher and lecturer, having delivered lectures all ove.
the coast of Alaska to pay his way up there.   He spent three
years in the territory before his great opportunity came.   He
was at Glacier Creek when the news was brought down of the
immense strike in Bonanza Creek.   Here is his story in his own
words, which give an admirable idea of the way the mines are
worked :

"With hundreds I rushed to the new fields.   After a few day·
I became disgusted and started to leave the country.   I had
gone only a short distance down the river when my boat go
stuck in the ice and I was forced to foot it back to Dawson
City.

"Well, it was Providence that did that.   I purchased claim
No. 60, below Discovery claim, and it proved one of the riches
pieces of ground in the district.   My claim will average 16 or 1·
dollars to the pan, and in addition to what I have already take·

GLACIER AND FLOATING ICE IN GLACIER BAY, ALASKA

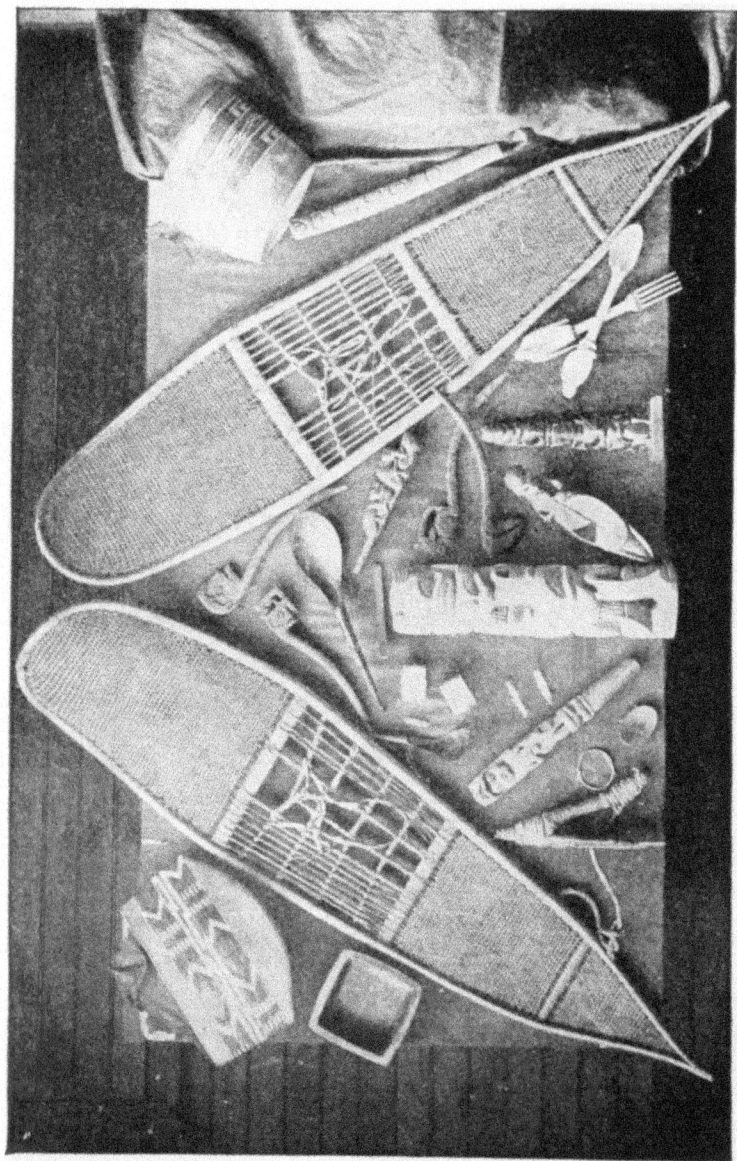

SNOW SHOES AND INDIAN CURIOS OF ALASKA

out, there is at least $250,000 in sight. Last season I worked thirty men, and I intend to employ more next year."

· B. W. Shaw, a former insurance man of Seattle, writing from Klondike, says he does not expect to be believed when he says he counted five five-gallon oil cans full of gold dust in one cabin, the result of a winter's work by two men. He adds that 100 ounces have been taken out of a single pan.

William Kulju sold his claim for $25,000, brought down 1000 ounces of dust and started home for Finland.

Fred. Lendeseen went to Alaska two years ago, and in July brought down $13,000 in dust, besides having an interest in a claim.

Greg Stewart sold his share in a claim for $45,000.

ONE OF THE FIRST SETTLERS.

Thomas Flack brought along $6000 in dust - for expenses, and said he had refused $50,000 for his share of a claim, out of which his partners realized, respectively, $50,000 and $55,000.

J. B. Hollingshead had $25,000 in dust to show for two years' work.

M. S. Norcross said: "I was sick and couldn't work, so I cooked for Mr. McNamee. Still I had a claim on the Bonanza,

but didn't know what was in it because I couldn't work it. I
sold out last spring for $10,000, and was satisfied to get a chance
to return to my home in Los Angeles."

John Marks reported thus about his "pile:" "I brought
$11,500 in gold dust with me, but I had to work for every bit
of it. There is plenty of gold in Alaska—more, I believe, than
the most sanguine imagine—but it cannot be obtained without
great effort and endurance."

This is Talbot Fox's story: "I and my partner went into the
district in 1895 and secured two claims. We sold one for
$45,000. I brought 300 ounces, which netted $5000. Everybody is at Dawson for the present. The district is apt to be
overrun. I wouldn't advise anyone to go there this fall, for
people are liable to go hungry before spring. About 800 went
over the summit from Juneau, 600 miles, so there may not
be food enough for all."

Riches on the American Side.

F. G. H. Bowker, a Yukoner of six months' standing, brought
out $40,000 and the information that the placers were richer on
the American than on the Canadian side of the boundary line.

Wonderful tales are told of the great richness of the Klondike
placers. More than one man reports having obtained $1000
from a single pan washing, while reports of yields of $500 and
$600 to the pan are numerous. An ordinary pan of gravel will
weigh twenty-five pounds and a yield of $1000 worth of gold
means sixty-two ounces, or nearly one-sixth of the entire bulk
in precious metal. The average is said to be fifty dollars to the
pan, and this is phenomenal when it is taken into consideration
that the California pan washer was well pleased with a uniform
product of three dollars to a washing, and could make money
with a yield running as low as fifty cents. With this kind of

field to work in, it is small wonder that claim-holders gladly pay
fifteen dollars a day for common labor, and are unable to get
anything like a fair supply at that.   It is only men who are
" broke" who are willing to work for wages.

Fever Strikes the Navy.

Lieutenant John Bryan, of Lexington, who is on the revenue
cutter Rush, stationed at Unalaska, Alaska, watching the seal
fisheries, writes under date of July 9th to relatives in Kentucky
that the Alaska gold fields are not overestimated.   He says the
placer mining is in the old bed of the Yukon River.   He says :

" You dig no deeper than fifteen feet into the river bed when
you strike a strata of pure gold nuggets among the stones.
There are eighty claims already taken, each 5,000 feet long and
the width of the river bed.

" The great obstacle in reaching the gold fields is the uncomfortable mode of travel.   Steamers go no further than the mouth
of the Yukon, and you have to walk the 1000 miles or pay the
extravagant fare asked by the company, which runs a small boat
up the river and finally lands you near the gold fields.

" All who are fortunate enough to reach the country are certain to find employment, even if they do not strike a claim,
which at present they could avoid only by not looking for it.
The poorest miners will pay fifteen dollars a day for help on
their claims, but it will cost five dollars per day to live unless
you take your provisions with you."

The lieutenant says he has the gold fever badly, and if it were
not for the fact that he is in the government service he would go
to the new Eldorado.

The *Toronto Globe* says editorially of the Klondike situation :

" While there is probably much exaggeration in the stories
that are brought back from the Yukon, it is only necessary to

read the calm official reports of Mr. Ogilvie, the well-known officer of the Geological Survey, to realize that it is equally possible that there is no exaggeration in them at all.   Mr. Ogilvie's notes read like passages from Monte Cristo.   Writing on December 9, 1896, he said : ' Bonanza Creek and tributaries are increasing in richness and extent until now it is certain that millions will be taken out of the district in the next few years.   On some of the claims prospected the pay dirt is of great extent and very rich.   One man told me yesterday that he had washed out a single pan of dirt on one of the claims on Bonanza and found fourteen dollars and twenty-five cents.   Of course that may be an exceptionally rich pan, but five to seven dollars per pan is the average on that claim it is reported, with five feet of pay dirt and the width yet undetermined ; but it is known to be thirty feet even at that ; figure the result at nine to ten pans to the cubic foot, and 500 feet long—nearly $4,000,000 at five dollars per pan.   One-fourth of this would be enormous. Another claim has been prospected to such an extent that it is known there is about five feet pay dirt averaging two dollars per pan, and width not less than thirty feet.   Enough prospecting has been done to show that there are at least fifteen miles of this extraordinary richness, and the indications are that we will have three or four times that extent, if not all equal to the above, at least very rich.' "

Captain McGregor's Story.

Captain John G. McGregor, of Minnesota, went into Alaska last March, and the last of letters to his relatives came from the land of gold June 14th.   This was before the rush of the fortune hunters had begun or before, in fact, much was known of the Dawson City diggings.   Notwithstanding that fact, the letter contains estimates of wealth which distance far and away any of the hitherto published accounts of the yield from Alaska's glittering sands.

"We have washed $3000 to a single pan," says the captain, in one of his letters. This is almost incredible. It would be quite so in fact were it not for his well-known reputation. He has been a mining expert for thirty years, and much of that time has been engaged in the very work he is now doing—placer mining.

Up to date the world's record has been $1000 a pan. This was in Montana at Montana Bar. There was a group of properties in what was known as the Confederate Gulch, and every 100 feet for half a mile along the shore produced $1000 a pan for every washing. The year was 1868. Captain McGregor owned those properties then, and does now, so that in the present instance his word must command a good deal of respect on that ground alone.

Results of Prospects.

His attention was directed to the Yukon valley basin some time ago, and a year ago last March he sent two men who had been in his pay for a number of years out to prospect. He heard from them from time to time, but the message he waited for did not come until last March. Then the word he received caused him to form a party immediately. He had had his preparations all planned, and within a very short time was breasting the mountain snows in the Chilkoot pass. He could not wait for the warm season, and made the trip successfully, though at the expense of considerable suffering by members of his expedition. On his arrival he immediately assumed charge at the claims which had been located and staked out by his men, with the result that he uncovered the tremendously rich find he reports.

Captain McGregor began his prospecting immediately after the war. He came into control of the Confederate Gulch properties shortly after his start, and most of the gold taken out was washed under his direct management. The gulch was then 500 miles from the borders of civilization, and each installment of the yellow stuff had to be escorted down to the railroad by armed bodies of 200 or 300 men. The metal was packed in beer kegs and so carried without trouble.

The captain is a Scotchman and has all the caution and conservatism characteristic of the nationality. Coming from such a source, the character of his statement is far superior to the report which might be brought from some prospector or from entirely irresponsible parties. Captain McGregor has had men in his employ and prospecting various regions since the seventies. He is now looking for quartz, and will undoubtedly, later on, place himself at the head of some very important deep-earth operations.

Placer mining will pay when not more than twenty-five cents is realized on a pan. The operation is very generally familiar, even to those who know nothing about mining. The earth washed in the Confederate Gulch was so dazzlingly heavy with gold that it seemed as if it were nearly pure, so it can be imagined what description the wash from the Klondike soil must take on.

How Berry Got His Stake.

Clarence Berry, the "Barney Barnato" of the Klondike, tells a thrilling story of his experience.

Berry was a fruit raiser in the southern part of California. He did not have any money. There was no particular prospect that he would ever have any. He saw a life of hard plodding for a bare living. There was no opportunity at home of getting ahead, and, like other men of the far West, he only dreamed of the day when he would make a strike and get his million. This was three years ago. There had then come down from the

PROSPECTORS IN CAMP.

99

frozen lands of Alaska wonderful stories of rewards for men brave enough to run a fierce ride with death from starvation and cold. He had nothing to lose and all to gain. He concluded to face the danger. His capital was forty dollars. He proposed to risk it all—not very much to him now, but a mighty sight three years ago. It took all but five dollars to get him to Juneau. He had two big arms, the physique of a giant and the courage of an explorer. Presenting all these as his only collaterals, he managed to squeeze a loan of sixty dollars from a man who was afraid to go with him, but was willing to risk a little in return for a promise to pay back the advance at a fabulous rate of interest.

Juneau was alive with men three years ago who had heard from the Indians the yarns of gold without limit. The Indians brought samples of the rock and sand and did well in trading them. A party of forty men banded to go back with the Indians. Berry was one of the forty. Each had an outfit—a year's mess of frozen meat and furs. It was early spring when the first batch of prospectors started out over the mountains, and the snow was as deep as the cuts in the sides of the hills, the natives packed the stuff to the top of Chilkoot pass. It was life and death every day. The men were left one by one along the cliffs.

Disaster to the Outfit.

The timid turned back. The whole outfit of supplies went down in Lake Bennett. The forty men had dwindled to three —Berry and two others. The others chose to make the return trip for more food. Berry wanted gold. He borrowed a chunk of bacon and pushed on. He reached Forty Mile Creek within a month. There was not a cent in his pocket. The single chance for him was work with those more prosperous. His pay was $100 a month. It was not enough, and, looking for better

INTERIOR OF HOME OF ALASKA INDIANS

DOGS AND SLEDGE IN NORTHERN ALASKA.

pay, he drifted from one end of the gulch to the other, always keeping his shrewd eye open for a chance to fix a claim of his own.   There was a slump in the prospects of the district and he concluded to go back to the world.

The slump was not the only reason.   There was a young woman back in Fresno who had promised to be his wife. Berry came from the hidden world without injury and Miss Ethel D. Bush kept her pledge.   They were married.

Berry told his bride about the possibilities of Alaska. She was a girl of the mountains. She said she had not married him to be a drawback, but a companion.   If he intended or wanted to go back to the Eldorado, she proposed to go with him.   She reasoned that he would do better to have her at his side.   His pictures

A MINER IN HARD LUCK.

of the dangers and hardships had no effect upon her. It was her duty to face as much as he was willing to face. They both decided it was worth the try—success at a bound rather than years of common toil.   Berry declared he knew exactly where he could find a fortune. Mrs. Berry convinced him that she would be worth more to him in his venture than any man that ever lived. Furthermore, the trip would be a bridal tour which would certainly be new and far from the beaten tracks of sighing lovers.

Mr. and Mrs. Berry reached Juneau in May, 1896. They had little capital but lots of determination. They took the boat to Dyea, and the rest of the journey was made with dogs. They slept on a bed of boughs under a tent. They reached Forty-Mile Creek a year ago in June, three months after they were married. They called it their wedding trip.

Off for the Discovery.

Klondike was still a good way off, and it was thought at first that the claims closer at hand would pay. One day a miner came tearing into the settlement with most wonderful tales of the region further on. His descriptions were like fairy tales from "Arabian Nights"—accounts fitting accurately the scenes in spectacular plays, where the nymph or queen of fairy land bids her slaves to pick up chunks of gold as big as the crown of a hat. Berry told the tale to his wife. She said she would stay at the post while he went to the front. There was no rest that night in the camp. Men were rushing out pellmell, bent on nothing but getting first into the valley of the Klondike and establishing claims. Mrs. Berry worked with her husband with might and main, and before daylight he was on the road over the pass. There were fifty long miles between him and fortune, and he worked without sleep or rest to beat the great field which started with him. He made the track in two days. He was among the first in. He staked Claim 40, above the Discovery; which means that his property was the fortieth one above the first Aladdin. It was agreed that each claim should have 500 feet on the river—the Bonanza. This was the beginning of Berry's fortune. He then began to trade for interests in other sites. He secured a share in three of the best on Eldorado Creek. There is no one living who can tell how much this property is worth. It has only been worked in the crudest way, yet five months netted him enough to make him a rich man the rest of his life. There are untold and inestimable millions where the small sum from the top was taken.

Berry gives all the credit of his fortune to his young wife. It was possible for her to have kept him at home, after the first trip. She told him to return—and she returned with him. It was an exhibition of rare courage, but rare courage rarely fails. The wedding trip lasted about fifteen months. Berry says it was worth $1,000,000 a month. This estimate is one measured in cold cash—not sentiment.

One day while they were working the claim on Eldorado Creek, Mr. and Mrs. Berry gathered $595 from a single pan of dirt. This dust they have saved in a pan by itself.

Mrs. C. C. Adams' Letter.

Mrs. Chester C. Adams, who went from Tacoma to Dawson City last April, writing under date of June 17th, says that miners were then coming into Dawson City daily with all the gold dust they could carry. It was considered a small matter to have 100 pounds. Many were bringing this amount in as a result of seven or eight months' working of claims on shares.

Other men brought to Dawson from 200 to 500 pounds of gold dust, and Mrs. Adams makes the startling statement that one man had brought in 1300 pounds, which would amount to over $250,000.

Her husband estimated that the steamer then loading at Dawson would take over $2,000,000 to St. Michael's, from which point it will be brought out by the steamers Portland and Excelsior on their next trips down. They are due between August 15th and September 1st.

Mrs. Adams declares the whole truth regarding Klondike has not been told and cannot be, because people would not believe it.   She tells of new discoveries this spring on the Stewart River and Henderson Creek and the creeks emptying into them.

High water had prevented complete prospecting, but when she wrote it was known that some dirt considerably above bed rock would run $10 and $12 per pan.   Bed rock cannot be reached until winter.

Miners are also preparing to do more thorough work on Chicken, Mastodon, Miller, American, Last Chance and other creeks, on which men formerly took out as high as $30 per day each.   These creeks were deserted by last fall's rush to the Klondike.

When she wrote new creeks were being found and prospected in all directions from Dawson, and every day witnessed a stampede of men to one or another of them.

She speaks of an overland trip as one of pleasure rather than hardship when properly made.

Ship Gold in Barrels.

Warren Shea, of New Whatcom, Wash., a reputable and reliable man, writes from Klondike to his brother, S. Shea, of New Whatcom, and says the next boat to leave the gold field will bring out dust and nuggets in barrels.

Two days after the boat that brought out the miners, who arrived on Puget Sound aboard the steamer Portland, left Dawson City one of the largest stores at that place was closed and the building was turned into a gold packing warehouse.   So great a quantity of gold was offered for shipment that it was decided to pack it in barrels holding about twenty-two gallons.

The barrels have heretofore been used for packing salt fish.

An interesting letter from Captain J. F. Higgins, of the steamer Excelsior, describing his last voyage to Alaska, is as follows:

" Bonanza Creek dumps into Klondike about two miles above the Yukon.

" Eldorado is a tributary of the Bonanza.   There are numerous other creeks and tributaries, the main river being 300 miles long.

PUGET SOUND AND MT. RAINER.

" The gold so far has been taken from Bonanza and Eldorado creeks, both well named, for the richness of the placers is truly marvelous.

" The Eldorado, thirty miles long, is staked the whole length, and as far as worked has paid.

" Each claim is 500 feet long and is worth half a million.

" So uniform has the output been that one miner, who has an interest in three claims, told me that if offered his choice he would toss up to decide.   One of our passengers, who is taking $1000 with him, has worked 100 feet of his ground and refused

$200,000 for the remainder, and confidently expects to clean up $400,000 and more.

" He has in a bottle $212 from one pan of dirt.

" His pay dirt while being washed averaged $250 an hour to each man shoveling in.

" Two others of our miners who worked their own claims cleaned up $6000 from the day's washing.

" There is about fifteen feet of dirt above bed rock, the pay streak averaging from four to six feet, which is tunneled out while the ground is frozen.

" Of course the ground taken out is thawed by building fires, and when the thaw comes and water rushes in they set their sluices and wash the dirt.

Sold Out for $45,000.

" Two of our fellows thought a small bird in the hand worth a large one in the bush and sold their claims for $45,000, getting $4500 down, the remainder to be paid in monthly installments of $10,000 each.

" The purchasers had no more than $5000 paid. They were twenty days thawing and getting out dirt.

" Then there was no water to sluice with, but one fellow made a rocker, and in ten days took out the $10,000 for the first installments. So, tunneling and rocking, they took out $40,00c before there was water to sluice with.

" Of course these things read like the story of Aladdin, but fiction is not at all in it with facts at Klondike.

" The ground located and prospected can be worked out in a few years, but there is still an immense territory untouched, and the laboring man who can get there with one year's provisions will have a better chance to make a stake than in any other part of the world."

W. F. Parish, of Chicago, has received from a business associate in Spokane, Wash., H. D. Heacock, a letter written to the latter by J. F. Wallace, dated Klondike, Northwest Territory, May 14th. It is as follows :

" I have been here a month or so. There is a placer mining camp, discovered last summer and supposed to be as rich as Alder Gulch in Montana. They have got as much as $800 to a pan, and will have out over $2,000,000 this winter. There are three creeks known to be good. Eldorado is the richest, there being four miles without a blank claim, and all selling from $50,000 to $100,000 each. Some will not sell at any price. It is in British territory, fifty miles above Forty Mile Post, on the bank of the Yukon River. Mostly every one has left Circle City and come up on the ice. During the winter provisions were scarce. Boats did not get up here last fall on account of the ice. Flour was $1.30 per pound, bacon $1.50 per pound, shovels, $20 each. Dogs sold for $200 and $300 each for freighting. Freight cost $1 per pound from Circle City here. Wages are $15 per day. Lumber is $600 per 1000 feet at the mines. Mines are from five to twenty miles from Dawson City, situated at the mouth of the Klondike. Claims are 500 feet in length. Ground frozen from top to bottom and has to be thawed with fire. Mostly drifting diggings about twenty feet deep. Some twenty or thirty claims will open from top. I did not get here in time to locate, so I am still a prospector. Very mild winter ; only seventy-four below zero the coldest. River frozen yet, but expect it to break almost any day."

Inspector Strickland's Report.

A special from Regina, Northwest Territory, says : " Inspector Strickland, of the Northwest mounted police arrived here last night from the Yukon.

" Mr. Strickland does not believe the story of $250,000 having been made there by any one man, but says the most liberal truths read like fairy tales. It is hard to say just what is being made. The miners are reticent about their earnings. He says that miners who have come out and staked claims this year, number- ing about 100, have taken or sent away sums varying from $5000 to $50,000 each, and have kept back considerable sums for development and other investments. Miners from California, Australia and South Africa say that nothing in the world has been struck as rich.

" Inspector Strickland says that if the country fills up as rapidly as it is doing, the two trading companies will not be able to supply food for the inhabitants. Provisions are not so dear as might be expected : Flour is $12 a hundred ; bacon 40 cents a pound ; canned meats 75 cents and $1, and cariboo and moose flesh is sold by the Indians at 50 cents a pound. Inspector Strickland strongly recommends that no person should go out to the Yukon district without taking with him a year's food, as well as some money, because paying claims are not always found immediately, and there is the long and hard work of building a home. He says that mining is not a picnic. All is hard work. Wood is scarce and requires a great deal of labor. The climate is healthy and there is very little sickness. The chief complaints are scurvy, kidney trouble, and rheumatism.

" Though the winter is eight months long, it is only three weeks that the sun is not seen. Miners' wages are fifteen dollars a day, but this rate will fall soon if the present rush continues from the Pacific coast."

Finds No Hard Times.

J. P. Staley, who is working a claim on Bonanza Creek, wrote to C. P. Enright, of Gilman, Ills., as follows :

"There is no doubt this is the best place to make money in

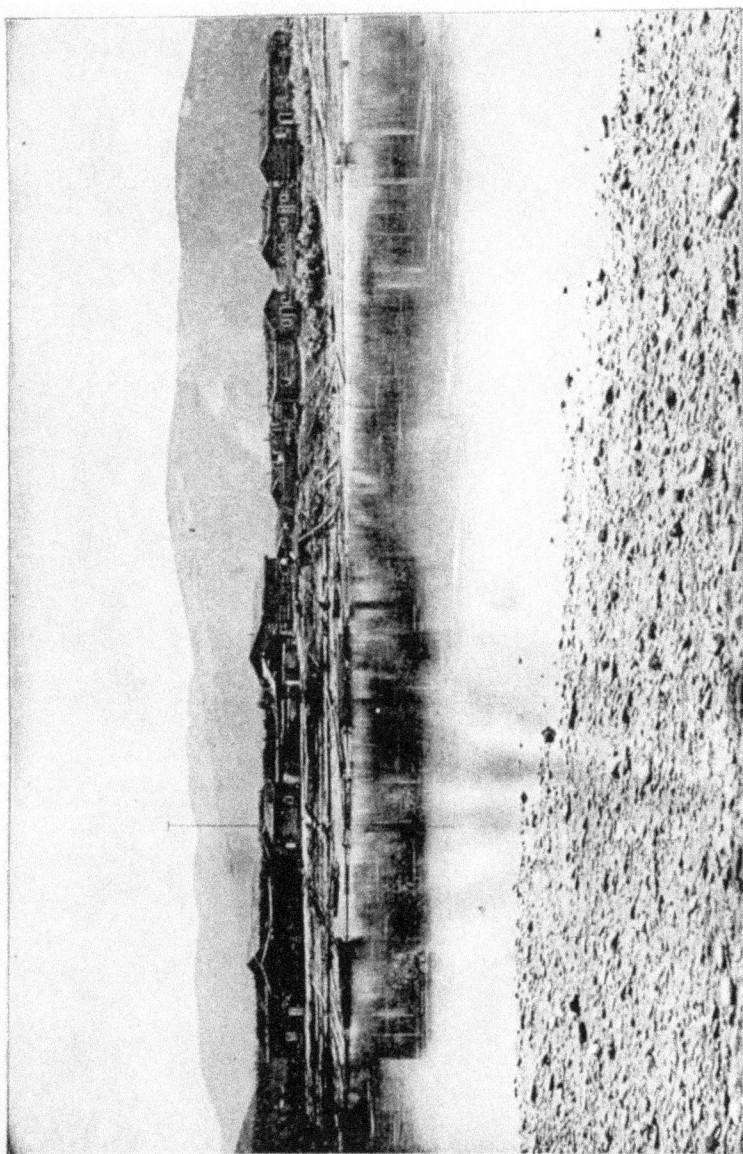

FORTY MILE POST

BIRDS-EYE VIEW OF JUNEAU CITY

the world. Sell out and come here. We need live business men. Flour is $12 a hundred, bacon 40 cents a pound, sugar 25 cents a pound, rice 25 cents a pound, any kind of dried fruit 25 cents a pound. All kinds of canned fruit, 75 cents a can. Bring fur moccasins with you. They will fetch from $15 to $25 a pair.

"Brother Dan and I are working in a mine, or rather in a bed of a creek. We are getting $15 a day each for ten hours, and it is thought wages will be $25 a day during the winter. It takes about $600 a year each for provisions, blankets, gloves, moccasins, etc. We expect to remain here all winter. It is too long a trip to lose the chance of making a stake by refusing to stay.

"Everbody is pleased with the country. There are no hard times. All have buckskin socks, containing more or less gold dust. There is no other kind of money.

"During June and the first days of July it was very hot, but under the moss, which is eight inches thick, solid ice is encountered. It has not been dark for over a month, and will not be until the last of September. It is possible to read any time during the twenty-four hours. The sun goes behind the mountains about 10.30 p. m. and comes up about 1 a. m. Old-timers say the winters are not so bad even if the thermometer goes down to 70 degrees below zero. There is no wind. All dress in fur clothing.

"I expect to work a claim on shares this week and will make plenty of money. No matter how big the stories are you hear of this place they are not big enough. I have received but one letter from home. It was forty-three days on the way."

Go to Work for Wages.

Two other letters from men who found it necessary to resort to day labor at the start are interesting reading.

Hart Humber, a young man who left Rossland, B. C., early last spring and arrived at Dawson City, Northwest Territory, on June 9th, over the Chilkoot Pass route, writes the following:

"Dawson City, N. W. T., June 18, 1897.—Friend Charlie: After leaving Dyea we had a trip full of hairbreadth escapes and arrived at Dawson City on June 9th.

"I will start to work to-morrow morning at $1.50 per hour. I will work with pick and shovel about three weeks, and will then have a better job with the same outfit and will get an ounce of gold per day ($17).

"There are at least fifty people going out on the boat to-morrow, who are taking out all the way from $10,000 to $100,000.

"This is undoubtedly the richest placer camp ever struck. The diggings are fifteen miles from Dawson. One Montana man took $96,000 out of forty-five square feet, another took $130,000 out of eighty-five square feet, and there are many more strikes equally as rich."

Klondike Will Kill Bryan.

Lewis W. Anderson, a Tacoma machinist, wrote this to his wife:

"I have been here a little more than two months and have already secured a quarter interest in a claim for which I have been offered $26,000, but out of which I expect to make as my part more than $100,000 in the next year. This for us, you know, is a big thing, and yet there are dozens of men who are making ten times as much.

"When I arrived my money had almost given out. I had only $31 left, so I worked ten days at sawing lumber at $15 per day to get a start. Nothing like this has ever been heard of in the world. Money, that is gold dust, is almost as plentiful as water. There are many hardships to be endured, but I expect to return to Tacoma next year safe and sound with lots of money.

" Tell Henry that we will have to change our politics, because the Klondike will kill Bryan and the silver question and the money power of Wall Street will try to demonetize gold. The gold that will come out of here inside of two or three years will make Wall Street more anxious to demonetize gold than it ever was to demonetize silver."

But in spite of this long list, at best only partial, of men and women who have "struck it rich," there is another side to the question, and fairness towards the reader demands it to have a showing. Let it speak for itself.

Hestwood Tells of Drawbacks.

J. O. Hestwood, who brought a small fortune with him to Seattle, in an article telegraphed from Seattle to the *New York World*, says :

" Modern or ancient history records nothing so rich in extent as the recent discoveries of gold on the tributaries of the Yukon River. The few millions of dollars recently turned into the banks and smelters of Seattle and San Francisco from the Klondike district is but a slight indication of what is to follow in the near future. When we consider the fact that there is scarcely a shovelful of soil in Alaska and the Northwest Territory that does not yield grains of gold in appreciable quantities, who can compute the value of the golden treasure that the great country will yield in the next few years?

" The Yukon River, which forms a great artery flowing through this frozen, rock-ribbed region for 2600 miles, seems to be a providential highway, opened up for the pioneer gold hunters and their followers, who are numbered by thousands yearly. There is room in that country for 100,000 miners for

100 years.   I do not make this statement from what some one
else has told me, or from what I have read.   I speak from
actual experience in that land of gold.   I have traveled over her
rivers of ice and mountains of snow in the springtime for three
years.

Perils of the Trail.

" Four years ago last May, when I first went into that country,
little was known of its wonderful possibilities.   With a heavy
outfit strapped to the backs of Indians, squaws and dogs, I
struggled over the trail from Dyea, on the southern coast of
Alaska, to Sheep camp, twelve miles distant, which was my first
camping place.

" The softening snow, under the sun's hot rays, rendered
traveling difficult, and it was a pitiable sight to watch the half-
starved, half-clothed Indians struggling along with their heavy
burdens on their backs, climbing the mountain side, frequently
breaking through drifted snow and being buried almost out of
sight ; wading icy streams, falling trom foot logs and enduring
hardships from which death would seem a welcome relief.

" The endurance of these Indians, or human beasts of burden,
was a constant surprise to me.   I remember one young buck
whose smallest load was 150 pounds.   His wife was a young
squaw, who, with seventy-five pounds strapped to her back and
a four-weeks-old child in her arms, struggled up the Chilkoot
Pass, where the declivity was so steep that we were compelled
to dig steps in the ice and snow in order to make the ascent.
One poor old Indian, I remember, had but half a dozen small
cawdle fish and one grouse to subsist on for three days.

" We were landed on the summit of Chilkoot Pass, 4100 feet
above the sea level, at Dyea, in the midst of a terrific snow storm,
such as takes place frequently in this pass in the spring of the
year, endangering the lives of many who attempt going over

it.   The blinding snow rendered it dangerous in the extreme to attempt the descent from the mountain toward Lake
Linderman, the headwaters of the Yukon River.   To make
matters worse, the clouds settled down on the mountain top,
and we dared not leave the camp for more than a few hundred
feet for fear we might lose our footing and be plunged over a
precipice or into some yawning chasm in the mountain.   A misstep meant death.

Among the Awful Glaciers.

" We took shovels and dug a hole in the ice and snow and
spread a tent over it, placing sacks of provisions on the tent to
weigh it down so the fierce wind would not carry it away.   Our
supper consisted of a cup of tea and a few crumbs of bread.
Great glaciers were sleeping all around us, but there was little
sleep for the weary travelers that night.   The glaciers, however,
seemed to be endowed with life and fits of wakefulness, for every
now and then we would hear a crackling sound, followed by a
noise as of crashing thunder, and 10,000 tons of sleeping giants
would be precipitated from the mountain heights and shattered
into icy diamonds to feed the roaring torrents in the chasm
below.

" Morning broke bright and clear.   There was no wood on
the mountain top, and we were compelled to chop up a sled for
fuel.   This was expensive.   We tried to breakfast on a pot of
half-cooked beans and a little coffee, which would freeze at the
slightest provocation.   Two sleds were then loaded with provisions and started down the mountain.   They went with a
velocity as if fired from a cannon until they struck the ice in
Crater Lake, three-quarters of a mile below.   After that every
foot of the ground we gained was by the most excruciating
labor a human being can be subjected to.

" Two weeks were consumed in reaching Lake Linderman,
8

~leven miles farther on.   Another week had passed before a boat was completed with which we could make our way down the river.   While in camp at Lake Linderman one of the party injured his knee, and three times a hunting knife had to be brought into requisition and incisions made.   Only after the most careful nursing was he able to proceed on the journey. Men are often taken with snow blindness in that country and lie helpless for days in their tents, unable to cook enough to sustain life.   If deserted by their companions in this condition their fate is sealed.

On to Forty Mile.

" From this point we encountered few difficulties in the way of river transportation until we reached Forty Mile, which is located where the 141st meridian crosses the Yukon.   Between Marsh Lake and Lake Lebarge there is sixty miles of river, in which occur the Grand Cañon and the White Horse Rapids.   Before reaching Grand Cañon the river is wide and smooth, when all at once the water is forced through the cañon at incredible speed. The cañon is a crevice where the mountain has been split in twain, apparently, to make an outlet for the water.   The walls are perpendicular on either side, rising to a height of 100 feet. Three miles below is the White Horse Rapids ; the most dangerous portion of the Yukon River.

" I simply mention these facts in order that any one who thinks of going into that country may know before hand that the search for gold there is preceded by hardships and privations which they little dream of unless they have penetrated the American land of the midnight sun.   But after the dangers are passed the adventurer finds himself in a country rich in mineral resources.

" Mark you, the country has yet given but a faint indication of its real wealth.   The gold that has been found only points the way to the true deposits, which will prove to be the wonder of the world."

John Welch, a former employe in an Indianapolis iron foundry, has written to his mother from Circle City, saying he has been in the Alaskan gold fields for fifteen months and could come home at any time with a few thousand dollars, but he prefers to remain a while longer and return rich.   He says that gold nuggets worth from twenty to fifty dollars are being found daily, but many men have become insane from hardships and from disappointment.   Successful miners are squandering fortunes in reckless extravagance.

Says Lucky Ones Are Few.

William Ireland has sent a letter from Alaska which ought to be a warning to men who are hastening to the field without due deliberation. He says:

"Undoubtedly it is true that some very rich discoveries have been made on the Klondike in the last year or so. I have been in the midst of the excitement and know that a large amount of gold has been taken out. As in California, a few lucky ones have made the killing.

"Of the 200 miners working near where I am located thirty-one are mine owners and the others laborers. I receive $10 a day, and I can work about 165 days during the year. The cost of living, I should say, would average about $2 per day per year, and at this price I enjoy none of the luxuries. I am on an equality with the rest of the workers, only three of whom receive higher wages.

"The mine-owners are making fortunes. Just how much money has been taken out can only be roughly guessed at, but it is certain that the placers here are exceedingly rich. Those who come from California, if they possess money enough, may succeed in making a strike, but I would not advise anyone to come up here without a sufficient supply of money to carry him over a year. There is plenty of country to prospect in, and the summers are delightful, so that for about five and a half months in the year a miner can work out of doors as well here as in California. Be sure and send a big supply of papers. If I were starting out again, I would carry at least one-third of my load in reading matter. Life in the long months of winter is unbearably dull without something to read."

Kills Himself on the Road.

There is a story of despair and death from the rush into Alaska gold fields. It comes from Lake Linderman on the Dyea route, and the victim was Frank Matthews, of Seattle.

Matthews and his partner, George Folsom, had safely crossed the divide, and were rafting their supplies along the lakes toward the Yukon. In the rapids between Lakes Linderman and Bennett the raft went to pieces, the supplies were scattered along the river, and Matthews was rescued after a severe injury to his leg. His partner placed him in a comfortable position and started back for help. Before going a hundred yards he heard the report of a rifle and was horrified to find Matthews shot dead. Undoubtedly he committed suicide.

Miss Mary E. Mellor, superintendent of the United States Indian Training School at Unalaska, who came on the Portland, July 17th, said the hardships in the Northwestern gold region are terrible. Summers are short, winters long and the supply of food and clothing inadequate.

"When I left flour was selling at the rate of $50 a sack, and if the luxury of eggs was indulged in, the consumers paid $4 per dozen. Then it must be remembered that each egg of the twelve was not what a Pennsylvania farmer would consider

PLACER MINING—HYDRAULIC SYSTEM

OLD BLOCK HOUSE BUILT IN 1805 FOR PROTECTION
AGAINST THE INDIANS

freshly laid. Clothing is also hard to obtain and is high in price, the majority of the gold seekers wearing clothes made of coarse woolen blankets."

Fred. Moss returned from Klondike to Great Falls, Mont., and said the upper Yukon was a country of starvation, outlawry and death. He had no story about how much he was worth and exhibited no dust.

J. D. Clements, of Seneca Falls, N. Y., told a story something like Moss'. He said he almost starved to death while prospecting. But he brought back $40,000 and said he would return to Klondike in the spring.

Mrs. Poppy Calls on Mrs. Gage.

Among the many women who called on Mrs. Eli Gage in Chicago before she started for Dawson City was a Mrs. Poppy, whose husband had spent fifteen years in Alaska. Mrs. Gage told her that if her husband had been long in the gold fields, he could probably give her more information than she could. According to Mrs. Poppy, the stories her husband tells indicate that there are some things in Alaska that are quite as valuable as gold, and his experience has demonstrated that some of them are really " worth their weight in gold." At one time when he was in the gold fields he had in his possession 300 ounces of virgin yellow metal, but not enough food to maintain the spark of life in a rabbit.

E. W. Egalbrecht, who went over Chilkoot Pass in February, wrote back from Dawson City in June, as follows:

"If I and many another had known anything about the hard-ships and exposures of this trip we would not have gone. It took me three days and half of the nights to reach Pleasant Camp with my outfit, and I will only add that when I slept at the foot of the cañon during the last night I awoke to find my camp six inches under water. All my clothes were soaked and my misery was indescribable. My feet especially suffered, because the skin had become very soft from perspiring in the rubber boots, and sore from walking, so that I suffered excruciating pain at times. I also suffered much from nausea, not being able to accustom myself to the food. The everlasting odor of bacon and beans that clung to everything took away my appetite. The poorest hut in civilization seems like a palace, but people never know when they are well off.

"I have worked hard all my life, but it is nothing compared to what one has to accomplish on a trip like this. Snow and ice all around wherever one looks, and one's face feels as though he was being whipped, but we had to push on if we did not want to perish.

"At Sheep Camp we found about 200 miners, mostly from the Mexico and Al-Ki, all of whom were unable to proceed to Stone House, owing to the stormy weather. However, the wind died out, and now began some climbing up a steep mountain trail, with 100 pounds on the sled, as much as the strongest man could pull, otherwise he would be dragged backward. I tell you one's limbs tremble with the horrible exertion. Such a trip takes from two to three hours, and we made three of them.

No Laughter in the Camps.

" We were allowed thirty minutes for a lunch of frozen beans and a pipe of tobacco, and then forward again. If after such a day's work you pass through a camp you hear no laughter, but see only pale, tired faces. Everthing is quiet, and you might kick their hands and they would not move out of your way.

"Fourteen hundred feet up a steep incline, step by step, with your feet firmly planted down and your pack on your back, you push on. If you slipped there would be no stop until you reached the bottom. In this way our journey continued for some time. We had many narrow escapes, and suffered severely from cold, but arrived eventually at our journey's end—Klondike, the land of promise and of gold."

Mrs. Julia Cook, of San Francisco, received the following letter, via the Portland, from her husband, at Dawson City:

"At last, at last, we reached here to-day. What we have lived through I will not trust to pen and paper; the many little crosses on the road here—they count up over a hundred—speak only too plainly of the innumerable dangers of this terrible journey. Let us rather pass over our experiences in silence, for surely we are fortunate to have reached here. Now we must get to work.

"The news of the gold strike, though I feared it might be, is not exaggerated. On the contrary, all the stories are surpassed by the facts. There are fellows here of doubtful calling who since last fall have gathered in over $100,000. Two brothers have over $150,000.

" We were in a great hurry to get here, and now learn that for a month work cannot be begun in the mines, although the roses and the most beautiful flowers are blooming. Still we can dig down but a few inches without striking ground frozen hard as rock. There is all kinds of work going on in this mushroom city, still there are plenty of idle men."

Hurley's Pay-Dirt Swept Away.

James Young, General Agent at Milwaukee for the Great Northern Railroad, received a Klondike nugget one day in August from James Hurley, a well-known mining promoter, who was active in operations on the Gogebic iron range during its palmy days.

Mr. Hurley has had an interesting experience in Alaska. Mr.

Young sold him a ticket to that region some months ago, and was surprised to hear from him.

Accompanying the package containing the piece of metal was a letter from Mr. Hurley which stated that he had not become very rich, although he had acquired more money in Alaska than he ever had before.

This is not Mr. Hurley's first experience in gold mining in Alaska. He went to that country with several friends as long ago as the 70's.

Most of the miners at that time were so poor they were compelled to wash the dirt as fast as possible, that they might get enough gold to exchange at the store for the necessaries of life.

Hurley and his companions had plenty of money, and they conceived and partly carried out the idea of digging out a pile of the pay dirt, building their cabin up against it and washing it out during the winter, alongside of the fire in the cabin.

By this plan they expected to keep themselves employed all winter, whereas by the ordinary method they would have to discontinue operations all through the long winter.

Just before the winter set in there was a big freshet that washed away the pile of pay dirt that they had been working all summer to secure.

They were nearly out of money and lost courage. They made their way back to their homes, and Hurley did not return until about a year ago.

Jerseymen Have Good Luck.

W. J. Hibbert, one of a party of seven from Trenton, N. J., who went to the Yukon late in 1896, grubstaked by some Philadelphia and Trenton merchants, has written back to his "angels" that the seven prospectors have laid claim to a large tract of rich dredger land, and that they will add to that area twenty-one placer claims.

He tells some big stories about the luck of the prospectors in that country. One man worked five days, at the end of which time he cleaned up $40,000. Another man who had worked industriously two months found at the end of that time that he was $150,000 ahead of the game.

J. R. Fitzgerald, of Springfield, O., wrote that a boat which he and his two companions had built was wrecked on the trip to Dawson City, and they lost everything they had; but he had some friends connected with the Alaska Commercial Company and went to work at ten dollars a day as soon as he got there. He said the most dangerous places are the cañon, White Horse Rapids, and Leads River, many people being drowned at those three places.

Fitzgerald said that reports as to the richness of the Klondike fields have not been exaggerated, and he knows of as high as $1000 worth of dust being taken out of a single pan, while some claims now pay as high as $12,000 to $15,000 a day.

The prospectors are locating new claims every day, which seem to be paying as well as the old. He said that miners frequently came down from the diggings loaded with sacks of dust weighing from 100 to 300 pounds. He said that one eastern young man sold his claim for $30,000 and died of heart disease just as he was about to board the steamer on the return trip.

Perish on the Glacier.

Few of the tales of hardship endured by gold seekers in the Arctic surpass in thrilling sadness the story of the deaths of Charles A. Blackstone, George Botcher and J. W. Malinque, expert miners from Seattle, who were killed on the glacier last April. The three men went north on the steamer Lakme in March, 1896. For a time they were at Cook's Inlet, and later they went to Circle City. They remained in the district until

March of this year, but fortune did not favor them, and March 25th they started back to Seattle, intending to go to Portage Bay, an arm of Prince William Sound.   March 27th they were seen on the glacier by a Mr. Gladhouse and by a Swede named Peterson.   They were never seen alive afterwards.

Before Blackstone left this city he asked a friend, George Hall, to look out for his wife and family should anything happen him. When word reached this city that the three men had left Circle City and had not made connections with the steamer at Portage Bay Hall went to Alaska to investigate.   He easily found traces of the men.   They had lost their way and had ascended that terrible mountain, coming out on the wrong side of the glacier. Mr. Hall found how Blackstone, Botcher and Malinque, after searching the top of this perpendicular cliff, had crawled under a ledge of ice.

Miners Frozen to Death.

The following statement was found on Blackstone's body:

"Saturday, April 4th 1897.—This is to certify that Botcher froze to death on Tuesday night.   J. W. Malinque died on Wednesday forenoon, being frozen so badly.   G. A. Blackstone had his ears, nose and four fingers on his right hand and two on his left hand frozen an inch back.   The storm drove us on before it. It overtook us within an hour of the summit and drove us before it.   It drove everything we had over the cliff except blankets and moose hide, which we all crawled under.   Supposed to have been 40 degrees below zero.   On Friday I started for Salt Water.   I don't know how I got there without outfit.   On Saturday afternoon I gathered up everything.   Have enough grub for ten days, providing bad weather does not set in.   Sport was blown over the cliff.   I think I can hear him howl once in awhile."

The bodies of Malinque and Botcher were never found.

H. Juneau, of Dodge City, Kansas, who was one of the

founders of the town of Juneau, had something to say of the dark side of life in Alaska, in these words:

"I have found the country full of disappointments, and I don't want to paint the picture too bright. Enough has not been said of the dark side.

"It is no place for men of weak constitution. The hardships to be encountered require the strongest hearts and sinews as well.

"I have seen nothing published of the fact that a large portion of the country is covered with a moss and vine which contains sharp thorns, like porcupine quills, with saw edges. These will penetrate leather boots, and when once in the flesh nothing but a knife will remove them. These are worse than the mosquito pest.

"Another thing which must not be overlooked is the total lack of law in the interior. When only Indians and a few prospectors were in the country there was little need of courts, but with the great influx of mixed humanity lawlessness is almost sure to break out.

"Alaska is a country on edge. It is so mountainous. Basins are mainly filled with ice. The weather is always hard in great extremes. Where there is no ice there is moss and devil's club, the latter a vine that winds around everything it can clutch. Persons walking become entwined in a network of moss and devil's club, and passage is extremely difficult and 'torturous' as well as tortuous."

Leave Good Claims for Better.

The opinion of Mrs. Eli Gage on the Klondike situation is interesting reading, for her opportunities to know have been exceptional. She says:

"There are many claims along the best known creeks that

have been abandoned. The prospectors would be digging on them contentedly, earning big money every day. There would then come a report from some neighboring place of fabulously rich finds, and there would follow at once a wild rush. In this way sites that paid moderately were passed in the search of others that would banish poverty in a month. The two kings of the region were wise enough to profit by the craze which carried

VERTICAL SECTION OF A QUARTZ MINE.

the men along, and they bought claim after claim along the Bonanza and the Eldorado. I do not think any man on earth can guess how much these men are worth to-day. They would be millionaires to stay at home the balance of their lives and sell interests in the mines they now have in operation.

"Experts say that the best mines are still to be found. It is an old saying that the existence of the placer mine merely shows that not far away the mother rock must be found. It

TOTEM POLE, FORT WRANGEL, ALASKA.

THOUSANDS OF SEALS—ST. PAUL ISLAND, ALASKA

looks as if the gold in the loose dirt about the creeks had been brought down from the mountains by some great glacier. The men who have gone in, and are going in, have no capital for machinery and the placer mining is the only kind they can undertake. The late comers and the men with money for machinery will probably search for quartz veins and get bigger fortunes with but comparatively small expenditures. It is reported by government officials and everybody else that the whole country is gold producing, and the work of 10,000 men who will be able to get there within the next twelve months will not begin to exhaust the resources.

Advice of a '49er.

No better words to close a chapter on the " luck " and experiences of the Klondike argonauts have been written than these from a '49er who " made his pile" before California was a State, and who still sympathizes with each one of the "thousand " gold seekers in the Arctic wilds who believes he is the " one " who is predestined to have fortune thrust upon him in the Yukon valley. He says, this snow-capped veteran of the early placers:

" It was this belief that encouraged the multitude of '49, and populated California with refugees from every quarter of the globe ; it was the same idea that sent the tide of a tumultuous humanity into the deserts of Nevada to hunt for silver; it was the same egotism that starved on Fraser River and shivered in the blizzards of Cariboo ; it was the same spirit that went up against the false hope of Panamint, and wandered helplessly across the hot sands of Lower California.

" So it will be this time ; so it has ever been from the going out of Ishmael ; and so it will ever be until men cease to care for gold—subduing the love of riches, which the wise man has said is the root of evil.

"Of course, the effort to deter these men from hazarding their lives and risking their fortunes in the Arctic is merely perfunctory. Even those who are advising that the wolf of Unalaska be permitted to howl undisturbed do not expect that the beast will long enjoy that privilege.

Survival of the Fittest.

"The weaklings may perish, as the advisory board of editors predicts, but the strength, the bone and sinew and the brawn of this movement will pull through, barring the accident that the litany refers to as ' battle, murder and sudden death.'

"These are of the stuff that builds commonwealths and perpetuates races of men. These are of the lineage that followed the Vikings; the ancestors of these conquered with William and crossed the storm-lashed Atlantic to subdue a wilderness and found an empire.

"These are the kind of men they want, whether they return from the Yukon burdened with wealth or as poor as they went. There's good leather in the stock that will come out of that frozen desolation, and it will work up into excellent material in a land where energy compels prosperity, and industry is rewarded with contentment.

"Suppose it is true that hardships must be endured in this quest? Are they any more disheartening than those which the poor man faces in the overcrowded cities ?

"Let it be conceded that the climate is rigorous. The winters of Minnesota are almost as severe, and the thermometer often registers as low in Quebec and the northern cities of Europe.

"The climate of Alaska may be deadly at certain seasons of the year if the inhabitant exposes himself to its clemency, but the mortality resulting from such foolishness will not, under the most favorable circumstances, equal the record of the recent "hot spell" in New York, Chicago, St. Louis and throughout the Middle West.

"As for starvation, there is less danger of that unhappy consummation in a mining camp than there is in the most opulent ' centre of civilization.'

Makes Light of Journey.

"The distance and the difficulty of reaching the mines of Alaska have been urged as an obstacle to be seriously considered by those who contemplate this adventure.

"As a matter of fact, it is a less arduous journey from New York to Dawson City than from Sandy Hook to Johannesburg. Steamers comfortably fitted are plying between San Francisco and St. Michael's, at the mouth of the Yukon, and thence to Klondike.

"The voyage is long, true, and somewhat expensive; but, aside from these natural consequences of a trip to the Arctic, there is no valid reason why anyone who wishes to go there should be discouraged.

"As for the tedium of the voyage, that can be endured in anticipation of the varied excitement that awaits the traveler at the end of his journey, and the expense that may attend the trip must be hopefully borne in the certainty of a manifold return when the industry and ability of the adventurer is put to the test in the land of the long twilight.

"The most encouraging information that has come out of the north with the homing millionaires is the assertion that a miner in Alaska does not need to know anything about mining. If all accounts are accurate, in fact, the less a man knows about 'formations,' 'strata,' 'deposits,' or 'dips, spurs, and angles,' the more likely he will be to strike it rich."

"It is the tenderfoot who finds the plethoric ' pockets' of the Klondike placers. As soon as he has been in the country long enough to think he knows all about it, his 'luck' forsakes him and it is time for him to come home. The 'tip' of a Freiberg expert on the Yukon isn't worth the icicles on his Vandyke Touting on the sixty-fourth degree of north latitude is not as absolute as it is at Ingleside.

"A great many people are encouraged to believe that the stories of hardships and privation in the diggings are exaggerated because several women have weathered an Arctic winter—some of them have lived for two and three years in Circle City and St. Michael's. But this is no criterion of a possible mildness of climate in that region.

"Last season a woman old enough to admit her age climbed Mount Shasta, and, within a thousand feet of the apex, was compelled to shame the young men of the party into renewed exertion by guying them on their lack of pluck and endurance. The circumstance that women can withstand the rigor of the Arctic is no evidence that a man would not succumb to it, for it is a physiological fact that women may display a more commendable fortitude under stress than her masculine congener.

CHAPTER IV.

How To Get There.

Main Routes to the Klondike—By Water and Land—Voyage via St. Michael's —Trip Up the Yukon—Choice of Trails via Juneau and Dyea—In by Chilkoot Pass—Over the Chilkat—The White Pass Route—Lieutenant Schwatka's Trail via Taku—By Way of Fort Wrangel and Lake Teslin —Railroads Suggested—The "Back Door" Route—Up the Copper River—By Moose Factory and Chesterfield Inlet—Other Trails—Telegraph and Telephone—Postal Service—Outfits for Miners—List of Necessaries.

THOUGH in a sense all roads lead to the Klondike, the gold-seeker does not become especially interested in a choice of routes until he reaches the Pacific seaboard. Then, whether he be at San Frascisco, Portland, Seattle, Tacoma or Victoria, the problem of "how to get there" becomes an engrossing one. Time, money and danger and the season of the year must all be considered, and the question is too often more perplexing than the unposted traveler can successfully grapple alone and hope to get the best solution. At the present time, in addition to the established routes, there are dozens of projected transportation schemes in the air, all possible to develop into untility on short notice. The wise argonaut, then, when settling upon his itinerary, will consult the latest sources of information—railroad and steamship literature and the folders and guides of land transportation concerns—and make up his mind accordingly.

Two Main Routes.

In a general way there are two main routes into the gold fields —the one entirely by water, via St. Michael's and the Yukon;

9   129

the other by water and land, via steamer to Fort Wrangel or Juneau, and then over the passes and down the rivers to Dawson City.

The former is only available during the "open" season, for the Yukon River, throughout the greater portion of its course, is closed by ice from September to May.   When the river is open, however, this route, though the longest in point of time and distance, has certain advantages, especially in the line of comforts, for it avoids the hazards of the mountain passes and the perils of the inland rapids, as well as the arduous labor of the portages as yet inseparable from the overland routes; and the traveler is reasonably sure of three "square" meals daily and a warm, dry bed at night.   To people who have money and reasonable leisure, and who are not used to roughing it, these are advantages not to be lightly foregone.

On The Overland.

The latter, the overland route, is shorter in time and distance, but more laborious, and, if the traveler has much of an outfit, and the "boom" prices for "packing" keep up, not less expensive than the water way.   It has the somewhat dubious advantage, as things are now, of being measurably "open" all the year round.   But to those who know what a mountain pass in Arctic weather means—rain, snow, hail, mud, ice, glaciers, fords, upsets, wrecks, perilous days of Sisyphean toil and deadly nights in sodden clothing on frosty beds—there will easily be apparent the dark side of the overland route.   By St. Michael's and the Yukon, the traveler will find most things done for him; by the mountain passes and the upper rivers he will have to do most things for himself and the "tenderfoot" is apt to find his troubles multiply as he presses forward, till only the most stalwart and the stoutest hearted will get through to the modern Ophir with heart or health to seek the fortunes hidden in the gravel.

There is still another overland route than those via Juneau, Dyea, or Wrangel. It is termed expressively the " back-door " route or " inside track," and is simply the old Hudson Bay trunk line to the North. It goes from Calgary, in Alberta, by railroad, stage or wagon, and cañon to Fort Macpherson at the mouth of the Mackenzie River, and then by the Peel River, leading south-ward to the gold fields.

The time via St. Michael's is from thirty-five to sixty days in the summer season ; via Juneau, Dyea or Wrangel, from sixty days upward according to the season ; by the "back door " route from sixty to ninety days.

Sailing to St. Michael's.

St. Michael's may be reached by the steamers of any of the great commercial companies from San Francisco or Seattle, though up to the present time the bulk of the transportation business has been in the hands of the North American and the Alaska companies, the old-time rivals for the trade of the Yukon country. The former owns the stores along the Yukon River, and has been a practical monopoly except where it has come in contact with the agents of the Alaska Commercial Company.

Dutch Harbor, in the Aleutian Archipelago, is the first port made on he outward trip to St. Michael's. Here the company owning the sealing privilege on the Pribyloff Islands has a coal-ing and supply station. It is 1800 miles on the way to the gold fields. Then away to the north, 800 miles through Behring Sea and past the seal islands to St. Michael's. The journey has so far been a pleasant one, unless the weather has been stormy. The one great peril of this route lies in that portion of the sea known as " the Boneyard of the Pacific," from the vast number of ships which have gone down beneath its treacherous surface, and which is still one of the most dangerous spots known to northern navigators. This once passed, the other hazards of the long voyage can happily be made light of.

On St. Michael's Island.

St. Michael's, on the island of the same name, near the mouth of the Yukon, used to be a Russian fortification, and some of the old Russian buildings are still standing; but for many years it has been the transfer and forwarding point for all goods going into or coming out of the interior. Both the commercial companies doing business on the river have warehouses here. During the two or three months of open navigation it is a place of considerable activity. Then communication is cut off, and it goes into the long, uneventful night of winter.

The inhabitants of St. Michael's are the white resident employes of the companies, the collector of customs, several missionaries, and a number of traders. There are several hundred Eskimos on the island. The surface of the country immediately surrounding St. Michael's is gently rolling, and in summer it is covered with a great growth of grass, having more the appearance of Nebraska prairies than of an Arctic region. A series of six or seven low, cone-shaped hills across the shallow estuary are extinct volcanoes. In all the landscape there is no timber, nor are there trees anywhere near Behring Sea.

At St. Michael's passengers and freight are transferred from the ocean liners to the river steamers. These run down the coast sixty miles to the north mouth of the great Yukon, a river larger than the Mississippi and navigable for boats of light draught for 2300 miles above its mouth, and there begins the long journey up stream to Dawson City and the golden placers.

The source of the Yukon is in the Rocky Mountains and in British

PROSPECTORS IN CAMP

SAW MILL NEAR THE HEAD OF LAKE BENNETT

territory, at a point northeast of Sitka. The river drains practically the same territory in its headwaters as the Stickine, Peace, Columbia and Frazer rivers, all well known for many years to treasure-hunters because of the great placers in their valleys. It was natural, therefore, to expect that gold would be found along the main channel of the Yukon or some of its tributaries. Explorers were sent out from two bases. One set went up the river from its mouth, traversing the whole of Alaska from the west to east.

Fine gold dust, in small quantities, was found at the mouth of the Porcupine River, a stream that joins the Yukon about 100 miles west of the boundary, and also near the mouth of Forty-Mile Creek, most of whose course lies in Alaska, but which crosses into British territory before emptying into the big river. Fort Cudahy is situated here, and Circle City, where there were other mining camps, is about fifty miles further west. These places are about 800 or 900 miles from the sea, if one travels by steamboat, and in the winter are completely cut off from the outer world. The discoveries above the Porcupine are the cause of the present rush of gold hunters—they are the richest placers in the world.

Stop at Fort Yukon.

The first point of more than passing importance on the journey up the river is Fort Yukon, a misnomer as to the " Fort," as is the case with all the stations on the lower river. As stations in the wilderness, most trading posts were fortified after a fashion in the early days, and this custom led to dignifying them by the term " fort." Fort Yukon was established by Robert Bell as a post of the Hudson Bay Company, he assuming that it was in Canadian territory. He made a mistake of 300 miles, measured by the river. Hudson Bay Company held the post until it was warned away by an American officer.

Here the argonaut finds himself fairly under the Arctic circle. In June and July he will see the sun twenty-four hours without a break, and all along the river at this time he can read a paper at any time of day or night without a lamp.

Above Fort Yukon is the once important town of Circle City, formerly a mail station and a thriving post, but now practically depopulated by the stampede to the Klondike gold fields, higher up the stream. Circle City stands on a dead-level plain, twenty feet higher than the river at the ordinary stage of water. In the distant background is a low range of purple hills, which marks the dividing line between the Birch Creek district and the river. On the opposite side from the town the river runs away into space, with no very well defined shore line.

It is a town of log huts, square and low, with wide projecting eaves and dirt roofs. Two men would get out the logs, build the cabin and " chink " it with the abundant moss in two weeks ; and before the Klondike fever such a house would rent for fifteen dollars a month (in gold dust) or sell for $500. But the inhabitants have fled and most of the cabins are empty. From the present outlook hardly a dozen white persons, and perhaps a dozen Indians, will be left in the town during the coming winter. In April it had 1500 white residents. It also had dogs, unlimited quantities of them, worse pests than mosquitos, but the call for dogs in " packing " miners' outfits over the southeastern passes materially reduced the supply. A good dog is worth $100 in dust in Circle City.

Gold on Birch Creek Claim.

The rich discoveries of gold on Mammoth and Mastodon Creeks and many gulches which terminate in these creeks all tributaries of Birch Creek, " just over the divide," gave Circle City its first boom. Many wise men among the miners prophesy that when the surrounding country is carefully prospected, its diggings will be found equal to the Klondike, and Circle City will again become a formidable rival of Dawson City.

At Forty-Mile, or Fort Cudahy, across the boundary line in the British territory, the next important stop, some gold was found by the expedition mentioned heretofore. This place was named for John Cudahy, of Chicago, of the North American Transportation and Trading Company, and was for years the company's headquarters on the upper river. It contains about 200 log cabins of the prevailing Yukon style—square, low, flat, and dirt-roofed—the companies offices, a few stores and saloons, and a hotel or two. Whiskey is worth ten dollars a quart, or fifty cents a drink, and half a dollar will buy three loaves of Yukon bread.

Arrive At Dawson City.

Passing Fort Reliance, the next stop is Dawson City, the metropolis of the gold fields, the Mecca of the '97er, the threshold of the Klondike treasure house. This new town and trading post, though barely six months old, is already the busiest town on the river. " Old Joe " Ladue, as he is locally and unappropriately named, for he is not old at all, the owner of the town site, was being kept busy selling town lots at $5000 each when he made up his mind last summer to run back to New York and claim for his bride the sweetheart who had been waiting for him to " make a stake " under the Midnight Sun.

There were said to be 3000 people in Dawson City in July and that number has been greatly increased since by the influx of men with the gold fever who had had prescribed " Klondike refrigeration " as a remedy for the almost hopeless malady, Dawson City will probably have to winter 12,000 to 16,000 people, and there has been general fear that there would be great suffering there this winter in consequence of lack of supplies and shelter for the great rush of unprepared prospectors. And winter at Dawson City begins in September. However, strenuous effort was made up to the last moment by the commercial companies to get in provisions against a possible famine, and as many of the later argonauts carried in fairly good and liberal outfits, it is hoped the long season of cold may pass without general disaster.

A miner who came in on one of the late steamers, described Dawson City as wild with speculation. He said:

"Speculation is already the ruling idea. A purchaser inspects a claim that he thinks he would like to buy. He offers just what he thinks it is worth. There is no skirmishing over figures; the owner accepts or refuses, and that is the end of it. With this claim goes the season's work. By that I mean the great pile of earth that may contain thousands or may not be worth the expense necessary to run it through the sluice. That is a chance one must take, however, and few have lost anything by it this season.

"It may be said with absolute truth that Dawson City is one of the most moral towns of its kind in the world. There is little or no quarreling, and no brawls of any kind, though there is considerable drinking and gambling. Every man carries a pistol if he wishes to, yet few do, and it is a rare occurrence when one is displayed.

Around The Gaming Table.

"The principal sport with the mining men is found around the gambling table. There they gather after nightfall and play until late hours in the morning. They have some big games, too, it sometimes costing as much as fifty dollars to draw a card. A game of $2000 as the stakes is an ordinary event. But with all that, there has not been any decided trouble. If a man is fussy

AMUSEMENTS IN DAWSON CITY.

137

and quarrelsome, he is quietly told to get out of the game, and
that is the end of it.

"Many people have an idea that Dawson City is completely
isolated, and can communicate with the outside world only once
every twelve months. That is a mistake. Circle City, only a few
miles away, has a mail once each month, and there we have our
mail addressed. It is true, the cost is pretty high—a dollar a letter and two dollars for paper—yet by that expenditure of money
we are able to keep in direct communication with our friends on
the outside.

[The Canadian authorities have since established a post-office
at Dawson City, with regular service.—ED.]

In the way of public institutions, our camp is at present without any, but by the next season we will have a church, a music
hall, school-house and hospital. This last institution will be
under the direct control of the Sisters of Mercy, who have
already been stationed for a long time at Circle City and Forty-
Mile Camp."

Mines Not At Dawson.

The general impression that the mines are at Dawson City is
erroneous. They are twelve to fifteen miles up the Klondike
River, and are easily reached by poling up the stream in summer
or sledding over its frozen surface in winter.

Dawson City is under the British Government, and its laws
are enforced by the famous mounted police.

Inspector Strickland, of the Canadian mounted police, who
came down from Alaska on the Portland, said :

"When I left Dawson City there were 800 claims staked out.
We can safely say that there was about $1,500,000 in gold
mined last winter. The wages in the mines were fifteen dollars
a day, and the saw mill paid laborers ten dollars a day.

"The claims now staked out will afford employment to about

5000 men, I believe. If a man is strong, healthy and wants work he can find employment at good wages. Several men worked on an interest, or what is termed a " lay," and during the winter realized $5000 to $10,000 each. The mines are from thirty-five to 100 miles from the Alaska boundary."

Inspector Strickland paid the miners at Dawson City a compliment, saying "they do not act like people who have suddenly jumped from poverty to comparative wealth. They are very level headed. They go to the best hotels and live on the fat of the land, but they do not throw money away, and no one starts in to paint the town red."

Price List at Dawson.

He gave the following price list as a sample of the cost of living in Dawson City : Flour, $12 per hundredweight. Following are prices per pound: Moose ham, $1 ; caribou meat, 65 cents ; beans, 10 cents ; rice, 25 cents ; sugar, 25 cents ; bacon, 40 cents ; potatoes, 25 cents ; turnips, 15 cents ; coffee, 50 cents ; dried fruits, 35 cents ; tea, $1 ; tobacco, $1.50 ; butter, a roll, $1.50 ; eggs, a dozen, $1.50 ; salmon, each, $1 to $1.50 ; canned fruits, 50 cents ; canned meats, 75 cents ; liquors, per drink, 50 cents ; shovels, $2.50 ; picks, $5 ; coal oil, per gallon, $1 ; overalls, $1.50 ; underwear, per suit, $5 to $7.50 ; shoes, $5 ; rubber boots, $10 to $15.

The latest reports are that these figures are still maintained, despite the great amount of supplies brought in by the commercial companies, and it is expected they will go higher rather than lower before spring comes around again.

Whisky is fifty cents a drink, and some of the saloons are said to be making $6000 to $8000 a day. There is some gambling, though not of a bloodthirsty kind, and chips are commonly $500 a " stack."

Should the argonaut decide to go in by the Juneau and Dyea, or "mountain" routes, he will find the trail by Chilkoot Pass the one most talked of, and will probably this fall decide to try his fortunes by that way, though the spring and perhaps the winter even may find the Chilkat, the Taku and the White Pass routes, or even the Lake Teslin trail, becoming favorites.

Right here the gold-hunter, having fixed on his route, needs to make very sure of one other thing—his "outfit." When he leaves Dyea or Juncau he leaves civilization and all its adjuncts of stores and traders behind him. From Dyea to Dawson he must depend on his outfit for practically everything he has to eat, drink and wear and for every tool and appliance with which to build or repair any article needed for the long journey by trail and stream, 700 miles, to Dawson.

Via Chilkoot Pass.

If the "outfit" is all right, the prospector engages Indians at Dyea to pack his goods in a dugout and tow them to the head of canoe navigation on the Dyea River which is about six miles. If possible the Indians should be hired to pack the goods over the Chilkoot Pass to Lake Linderman, about twenty-two or twenty-three miles. The old rate for this work was from five to sixteen cents a pound, but the great stampede of prospectors has caused the price to rise to twenty-one and even twenty-two cents, and even at that almost prohibitive figure it is often impossible for prospectors to hire native carriers, and as a result they have to pack their outfits over themselves. A Chilkoot Indian will carry from 250 to 300 pounds over the pass, but even the strongest white man can "tote" little more than 100 pounds, and consequently when the Indians fail him, has to make "double trips," that is, take a pack a mile or two, caché it and return for another one, and keep this tedious and heart burning labor up

GROUP OF INDIAN CHILDREN

GROTESQUE ART OF ALASKA INDIANS

until the last article has been wearily dropped on the shores of Lake Linderman.

Many pack horses have been taken to Dyea for use on the Chilkoot Pass trail, and dogs are also to be experimented with this winter in hauling supplies.

From the head of canoe navigation a well-defined trail leads to the cañon at the summit. The first day's camp is made at the entrance to the cañon ; the next day's camp is well along in that formidable pass at a natural curiosity known as the " Stone House," a much frequented camping ground for packers. The place affords good shelter in stormy weather and, as it is very frequently impossible to cross the Divide on stormy days, packers have here a good place to wait for fair weather before attempting the fearful toil of the ascent.

An early start is necessary in crossing the Divide, the great Peraier Glacier, for it is urgent that the march should be made in one day in order to camp three or four miles beyond the Divide, where there are sticks and moss for a fire.

Passing the Divide.

Dr. E. O. Crewe describes the " passing " in these graphic words :

" Having arrived at the foot of the now almost perpendicular mountain of ice and half thawed snow, we struggle upwards, sometimes up to our knees in slush, sometimes clinging with hands and feet to the slippery mountain. Zigzagging from one side to the other until about half way up the ascent we drop our packs and survey the remainder of our journey up the glacier. On our left hand further progress is impossible ; a perpendicular wall of deep blue ice towers up a thousand feet above the actual pass ; on our right, we notice a pile of broken rocks that have crumbled from the cliff that forms the right hand side of the cañon. Towards these rocks we slowly pick our way, over which we slowly wend towards the base of the the cliff, and, having gained this comparatively comfortable foothold, our progress is quite easy and fairly rapid.   Ever keeping along the base of the cliff, ever getting nearer the crest of the ridge, we have little difficulty in managing our somewhat bulky pack, and almost before we are aware of it we have crossed the Divide and are over the most laborious part of our journey.

Off For Lake Linderman.

" Of course, if more than one trip is necessary the assent will consume much more time.   One should easily make the journey from Dyea to Lake Linderman in three days with an ordinary pack if ' double tripping ' is unnecessary.   After resting awhile on the summit of Chilkoot Pass, admiring the magnificent grandeur of the scene we begin our decent to the lake ; turning a little towards the left after coming over the divide we follow the trend of the hills which lead us down towards the North and we are very soon able to see Crater Lake (the actual source of the Yukon).   Skirting the right hand shore of this lake, we soon find ourselves in a well defined ravine, with a well worn trail running down the right hand side of the little stream that finds its way from Crater Lake and empties into Lake Linderman.   As soon as we find a convenient place to pitch our tent, we make ready for camping, and thoroughly enjoy a hearty meal followed by a well-earned refreshing sleep.   The following morning, as early as possible, we break camp and start with our pack toward Lake Linderman.   A few hours of easy walking will bring us to the lake, where we must at once break camp and prepare to go the balance of the way by water."

The next thing, after getting safely over the pass, is to build a boat.   Four men who are handy with tools can take a standing spruce, saw out lumber and build a boat large enough to carry them and their 4000 pounds of provisions all in a week. It should be a good, staunch boat, for there are storms to be encountered on the lakes, and rapids, moreover, that would shake a frail craft to pieces. The boat should have a sail that could be raised and lowered conveniently.

Some enterprising men have built a saw mill on the shores of Lake Linderman, and sell boats or lumber. A boat large enough for four men and their outfits costs $75. Lumber is worth $100 a 1000 feet, and 500 feet is enough for a boat.

From the end of navigation on Lake Linderman a trail leads over to Lake Bennett, making a portage of a mile and a half. There is a river between the lakes, but the rapids are so dangerous none but the most fool-hardy attempt to run them, and many lives and a great amount of property have been lost in the reckless ventures. Some gold-hunters who go in by Chilkoot Pass make a raft at Lake Linderman, sail it down to the portage and abandon it there, and carry their goods to Lake Bennett, where there is excellent timber for boat building.

Down Lake Bennett.

With boat built one starts from the head of Lake Bennett on the last stage of the trip—a sail of 600 miles down stream (not counting lakes) to Dawson City, at the mouth of the Klondike. With fair weather, at the evening of the second day, one reaches Miles cañon, the beginning of the worst piece of water on the trip. The voyager has passed through Lake Bennett and Takish and Marsh lakes. At the head of Miles cañon begins three miles of indiscribably rough water, which terminate in White Horse Rapids.

During the rush of gold-hunters it is probable there will be men at Miles cañon who will make a business of taking boats through the rapids, and unless one is an experienced river man it is economy to pay a few dollars for such service, rather than to take the greater chances of losing an outfit or even a life, for many have been drowned at this passage. Probably ten per cent. of the men who attempt the rapids are drowned.

Even lowering an empty boat through the rapids, with a rope fastened to each end of it, very often results in the loss of the boat, which is at this point of our journey exceedingly valuable.

In Miles Canon.

Miles Cañon, which is also called Grand Cañon, is the first dangerous water that the navigator encounters. Although this section of the river has a normal width of more than 200 yards, it is confined for a distance of three-quarters of a mile to a space hardly fifty feet across, with perpendicular walls of red volcanic rock. This cañon is broken in one place—about midway—by a circular enlargement of the channel, which causes a whirlpool of wonderful suction on each side of the river.

After the rapids comes Lake LaBarge, a beautiful sheet of water thirty-five miles long, and in this connection a suggestion is desirable. Near the foot of the lake, on the left side, is a creek coming in which marks a good game country. A year ago, and in previous seasons, moose, were plentiful there and in the rugged mountains near the head of the lake there always have been good hunting grounds for mountain sheep. A delay of a week either in this locality or almost any of the small streams that flow into the succeeding 200 miles of river, for the purpose of laying in a good supply of fresh meat, is worth considering. Moose meat that can be preserved until cold weather sets in will sell for a fancy price.

There is another suggestion to consider before arriving at Sixty-mile. All along that part of the river are many timbered islands, covered with tall, straight spruce. With such an influx of prospectors as is expected at Dawson City before winter begins, building logs will be in great demand. Cabin logs ten inches in diameter and twenty feet long sold at Circle City last year, in raft, at three dollars each. With an increased demand, and with better mines, the prices at Dawson City may be much higher. Four men can handle easily a raft of 500 or 600 such logs. Getting them out would be a matter of only a week or two.

From Lake LaBarge the journey is through Thirty-mile River, the Lewis River, 150 miles to Five-Finger Rapids, thence to the Yukon at Fort Selkirk and then down stream 250 miles to Dawson City.

Gold in Hootalinqua.

Within a few hours' run below Lake LaBarge is the Hootalinqua River, which drains Teslin Lake, the largest body of water in the Yukon basin. This river has long been a locality of great interest to prospectors because of the wide distribution of gold in its bars and tributaries. The metal is found everywhere on the whole length of the stream, but seems rather elusive when it comes to the test of actual mining. It has been prospected and worked sporadically for fifteen years, and in all that time the only Hootalinqua gold of any consequence taken out was found on Lewis River, a few miles below the mouth of the former stream, at Sassiar bar, where something like $150,000 was mined. It is deserted now for the better mines of the Alaskan side.

Five-Finger Rapids is one of the two or three obstructions that interfere with the free navigation of the river. A ledge of rock lies directly across the stream with four or five openings in it that afford a scanty outlet for the congested current. The largest passage and the one commonly used is the one at the

10

right shore. There is a considerable fall, but the water is not badly broken, the gateway being succeeded by several big waves, over which a boat glides with great rapidity, but with a smooth and even motion. Shooting this rapid is an exhilerating experience, but with careful management is not considered dangerous.

A few miles above Five-Finger Rapids is George McCormick's old Indian trading-post. This is now abandoned by the "venerable" George; he was the first man on the Klondike. A mile or so beyond McCormick's trading-post, (which by the way is very poorly stocked with anything, except Indian trading articles), on the right-hand side of the river, before turning to the Five-Finger Rapids, you see evidence of McCormick's shrewdness and enterprise. He has drifted a hole in the side of the mountain, and when prospectors last passed this point he was taking out good specimens of coal.

Next below Five-Finger Rapids are the Rink Rapids, so named by Lieutenant Schwatka, because of their musical rhythm. To run the Rink is mere child's play.

And now all the danger points in the Chilkoot Pass route are passed. It is clear sailing to Dawson City.

Past Fort Selkirk.

The first trading-post and settlement of white men to be encountered on the river is at Fort Selkirk, opposite the mouth of Pelly River. Thence, it is a little more than a day's run down to Sixty-Mile, and it takes less than a day to go from Sixty-Mile to Dawson City.

Dr. Crewe says of Pelly River:

"We will just run across the river and see how old man Harper is getting along at Fort Selkirk. He has been in the Yukon Valley, trading first with the Indians and then with the white men, ever since the Alaska Commercial Company estab-

lished trading-posts along the river. Before this time, I believe
he was employed by the Hudson Bay Company as a post-trader
at one of the northern stations. Wishing good-bye to our
Selkirk friends, a quick uneventful run of 120 miles brings us
to Stewart River. Gold was first discovered in the Yukon
Valley on this river. The prospects for the future of Stewart
River are as bright and hopeful as for any of the creeks that
are known to contain gold."

Colorado Miner's View.

The words of a Colorado miner, who went in by the Chilkoot Pass in the early summer and wrote back of his experiences, are worth reading as a practical man's summing up of the
case. He says:

" I think that the difficulties and dangers of the Yukon trip
have been much exaggerated. The cold up there is intense, but
is dry and a man does not suffer from it as would be supposed.
I spent one winter on the Yukon. The thermometer went down
to seventy-five degrees below zero, but the coldest day I ever
saw in my life was in Chicago last January.

" The Chilkoot Pass is only 3000 feet high, and that isn't any
height at all to a man used to mountains. With a good sleeping
bag a man may sleep out of doors there all of the winter. In
the interior there is very little snow. I did not find it over six
inches deep. In the dark part of the year there is almost always
enough of twilight to see by.

" Of course, a man who would kick about a crumpled rose
leaf on his couch would have a hard time in Alaska, but a man
who is a man could get along all right up there."

A company has been formed in Chicago which proposes to
build four or six small steamers of light draft which will be
launched in Lake Linderman, and will run in the chain of lakes,

the Lewis River and the upper Yukon River. The same company will build tramways, after the pattern of those in use by the Hudson Bay Company over the old route from the North, to overcome the difficulty of transportation at portage points.

The boats will go to their destination in parts, and will be put together on the waters of Lake Linderman. They will be provided with all the comforts that make steamboat traveling enjoyable, and will be of sufficent tonnage to carry a considerable amount of freight on each trip.

With the proposed wagon road that the Dominion Government and the Canadian Pacific Railway are figuring on, it is thought there will be little trouble in reaching the gold fields, and those who are caught on the Klondike when the lakes and rivers are frozen over can get out by way of the northern route, which is through Edmonton.

Over Chilkat Pass.

The Chilkat trail leads over the Chilkat Pass and is about 125 miles in length from the head of Chilkat Inlet to where it strikes the waters of Tahkeena River. This was the old trail used by the Indians to and from the interior, and leads all the way through to old Fort Selkirk by land. "Jack" Dalton has used this trail at times in taking horses and live stock to the mines, portaging to the Tahkeena, then by raft down that river to the Lewis, thus proving that the Tahkeena is navigable for a small stern wheel steamer for a distance of some seventy miles.

For the last three years several California and English companies have been studying the lay of the land between Chilkat and Circle City, with a view to establishing a quicker and more practicable way of transportation to the rich gold fields along the Yukon. Goodall, Perkins & Co. have made a thorough investigation of the matter. Captain Charles M. Goodall said:

SCENES ALONG THE ROAD TO THE KLONDIKE: BOAT-BUILDING AT LAKE LINDERMAN.

WEIGHING STATION AT DYEA.

It is at this point that the Indian packers weigh all the supplies that they pack over Chilkoot Pass.

"The rich find in the Klondike district will probably result in some better means of transportation, though the roughness of the country and the limited open season will not justify anybody in building a railroad for any distance. Recently we sent several hundred sheep and cattle to Juneau, and from there to the head of navigation by the steamer Alki. Dalton, the man who discovered the trail across the country from Chilkat River to Fort Selkirk, is taking the live stock to the mines. His route lies from the head of navigation through Chilkat Pass and across a route which is over a prairie several miles to the Yukon River, near Fort Selkirk. At this time of year the prairie is clear, and bunch grass grows on it in abundance.

"I believe this will ultimately be the popular route. People could go over it in wagons, as the prairie is level and the roads good. Stations could be established, as was done on the plains in 1849. It would be easy to go down the river in boats from where Dalton's trail strikes it to Dawson City and the other mining camps.

"The plan to build a traction road over Chilkat Pass from Dyea, the head of navigation after leaving Juneau, to Lake Linderman, is not a good business proposition. It has been talked of and the rest of the plan is to have steamers to ply from Lake Linderman through the other lakes to the Yukon. But to do this two portages would have to be made on account of the falls in the river, and these would be enormously expensive."

By the White Pass.

The White Pass is considered by many one of the best that cuts the mountains of the coast. It is at least 1000 feet lower than the Chilkoot and little higher than the Taku. It is reported timbered the entire length. Its salt water terminus is about eighty-five miles north of Juneau, and ocean steamers can run up to the landing at all times, where there is a good town site, well protected from storms. The pass lies through a box cañon surrounded by high granite peaks and is comparatively easy. The first seven miles from salt water lie up the bottom lands of the Skagway River through heavy timber. Then for about seven miles farther the way is over piles of boulders and moraines which would prove the most expensive part of the trail. This trail would not exceed thirty-two miles in length, and would strike Windy Arm of Tagish Lake or Taku Arm coming in farther up the lake. All of this part of the lake is well timbered and accessible to Lake Bennett and its connections. White Pass could be used as a mail route any month in the year.

Trail Open July 16th.

The Alaska *Searchlight* publishes a letter from William Moore, at Fourteen-Mile Camp, Skagway, Alaska, stating that the White Pass pack trail to the summit of the pass was opened for travel July 16th. On reaching the summit the traveler steps upon almost level country, the grade to the lakes being twenty feet to the mile. The distance from salt water to the Too-Chi Lake is thirty miles, and from salt water to the head of Lake Bennett, the distance is forty-five miles. Both routes from the summit are through rolling country, for the most part open, with plenty of grass for feeding stock, water and sufficient timber for all purposes. From salt water to the summit, stock and pack horses can be driven through easily.

C. H. Wilkinson, on behalf of the British-Yukon Company, has made an offer to the Minister of the Interior to build a wagon road through the White Pass for $2000 a mile. The distance is about fifty miles. About eight miles of the road would be very difficult to build. It would take $7000 a mile, being all rock excavation, to construct this eight miles.

At the rate the people are flocking into the new gold region of the Yukon country, something will have to be done soon to provide a way of getting provisions into the mining district.

If this road were built Victoria could be reached from the Yukon district in about fourteen days.   The Minister has taken the matter into consideration.

Mr. Wilkinson is also authority for the statement that the company has completed arrangements for placing a fleet of between ten and twenty steamers on the Yukon River in the spring, and will probably make an effort in the direction of a narrow guage railway over the pass.

Survey for Railroad.

George W. Garside, a well-known engineer, formerly in the employ of the Canadian Government, has recently completed the survey of sixty-two miles of railway running from Skagway Bay over the White Pass to Lake Tagish, and thence to the upper Hootalinqua River.   He is employed by the British-American Transportation Company, said to be amply supplied with funds with which to complete the undertaking.   It is said work will begin in the spring of 1898.   The new route will be 100 miles longer than that at present followed by miners going into the Yukon basin overland from Dyea.

The route surveyed leaves tide water at Skagway Bay, close to Dyea, and runs in a northerly direction over the summit by White's Pass, through which a trail has just been completed. The new trail is 1000 feet lower than Chilkat Pass, at which so much hardship is encountered by prospectors.   The route will eliminate all the danger of the White Horse Rapids and Miles Cañon, where now portages of from one to three miles are made, and where so many gold hunters have lost their all, in having their supplies turned out of the boat into the water by the bowlders

The report of the engineers on the project has been filed.
It endorses the plan as practical but costly. Skagway Bay has
a fine natural harbor, and is good anchorage for vessels of any
size. From the harbor the proposed railroad will follow the
Skagway River to its head, which is near the summit of the pass.

The grade is variable. The first four miles the ascent is
gradual. The next seven miles of the route is difficult and even
dangerous. In three more miles of easier grade the summit is
reached. The descent to Tagish Lake, about twenty miles, is
gradual and the total fall less than 400 feet. The surface of the
lake is 2200 feet above the sea.

Route by Taku Pass.

A new route to the Klondike (and it must not be forgotten
that " Klondike," as a destination, means anywhere in the great
gold-lined Yukon Basin) has been proposed by Mrs. Frederick
Schwatka, the widow of the great Northwestern Pathfinder. It
is by way of the Taku Inlet, River and Pass. Lieutenant
Schwatka discovered the pass and tried it.

According to Mrs. Schwatka, who has spent much of her
time in Alaska and who is familiar with a large part of the
country, the Taku Pass will prove to be a bonanza to the first
trading company that establishes a system of pack trains through
it to Juneau, the base of supplies for the mining region. It is
besides the easiest route for the miners themselves and a shal-
low-draft steamer that could be brought to run on Taku River
would leave only ninety miles of land to be crossed.

Mrs. Schwatka spoke of the Taku route in these words :

" Lieutenant Schwatka explored the Taku River and Pass
several years ago. He tried to get the people of Juneau to establish a pack train line through the pass, to connect with a
steamboat on the inlet. That was before there was much travel

to Juneau, and the people of the thriving village did not believe it would pay them. Now it certainly would, but I have not seen a word about the pass in any of the newspapers, and there appears to be almost no travel through it.

"In fact, the pass contains an excellent railroad grade, and it would cost a comparatively small sum to build and equip a road. The current of the river is strong and there are frequent floods, but a light draught steamer would have no difficulty in ascending it and making connections with the road to Juneau. It would be an easy matter to get supplies from Juneau then. The Canadian Pacific comes so near to that country it seems as if it could profitably build a line through the pass and connect the two branches by steamer.

"Lieutenant Schwatka made a map of the region, which I think I shall have published. He made the trip up the river by canoe and reported the current there very swift and strong. I am certain that the Taku route is the easiest for persons going from Juneau, however.

"From Taku to Lake Teslin it is ninety miles over level prairies, and the country from Lake Teslin is an open valley. With the aid of pack horses the Taku route is by far preferable."

Details of the Route.

The Taku Pass route may be briefly described as beginning at Juneau, thence up the Taku River to its end, where the portage of ninety miles is made by pack to the Teslin or Aklene Lake, the route through which is northwesterly. Arriving at the farther end of the last mentioned waterway the trip is by heavy canoes along the Hootalinqua or Teslin River to Lewis River, which joins the Yukon at Fort Selkirk. From the latter place Dawson City and other mining places are reached by the Yukon.

William A. Pratt, professor of electrical engineering at Delaware College, and P. I. Packard, of Wilmington, Del., are at the head of a party enroute to survey a line for a railroad to be built by an Eastern syndicate through Taku Pass to Lake Teslin.

Another route, whose promoters say is the best highway to

SCENE IN ALASKA NEAR THE COAST.

the gold fields from the coast yet discovered, is by way of the Lake Teslin, or Aklena Lake trail, and starts in American territory at Fort Wrangel. It leads up the Stickine River and Telegraph Creek from Wrangel to Glenora, a distance of 126 miles. The Stickine is navigable for stern-wheel steamers of four or five feet draught, and it is believed the channel of Telegraph Creek can easily be made ample for the same boats all the way to Glenora. The provincial government is at work improving the route.

The only point of peril in the water part of this route will be in the rapids in the Stickine River, but the trouble here is handily overcome at present by making fast heavy lines to trees on the banks and warping the boat up or down the dangerous passages.

From Glenora the route will traverse a newly-discovered pass and then straight across the smooth table land to Lake Teslin. Thence it is plain sailing down the Hootalinqua River, a tributary of the Lewis, by the Lewis to Fort Selkirk and thence on the broad Yukon to Dawson City.

Five-Finger Rapids.

The only danger on this part of the route is the Five-Finger Rapids, where so many prospectors and so much property have been lost. The Canadian Government will appropriate a sum of money to blow out the dangerous rocks at this point and clear the river of dangerous obstructions. This route avoids White Horse Rapids and Miles Cañon, the most dangerous spots in the river routes. The total distance to Dawson City via Telegraph Creek will be approximately 1780 miles.

John C. Galbreath, for many years a resident of Telegraph Creek, has been directed by the British Columbia government to open this new route and $2000 will be expended on it immediately. Even now the trip to the gold fields, it is said, can be made with less danger and more quickly by this route than by any other. It is open usually until the middle of October and sometimes as late as November.

It is also proposed to build a branch from Telegraph Creek to

Dease Lake, which connects with the upper waters of the Mackenzie River.

The "back door" route, or "inside track" from civilization to the Klondike diggings, is the old Hudson Bay Company's "trunk line," and has been in use nearly a century. It is said to possess many advantages, except perhaps in the matter of distance, over any of the other land and water trails.

Argonauts going in at the "back door" will go to Edmonton, in Alberta, 1772 miles from Chicago, via the Canadian Pacific Railroad, and thence by stage or wagon to Athabasca Landing. Edmonton is on the Saskatchewan River and the portage to the landing places the traveler on the banks of the great Athabasca River and at the head of a continuous waterway for canoe travel to Fort Macpherson, at the north mouth of the Mackenzie River, from which point the Peel River lies south to the gold regions. From Edmonton to Fort Macpherson is 1882 miles.

Only Two Big Portages.

There are only two portages of any size on the route—that from Edmonton to Athabasca Landing, over which there is a stage and wagon line, and at Smith Landing, sixteen miles, over which the Hudson Bay Company has a tramway. With the exception of five other portages of a few hundred yards there is a fine down-grade water route all the way. Wherever there is a lake or long stretch of deep water navigation, the Hudson Bay Company has small freight steamers which ply during the summer months between the portage points.

From Edmonton a party of three men with a canoe should reach Fort Macpherson within sixty days, provided they are strong and of some experience in that sort of travel.

Experienced travelers recommend that the canoe be bought at home unless it is intended to hire Indians with large bark canoes

VIEW OF A SHEEP CAMP ON THE DYEA TRAIL.

This is the camp which was wiped out of existence by a landslide.

ON THE WAY TO THE KLONDIKE—SCENE NEAR THE STEAMBOAT LANDING, DYEA.

for the trip. Birch-bark canoes can be purchased large enough to carry three tons, but are said to be unreliable unless Indians are taken along to doctor them and keep them from getting water-logged. The Hudson Bay Company will contract to take freight northward on their steamers until the close of navigation.

A recent letter from a missionary says the ice had only commenced to run on September 30, 1896, in the Peel River, the waterway from Fort Macpherson to the gold fields. If winter comes on the traveler can change his canoe for sleds and dog trains.

Advantages to Travelers.

The great advantage claimed for the "back door" route is that it is an organized line of communication. Travelers need not carry any more food than will take them from one Hudson Bay post to the next, and there is abundance of fish and wild fowl along the route. They can also get assistance at the posts in case of sickness or accident.

If lucky enough to make their "pile" in the Klondike, they can come back by the dog-sled route in the winter. There is one mail to Fort Macpherson in the winter. Dogs for teams can be bought at any of the Hudson Bay posts, which form a chain of roadhouses on the trip.

Parties traveling alone will need no guides until they get near Fort Macpherson, the route from Edmonton being so well defined.

It is estimated that a party of three could provide themselves with food for the canoe trip of two months for thirty-five dollars. Pork, tea, flour and baking powder would suffice.

Parties should consist of three men, as that is the crew of a canoe. It will take 600 pounds of food to carry three men over the route. The paddling is all done down stream except when they turn south up Peel River, and sails should be taken, as there is often a favorable wind for days. There are large scows on the

line manned by ten men each, and known as "sturgeon heads."
They are like canal boats, but are punted along, and are used by
the Hudson Bay people for taking supplies to the forts.

It is estimated $200 per man will be sufficient for expenses
via this route, and that two months, and possibly six weeks, will
be an ample estimate of time.

Another all-Canadian route to the Klondike is proposed, to
enable Eastern Canada to compete in transportation, traffic and
trade with the Pacific coast. It includes a railway to Moose
Factory, at the foot of James Bay, and a line of steamers thence
to the western end of Chesterfield Inlet, a distance of 1300 miles.
The rest of the journey would be mainly by the Mackenzie and
Yukon rivers, and it is estimated that in summer it could be
made in seven days from Toronto. Between Hudson Bay and
the Yukon it is believed the only piece of railway it will be necessary to construct is 200 miles or so between the head of Chesterfield Inlet and Great Slave Lake.

Offers Fine Steamers.

The late managing owner of a line of steamers on the great
lakes has examined the reports as to the waterways through
Great Slave Lake and the Mackenzie and Yukon rivers, and
offers to undertake to equip the route with a new style of steamers, which, while spacious and economical, would develop a
reliable speed of twenty miles an hour in slack water. A model
of an ice boat has been prepared for winter navigation of these
waters.

The plan for reaching Hudson Bay is the construction of a
railway from Missanabie to Moose Factory, to be operated by
electricity furnished by the water power of the Moose River.

The proposed route to Hudson Bay is disputed by Quebec,
which is desirous of securing the western connection for itself,

and having already constructed a railway to Lake St. John, to within 300 miles of James Bay, is ready, with a certain amount of Government aid, to extend it to Moose Factory by way of the valley of the Ashuamouchouan River.

J. M. C. Lewis, a civil engineer, has proposed to the Interior Department, at Washington, a route from the mouth of the Copper River, by which he says the Klondike may be reached by a journey of a little over 300 miles from the coast, a great saving in distance over the other mountain routes. He says the trail could be opened at small expense.

The route which he proposes will start inland from the mouth of the Copper River, near the Miles Glacier, twenty-five miles east of the entrance to Prince William Sound. He says the Copper River is navigable for small steamers for many miles beyond the mouth of its principal eastern tributary, called on the latest maps the Chillyna River, which is itself navigable for a considerable distance. From the head of navigation on the Chillyna, Mr. Lewis says, either a highway or a railroad could be constructed without great difficulty or very heavy grades, through what the natives call the "low pass," probably the Scoloi Pass. From the pass the road would follow the valley of the White River to the point where it empties into the Yukon, on the edge of the Klondike gold fields.

"Uncle Sam's" Survey.

"Uncle Sam" has had his eye on short routes to Alaska for some time. In 1886 a bill was introduced in Congress "to facilitate the settlement and develop the resources of the Territory of Alaska and to open an overland commercial route, between the United States, Asiatic Russia and Japan."

The Interior Department referred the subject to Director Powell of the Geological Survey for a report, which was made as comprehensive as the knowledge possessed by the survey of the topography of the country, through which the road would have to pass, would permit.

In the beginning of his report Director Powell says :

" Information on record bearing on the question does not indicate any greater obstacies to the construction of such a line than those already overcome in trans-continental railroad building, and the construction of the proposed line must be pronounced feasible.

" From the geographic knowledge available a tentative line may be indicated extending from the Northern Pacific Railroad in Montana northward to Behring Sea, about 2800 miles in length."

This tentative line, divided into three grand divisions, is as follows :

1. From some point on the Northern Pacific Railroad in Montana to the headwaters of the Peace River.

2. From the headwaters of the Peace River to the headwaters of the Yukon.

3. From the headwaters of the Yukon to some point on the shore of Behring Sea.

Straight to Klondike.

It will be observed that the proposed route would take the road right through the Klondike gold field.

From Montana northward through British Columbia as far as the Peace River, Director Powell considered two routes, which he calls plains and valleys, respectively, their names indicating their character.   His preference was for the valley route.

First, it would have a decided advantage in distance.

Second, it would afford easier grades.   He admitted the prospect for local business over the two routes appeared to be in favor of the plains route, "unless important mining districts should be developed on the other line."

From the Northern Pacific Railroad to the Canadian Pacific Railroad by the valley route is about 325 miles, and to connect Southern Alaska indirectly with the railway system of the United States via the Canadian Pacific Railroad would require the construction of only 840 miles of line, which is exactly the distance from Baltimore to Chicago by the Pennsylvania line.

One of the most perplexing problems of transportation to which the gold craze gave rise, in the first months of the epidemic, was to find steamers for the sea voyage either to Juneau or St. Michael's. The regular transportation companies used all their own boats and all that they could hire, and even then were unable to accommodate all who wanted passage, and private enterprise undertook the hazardous trips in almost any old tub that would float long enough to get out of the harbor.

The experiences of the season, however, and the demand for passage on the first boats to go North in 1898, which set in as early as the first week in August, set the steamship men hustling to be ready for the expected rush in the spring.

More Steamers Next Spring.

Manager C. H. Hamilton, of the North American Transportation and Trading Company, announce that his company has let a contract to Cramps, the Philadelphia shipbuilders, for the construction of two 2000-ton steel steamers. They will be the finest steamers on the Pacific coast, and will be used exclusively on the Seattle-St. Michael's run. They will have accommodations for 200 first-class and 500 second-class passengers.

The American Steel Barge Company, of West Superior, Wisconsin, arranged with a syndicate interested in the Alaskan gold fields to construct several small vessels on the whaleback plan to

11

navigate the Yukon. Arrangements are being made to open the shipyards of the company at Everett, Washington, and the plant at West Superior may be used to get out some of these little ships.

The whaleback steamer *Everett*, which carried the American contributions to the East Indian famine, one of the largest whaleback freighters afloat, will be remodeled to accommodate passengers and put on the San Francisco-Alaska route, making regular trips to the Yukon with gold-seekers who prefer the water route to the diggings.

Expert River Men.

In preparation for the spring rush up the Yukon River, and over the divide with supplies, a Canadian firm has been hiring lumbermen and river men from the Ottawa region. There is every indication that by the opening of navigation on the upper Yukon there will be abundant work for expert river men in transporting supplies to the Klondike.

A Seattle company has been organized to build a sea-going steamship, and also a light draft steamer for the river business between St. Michael's and Dawson City.

The Puget Sound Tugboat Company will put a steamer on the Yukon in the spring to carry freight and passengers from St. Michael's to the Klondike.

The Pacific Coast Steamship Company is arranging to use all its available boats on the northern route to Juneau in 1898, and may decide to make several additions to its fleet.

Both the North American and Alaska companies are adding to their facilities for taking care of traffic in the spring and expect to be fully equipped for the great rush of gold-hunters and supplies when the time comes. The North American has ordered several new ocean and river steamers.

Steamboat men in Seattle estimate that, beginning about the
first of April, a large steamer can leave Puget's Sound'for Alaska
daily with all the passenger and freight accommodations crowded.

Several new steamer companies are already in the field and
the promise has been made that next season will see a reduction
in the rate of fare.   But unless the reports received from the
gold fields during the winter indicate that the richness of the
placers has been exaggerated and that they give signs of peter-
ing out, the rush to the mines in the spring will surpass anything
the world has ever seen.

Transportation companies assert that those who are waiting
until spring to go North will be very much disappointed if they
expect a reduction in fares.   That some companies will be organized to make trips at reduced rates there is no doubt, but the
regular steamship lines say the fare will be the same.

Secretary Hamilton, of the North American Transportation and
Trading Company, spoke of the fares in the spring as follows :

" In my opinion the fare to St. Michael's will not be less than
$200 in the spring.   Transportation facilities will be improved,
but fares will not be less."

The Pacific Coast Steamship Company officials were equally
sure the fares would stay up.

Will Pay To Come Back.

The companies generally assert that in the early spring they
will be carrying to the sound hundreds of passengers who have
wintered in the vicinity of Dawson City.   All will have money
and will be in a position to pay the present fares, which
are considered reasonable.   The majority of the miners who
stay during the coming winter will undoubtedly come out by
way of St. Michael's.   They will not care to undergo the hardships of the trip over the pass.

The first ship from New York to Juneau with gold-hunters and supplies sailed late in August, going around the Horn. The fare to Juneau was $175. Several other sailing vessels are expected to leave New York for Juneau with miners during the winter.

A great demand for small boats arose on the Pacific Coast before the season closed, the argonauts thinking to save time on the overland journey by taking their boats with them. Several styles of boats that could be shipped "knocked down" at once came to the front, and several firms began making specialties of these handy craft. One that will carry a ton costs about $18, and weighs about 200 pounds. It is taken apart with no pieces more than six or seven feet long and packed for shipping. The principal objection to these boats is that the Indians and packers dislike to contract to carry them over the mountains on account of their awkward shape. One builder has worked out a model for a galvanized iron boat that can be carried in sections fitting together like a "nest" of custard dishes, and can be put together with small bolts. A canvas folding boat that would carry two tons would be available on the Yukon. A keel, mast and some additional bracing could be added after reaching the interior.

Wagon Road to Yukon.

The Canadian Pacific Railway and Dominion Government are conferring with a view to opening up a wagon road to the Yukon from Edmonton. Such a road is feasible, and would be only between 800 and 900 miles long, passing through a rich auriferous country. The object is to give a short and safe road for prospectors and to make it possible to maintain winter communication.

A joint resolution was reported favorably for the United States Senate Committee on Territories on July 22d, authorizing the construction of toll roads in Alaska. The resolution authorizes the Secretary of the Interior to grant right of way 200 feet wide. Franchises are to be limited to twenty years. The rates of toll are to be approved by the Secretary of the Interior.

One of the features of the stampede to Klondike via Dyea has been the number of burros, cayuses, mules and horses taken up to serve for packing over the Chilkoot, Chilkat and White Passes! Hundreds have been sent through, and their owners in many cases had contracts in their pockets for all the freight they could expect to handle at from thirteen to nineteen cents a pound. Old mountaineers, however, think the horses, and especially the mules, will prove a failure as a venture, for their hoofs will cut up the road, which has been barely good enough for human feet, so far, and this, in such a moist climate as that of autumn in Southeastern Alaska, will soon make the trails impassable for beast or even, perhaps, for man.

There are a few horses in the Yukon country, and one of the largest pack trains ever brought into Dawson City, Robert Krook, of Dawson City, says, was brought over the frozen river Yukon by thirteen horses and as many sleds all the way from Circle City. Feed, however, is expensive, and the horses are easily rendered useless. If water gets on the top of the ice and the horses or mules get wet feet, they are practically ruined for all time, as their hoofs split when the water freezes, crippling the animals. To avoid this, moccasins are used and have proved partially successful.

Dogs for Burdens.

Dogs are the choice beasts of burden on the overland routes during the long frozen season, and their points of merit have been recognized by a decided stiffening of prices in the canine market. Good dogs, are worth from $100 up, $200 for a fine brute not being an unusual price. There is not much danger of the supply running far behind the demand, however, even at Dyea, for if there is anything Alaska is "long" on besides winters and mosquitos, it is dogs.

Robert Krook says that Eskimo dogs will draw 200 pounds each on a sled, so that six dogs will draw a year's supplies for one man. He, however, puts in the proviso that the sleds should not have iron runners, because the snow sticks to the iron and increases the friction so much that the dogs cannot haul more than 100 pounds apiece. With brass runners this drawback is obviated.

Moccasins on Dogs.

Sometimes the feet of the dogs get sore, and then the Indians fit moccasins on them ; as soon, however, as the tenderness is gone from their feet the dogs will bite and tear the moccasins off. In speaking of the dogs, Mr. Krook said that they need no lines to guide them, and are very intelligent, learning readily to obey a command to turn in any direction or to stop. They have to be watched closely, as they will attack and devour stores left in their way, especially bacon, which must be hung up out of their reach. At night, when camp is pitched the moment a blanket is thrown upon the ground they will run into it and curl up, neither cuffs nor kicks sufficing to budge them. They lie as close up to the men who own them as possible, and the miner cannot wrap himself up so close that they won't get under his blanket with him. They are almost human, too, in their disinclinations to get out in the morning.

Where sleds cannot be used the dogs will carry fifty pounds apiece in saddlebags slung across their backs pannier fashion. Nature has fitted these dogs for their work, and so mastiffs and St. Bernards are not as serviceable. The two latter breeds cannot stand the intense cold so well, and, though at first they will draw the sleds cheerfully, their feet cannot resist the strain and

CAMP OF ESQUIMEAUX AND HERD OF REINDEER.

167

begin to bleed so freely that the dogs are useless. The pads under the feet of the Eskimo dogs are of tougher skin.

Reindeer are to be entered as rivals of the Esquimo dogs. Twenty sturdy bucks have been selected from the United States Government's reindeer herd at Teller's Station and will be taken to Circle City. The design is to materially decrease the cost of overland transportation in winter, for the benefit of the miner.

Much care has been exercised in the selection of the herd, and not one of its members is less than four and one-half feet in height and seven feet in length. The minimum weight of these bucks is 250 pounds, but some of them are twenty-five to fifty pounds heavier than the lightest. All are vigorous, healthy and in good working condition. Their antlers, which curve gracefully backward, are about two and one-half feet in length. Their general color is a soft seal brown, shading into black on the legs, which are covered with short, glossy hair, to which the snow does not adhere.

A prime advantage of the reindeer over the dog is the fact that he paws away the snow and secures his own food, instead of having to add his rations to the weight of his burden. Many a pioneer prospector, traveling by dog team, has been placed in a position in which his dogs have become useless from lack of provisions. Had these unfortunate pilgrims been provided with reindeer teams, such an emergency would not, in all probability, have arisen ; and in case of threatened starvation the traveler's means of transportation would have furnished him with a liberal quantity of meat.

Bicycles for Yukon.

One of the most novel and absurd of all the schemes of transportation fostered by the stampede to the Yukon diggings is the Klondike bicycle, theoretically adapted to carry one man and 500 pounds of outfit, but practically useless because there is not a piece of the wheelmen's "good roads" in the territory. Yet some "tenderfeet" have been seen in Seattle armed and equipped with just that thing. But it is to be hoped they were not typical "tenderfeet."

The Klondike is promised close communication with the world in a short time. The Alaska Telegraph and Telephone Company has been incorporated in San Francisco to construct a telegraph line from Juneau and Dyea to Dawson and Circle City. The capital stock is $100,000. The work of construction is to be pushed and it is hoped the line will be in working order before winter. The estimated length is 10,000 miles.

The line will be a novelty, as no poles will be used except in crossing cañons and rivers. The wire, which will be of large guage, pure copper, will be heavily coated with insulating substance and will be laid along the ground. Stations will be established at every fifty miles. It is thought that this line will answer perfectly for the present.

How it will be Built.

D. E. Bohannon, the chief of construction of the line, explained its details as follows:

"Our method is very simple. The line is to be constructed on the same plans as the ordinary military line used by armies for war purposes. We have a wire a quarter of an inch thick and covered with kerite insulation, which has proved able to stand the rigorous climatic conditions prevailing in Alaska.

"The wire is wound upon large reels, the same as an ordinary telegraph wire, and these coils are to be placed on dog sleds and dragged over the ice and snow. As we go along the reels will simply pay out the loose wire and run it along the ground, and thus our line will be through in something like six weeks, the time consumed in the ordinary tramp over the country."

The Dominon Government has made application to the United States Government to be permitted to build a telegraph line from a navigable point on Linn Canal, Alaska, to Tagish, across the summit, a distance of nearly 100 miles, so that communication may be had with the interior of the Yukon all the year around. It is said that the application will raise a new question only comparable to that which was involved in the establishment of the mixed mail route in Alaska, which gave rise to so much talk.

The Klondike will not be so badly off for mails this winter. The monthly letter mail which was started by the United States Government the first of July, 1897, will be continued, and there will be one round trip a month to Circle City until July 1, 1898. The Canadian Government has also arranged for postal service to Dawson City.

The scheme of the United States postal service is interesting. Between Seattle and Sitka the mail steamers ply regularly. Between Sitka and Juneau there is a closed pouch steamboat service. Seattle makes up closed pouches for Douglas, Fort Wrangel, Juneau, Killisnoo, Ketchikan, Mary Island, Sitka and Metlakatlah.

Service from Sitka.

Connecting at Sitka is another sea service between that point and Unalaska, 1400 miles to the west. This service consists of one trip a month between Sitka and Unalaska from April to October and leaves Sitka immediately upon arrival of the mails from Seattle. Captain J. E. Hanson is acting clerk. From Unalaska the mails are dispatched to St. Michael's and thence to points on the Yukon.

The Postoffice Department has perfected not only a summer but a winter star route service between Juneau and Circle City The route is overland and by boats and rafts over the lakes and down the Yukon, and is 900 miles long. A Chicago man

named Beddoe carries the summer mail, making five trips
between June and November, and is paid $500 a trip. Two
Juneau men, Frank Corwin and Albert Hayes, operate the
winter service, and draw for each round trip $1700 in gold.
About 1200 letters are carried on each trip.

FOREST SCENE NEAR SITKA.

The Canadian mail to Dawson City will be carried by the
mounted police from Dyea to Skagway.

In the expectation that the boom in Alaskan and North
British mining stocks will be one of the wildest in the history of

the world, and that the stock exchanges of London, New York, Chicago and San Francisco will be willing to pay handsomely for inside and speedy news from the centre of excitement on the Klondike, some capitalists have conceived the idea of establishing a carrier-pigeon service between Seattle or Victoria and Dawson and Circle Cities, with Juneau as the "way station" in the flight. The experience of Nansen, the Arctic explorer, with carrier pigeons in the ice fields surrounding the North Pole, has demonstrated the practicability of using these birds in Alaska during the coldest months.

Plan of the Service.

The idea is to transfer a number of "breeders" to Victoria, the nearest telegraphic station to the Klondike district, and also a number of them to Juneau and Dawson City, in the heart of the new Eldorado. It is believed that after the birds shall have been properly trained by frequent flights over the country between Dawson City and Juneau, they will be able to cover that extent of territory in about twenty-four hours. The birds, whose home cotes are located in Victoria, it is believed, can reach that place in less than thirty hours after being released at Juneau, a trip that is seldom made in less than three or four days by steamboat, although on one occasion it was made from Sitka in forty-nine hours. With such a line of communication opened up it ought to be possible for a message written in the frozen interior of Alaska to reach the most distant parts of the world within a few days.

A carrier pigeon, which was taken from Portland, Ore., on the steamer Elder, to Dyea, returned to Portland on August 9th with the following message:

"Dyea, Aug. 7th. Arrived safely here last night. All well on board. T. CAIN."

In preparing to make the long overland journey into the Klondike, one of the things of most importance to be considered and one in which the "tenderfoot" left to himself, is most apt to make a serious blunder of omission, is the "outfit."

There are all sorts of tastes and so there are all sorts of outfits, but the following table, prepared by a man of ample experience and good appetite, will serve as a sample for preparing a list of the articles necessary for a complete outfit for a year in the Klondike diggings:

CLOTHING:—3 suits heavy woolen underwear, 6 pairs heavy woolen stockings, 2 pairs blanket-lined mittens, 1 heavy Mackinaw coat, 2 pairs Mackinaw trousers, 2 dark woolen overshirts, 1 heavy sweater, 1 heavy rubber-lined top coat, 2 pairs heavy hip rubber boots, 2 pairs shoes, 1 Canadian toque, 2 pairs extra heavy blankets, 1 suit oil skins, 2 pairs heavy overalls, 1 suit buckskin underwear, towels, needles, thread, wax, buttons.

FOOD:—350 pounds flour, 200 pounds bacon, 150 pounds beans, 10 pounds tea, 75 pounds coffee (browned), 5 pounds baking powder, 25 pounds salt, 150 pounds assorted dried fruits, 100 pounds evaporated vegetables and dried meats, 10 pounds soap, 3 tins matches, 5 pounds saccharine, citric acid.

HARDWARE:—1 long-handled shovel, 1 pick, 1 ax, duplicate handles, 5 pounds wire nails, 5 pounds pitch, 3 pounds oakum, 2 large files, hammer, jackplane, brace and bits, large whipsaw, hand saw, 150 feet ⅝-inch rope, drawknife, chisel, jackknife, whetstone, hand ax, shaving outfit, frying pan, kettle, Yukon stove, bean pot, two plates, cup, teapot, knife, fork and six spoons, 2 buckets, 2 miners' gold pans.

ARMAMENT:—Repeating rifle, 40-82, reloading tools and 100 rounds brass shell cartridges, 1 large hunting knife, fishing tackle, snow spectacles.

CAMPING OUTFIT:—Heavy canvas tent, 8x10, pegs and guy ropes, 1 heavy-lined canvas sleeping-bag, rubber blanket, mosquito netting.

These supplies will weigh about 1350 pounds and will cost about $225 at Seattle, or at Juneau, if the rush of gold-hunters has not exhausted the supply.

It is important to pay attention to a sufficient stock of anti-scorbutics, for scurvy is the scourge of Arctic residence.

The shaving utensils listed may cause some to smile, as they think the Klondike is no place for " frills," but the experience of sojourners in those regions of long and intense cold is that a smooth face is a positive comfort. The breath's moisture congealing in moustache and beard is well nigh as painful a trial in winter in Alaska as the mosquitos in summer. It is comfort rather than style to shave.

In making purchases the argonaut should bear in mind that the very best of everything is none too good and will more than repay the outlay in the long run. The clothing and food in particular should be first quality throughout.

One of the most indispensable items in the list is the sleeping bag, with an outside covering of heavy duck and lined with warm lamb's wool. It is fixed up with handles, so that in case of necessity it can be swung up in trees.

Hip rubber boots are another necessary article, in addition to which a pair of heavy miner's boots is generally taken.

Native Costume.

Many miners adopt the native costume—and it is comfortable as well as highly serviceable and picturesque.

The boots, usually made by the coast Indians, are of several varieties. The water boot is of seal and walrus skin, while the dry weather or winter boot is of all varieties of styles and material. The more expensive have fur trimmed legs, elaborately designed. They cost from $2 to $5 a pair. Trousers are often made of Siberian fawn skin and the skin of the marmot, or ground squirrel. The parka, or upper garment, is usually of marmot skins, trimmed with wolverine around the hood and lower edge, the long hair from the sides of the wolverine being used for the hood. This hair is sometimes five or six inches in length and is useful in protecting the face of the wearer. Good, warm flannels can be worn under the parka, and the whole outfit will weigh less than the ordinary clothes worn in a country where the weather gets down to zero. The parka is almost cold proof. But it is expensive, ranging in price from $25 to $100. Blankets and fur robes are used for bedding. Lynx skins make the best robes. Good ones cost $100. But the cheaper robes can be made of the skins of bears, mink, red fox and the Arctic hare. The skins of the latter animal make warm socks to be worn with the skin boots.

A Cheap Outfit.

Outfits can be purchased more cheaply than the sample given heretofore, by lopping off some of the articles. Here is the bill of one in which each article was of first-class quality, no groceries nor armament being included :

3 suits heavy woolen underwear, at $4 50	$13 50
4 pairs heavy stockings, at 40 cents	1 60
2 pairs German socks, at $1.15	2 30
1 pair hunting stockings	1 25
1 heavy sweater	4 50
1 lighter sweater	2 35
1 leather fur-lined coat, short	7 00
1 pair jeans trousers, lined with flannel	3 00
1 Mackinaw coat	3 00
1 pair Mackinaw trousers	2 50
1 suit buckskin underwear	12 00
1 pair hip rubber boots	5 25
1 pair heavy miners' boots	5 00
1 pair heavy overshoes	2 10
4 blankets, at $2.40	9 60
1 pair leather-lined mitts	1 20
1 pair woolen mitts	1 00
1 sleeping bag	12 50
1 sleeping cap	75
4 canvas carrying bags	2 00
Tools, including two miners' pans, picks, shovels, axes, saw, file, knife	7 32
Total	$99 72

Some men buy sheepskin coats and vests, horsehide coats and trousers at $18 a suit and extra supplies of "jumbo" underclothing. Some other men, remembering only the outfits carried to Harqua Hala or Leadville, squeeze all their supplies into a $100 bill, but it is safe to say their frugality is " penny wise and pound foolish."

Here is a list of provisions sufficient for one man for a month, made by an expert. [He probably was not a heavy eater.—Ed.]

20 pounds flour, with baking powder, 12 pounds bacon, 6 pounds beans, 5 pounds desiccated vegetables, 4 pounds butter, 5 pounds sugar, 4 cans milk, 1 pound tea, 3 pounds coffee, 2 pounds salt, 5 pounds cornmeal, pepper, mustard.

One of the men who has " been there " has the following to say of the cost of the provisions a prospector should take with him :

" No one should venture into the region without some cash and a sufficient supply of provisions to last eight months. One should buy these things in Juneau, and he should start out with something like the following: 400 pounds of flour, 100 pounds of beans, 100 pounds of bacon, 100 pounds of sugar, 10 pounds of tea, 30 pounds of coffee, 150 pounds of mixed fruit, salt, pepper and cooking utensils. The whole outfit can be purchased well within $90. The cost of conveying this stock of provisions to the headwaters of Lake Linderman will average about $15 per 100, but even that makes it considerably cheaper than the same goods can be purchased in the mining camp.

Value of Salt.

Just how valuable salt sometimes becomes in the gold fields is illustrated in a story told by a miner who lately returned from there. His party ran out of that useful article, and it seemed that they would die without it. They came across another party that had salt, but refused to part with it. A pitched battle was about to begin for possession of the salt, when some one suggested that those who owned the salt were not overly flush with gold dust, while those who had no salt had plenty of gold. It was then arranged that gold should be weighed against the salt, and this was done.

And after this story of the salt, which needs not to be taken with a grain of that condiment, it is well to reiterate to every gold hunter going out to winter in the Klondike fields :

" Take your own grub—and plenty of it."

Food in Compact Form.

To those who find something terrifying about a heavy outfit, with packers' prices over the passes at twenty cents a pound and upwards, it may be suggested that many staple articles of food have been prepared in the utmost condensed or concentrated forms for the use of soldiers in the field, and will no doubt be equally as nourishing.to prospectors, while enabling them to carry extensive supplies in small bulk.

For instance, a cup of tea or coffee is crowded into the size and form of a quinine capsule, a mince pie is the size of a cough drop, and other delicacies are in proportion. Soup "buttons" are prepared in the same way, with meat, vegetables and seasoning all ready for hot water. A loaf of bread is compressed into the size and shape of a soda cracker, which swells up to normal size when put in hot water. Ten pounds of vegetables are put into one-pound can, and a cubic ounce of desiccated beef is equal in nourishment to several pounds of fresh meat.

Prospectors who go out by the St. Michael's route, if they purpose wintering on the Klondike, or in Upper Alaska, will not need to take so elaborate a provision supply by the amount of at least three months' consumption, but they had better keep

12

pretty close to the clothing, hardware, armament and camping schedules. They will find it advisable not to omit the food item altogether unless they have good assurance that the supplies brought in by the trading companies will be ample.

Robert Krook's Advice.

Lest any should think too much stress has been laid on the matter of supplies to be taken into the Klondike, these words of Robert Krook, the young Swedish miner, who returned from Dawson City during the summer, are given in full:

"Every one who goes to Alaska must rely mainly on two establishments for supplies. Even those who have a good outfit will find it often necessary to patronize one or other of the stores. Prices are on an average three times as high as at Juneau or St. Michael's, and four to five times as steep as in San Francisco. When the winter is nearly over and supplies begin to run short prices are, as a consequence, raised. Toward the close of last winter, before the new supplies came up the river, prices were doubled.

"All through the winter men arrive at such mining towns as Dawson City, bringing with them from one to two tons of food and clothing. They go up the streams and peddle their goods, taking care to lose nothing for their time and trouble.

"To one blacksmith shop all miners must go or send when they have tools to be repaired, or when they need anything made to order which the stores cannot supply.

"Dawson City can boast of two good practicing physicians— Police Surgeon Willis and another doctor who went from Circle City to Dawson last year. They carry their own supplies of staple drugs and medicines, so as to be able to compound their own prescriptions. Ordinary remedies are to be obtained at the two trading stores.

"I think it well to mention that the credit system has been entirely done away with in Dawson. No one can make a purchase of any kind without the necessary cash in the shape of dust. Of course it must not be understood that we would let a man starve, but on the other hand, no one must expect to be supported by the generosity of the people. We are all hard workers up there, and if any man will work he can always make a living.

"The impression seems to prevail that the mines are close to Dawson City. That is a mistake. The rich creeks are fifteen miles off, and it is a day's journey to reach them. The camp there is as pretty a place as one desires to see. The white tents and huts of the miners are scattered along the banks of the creeks or built on the mountain sides, as convenience or fancy dictated."

Medicine Chest.

Another thing which all prospectors should be careful to take along is a medicine chest. Doctors are few, distances long and emergencies of health or limb often most urgent in the Yukon mining camps.

Here is a list of contents of a medicine chest, whose cost is within $10, and every article of which is useful in the wilderness.

Quinine pills	50
Compound cathartic pills	50
Acetanilid tablets	3 dozen
Chlorate potash	1 box
Mustard plasters	6
Belladonna plasters	6
Carbolic salve	4 ounces
Chloroform liniment	8 ounces
Witch hazel	1 pint
Essence ginger	4 ounces
Paregoric	4 ounces
Laudanum	1 ounce
Borax	4 ounces
Tincture iodine	1 ounce

Spirits nitre	2 ounces
Tincture iron	1 ounce
Cough mixture	8 ounces
Toothache drops	1 bottle
Vaseline	1 bottle
Iodoform	2 drams
Lint	2 yards
Assorted bandages	½ dozen
Rubber adhesive plasters	2 feet
Absorbent cotton	4 ounces

Monsell's salts for hemorrhages—In quantities in accordance with the person's liability to attacks of the trouble.

Health and the Klondike.

As a rule, no one in ordinary health and strength need fear to winter in the Klondike or to risk the hardships incident to getting there, merely on account of the Arctic cold.   The bracing effect of the northern climate will probably prove beneficial to many.   Snow and ice are in themselves rather unpleasant than unhealthful.   Scientific records have well established that longevity increases as residence is advanced from the equator towards the poles.   There is more risk of disease in a voyage to Panama or India than in one to Behring Strait or Herschel Island.

But weak hearts and weak lungs cannot face northern blasts. Rheumatism unfits for such tests.   People of purely sedentary habits take big chances on the overland trails and in the gulches. Weak eyes would be severely tried and, perhaps, blinded by the glare of the snow-clad land.   Physical exhaustion, colds, scurvy, rheumastism and snow blindness are the ills chiefly to be dreaded by the Alaskan gold-hunters, and any who are subject to troubles of the heart, throat or lungs should stay religiously away from the Klondike.   The medicine chest would be a futile resort for them, and some volunteer sexton would likely do for them the last earthly office before the Alaskan spring bloomed in May.

But now that the daring prospector is in the Klondike and washing out the precious dust, his next thought will be, as his "pile" grows, to get out and back to the baked meats, and flesh pots of civilization.   Hear what Mrs. Frederick Schwatka, who had much personal experience in Alaska, and got the benefit of much more vicariously from pioneers returning from the wilderness, has to say about "how to get out :"

"This getting back is a formidable undertaking that appalls so many.   They choose rather to remain whole winters doing practically nothing that brings in more than a bare existence. In getting out it is necessary to make progress against the 600 miles of swift river current.   Rowing is out of the question, walking and poling being the methods used.   The poles are about twelve feet long and made of seasoned spruce saplings and sharpened at the butt end.   Sometimes an iron spike is put in, otherwise it must be sharpened two or three times a day. Two polemen stand in bow and stern.   To stand all day in a wabbling, cranky boat, and work like a beaver until six or seven hundred miles are traversed at about fifteen miles a day is in itself a formidable undertaking.   Then the great pass must be scaled without any assistance, for there are no Indians now to help.   Here it is that many a discouraged miner has given up all hope and found a grave in the ice-covered mountains.   It is the thought of again seeing something of civilization and the outside world that buoys up the traveler by this difficult trail."

CHAPTER V.

A Land of Wonders.

Land of the Midnight Sun—Great Distances—Primitive Conveyances—Terrors of the Arctic Regions—World of Wonders—Dangers of Travel—A Great Glacier—A Frozen Cataract—Beautiful Scenery—Rush of Torrents—Marvelous Sunsets—Great Yukon River—Cañon of Lewis River —Dominion of the Frost King—Towering Volcanoes—The Winter Moon—A Country of Romance—Totem Poles—Salmon Fisheries—Vast Solitudes—The Alaskan Natives.

THE man who goes from southern latitudes to seek his fortune in Alaska will leave familiar scenes for a land of wonders. His first experience will of necessity be one of surprise. He will find a country of new people, new scenes, and new modes of life.

Every one who has visited the land about which so much has been written and printed relative to the gold findings tells the same story of the matchless grandeur of the territory. With few exceptions all give the same report of the peoples and marvels there to be seen.

It is the land of the midnight sun and the mid-day moon; of salt water intruding hundreds of miles into the country, between mountains that overhang it in such a way as to excite a feeling of awe; of the Aurora Borealis, the displays of which are more magnificent than are ever witnessed in southern regions. It is a land of majestic mountains, of vast inland seas, of stupendous glaciers, compared with which those of the old world are but trifling affairs. It is a land from which thundering icebergs come plunging into the sea and float off in their glory of inimitable splendor. It is a land of exceeding wealth in fish, in timber, in minerals. And, above all, it is the land in which many think the

182

mother lode of the gold supply of the Western Continent is to be found.

One of the first things that will be forced upon the visitor will be the fact that Alaska is a country of magnificent distances. It is nine times the size of the New England states; twice the size of Texas and three times as large as California. It stretches more than 1000 miles from north to south, and extends west to the extreme limit of the Aleutian Islands.

Few people in the United States, probably, are aware of the fact that the gold fields which are attracting so much attention are pretty nearly on the medial line of the United States from east to west. From Sitka, for instance, westward to the limit of the Aleutian Islands, it is nearly 3000 miles; and eastward from the same point it is not over about 3500 miles to the most easterly coast of Maine.

The name of the country itself is simply a designation for the immensity of its territory—a wonder. It is a corruption of the Indian name Al–ay–ke–sa, which was given by the native islanders to the mainland, and which signifies "great country." As a matter of fact, the territory contains nearly 600,000 square miles; and it is thus nearly one-fifth as large as all the other states and territories of the Union combined. It would make more than twelve states the size of New York.

Poor Transit Conveyances.

These enormous distances soon impress themselves upon the traveler, and the sense of interminable space is accentuated by the lack of ordinary transit facilities. Alaska is a land in which the steam train is not known, and it may safely be said that a large proportion of the people living in the country have never heard of such a thing as a railroad. Even horses and wagons are virtually things unknown. The country has too rigorous a

climate for the successful use of any beasts of burden other than
dogs.   Hence, dogs as pack animals and as steeds for sledges
have become one of the chief possessions of the people.

These vast stretches of country are also observable in the
marked differences of climate.   Southern Alaska is really a different country than the more northern districts in which the gold
fields of the Yukon have been found.   William H. Seward some
years ago, writing from Berlin, makes use of these words: "We
have seen of Germany enough to show that its climate is neither
so genial, nor its soil so fertile, nor its resources of forests and
mines so rich as those of southern Alaska."

Akin to Norway.

In climate and all physical features southeastern Alaska is but
a repetition of southern Norway.   It enjoys, however, a greater
wealth of forests.   In latitude, configuration, temperature, rainfall and ocean currents it is identical.   Norwegians, therefore,
could be transplanted to Sitka and its neighborhood, and, barring
the lack of improvements of the old world, would scarcely realize that their location had been changed.   During the thirty-six
years that the Russians kept meteorological records in Sitka the
mercury went below zero but four times.

A comparison here may be of interest.   St. Johns, New-
foundland, is literally beset by icebergs in summer, and its harbor in the winter time is virtually frozen solid.   Yet Sitka, which
is ten degrees north of it has always an open roadstead, and it is
only the ends of the longer fiords that are ever covered by ice.

Again it is pointed out that Sitka Castle, which is three miles
north of Balmoral Castle, in Scotland, has a higher average winter temperature than the highland home.   In southern Alaska
the snow rarely lies on the ground at the sea level.   The mist
and rains reduce it to slush almost as quickly as in Kentucky or

the District of Columbia, the isothermal equals of this region. We hear much of snow shoes in connection with Alaskan life, and yet skating is one of the rarest of pleasures for the Sitkans.

It is a different matter, however, when one pierces the interior and wends his way over the mountain tops or through the valleys or along the mighty streams to the gold fields. As one ascends farther north, with the change of scenes comes a change of air, till in the neighborhood of Dawson City, Circle City, Klondike, and the other mining camps, it is no unusual thing for the mercury to fall from sixty to ninety degrees below zero.

Nine months of the year in these northern regions the ground is frozen to the depth of twenty-five or thirty feet as solid as a rock. Even in summer, which there is virtually but three months, the ground rarely thaws out more than from two to two and a half inches.

People who invade these northern districts find that a new mode of life is forced upon them. The clothing which would be comfortable even in Sitka no longer furnishes adequate warmth, and as a result, those who can do so, usually adopt the native costume, and dress largely in furs.

Wonders to Marvel At.

The voyager, be he excursionist or miner, thus finds an endless variety of things to admire, to wonder at and to ponder over. He will scarcely believe his senses or realize the fact that in sailing up the vast inland arms of the sea, which extend often hundreds of miles toward the interior, to which he is bound, he is really riding on salt water, mere inlets of the Pacific Ocean. It scarcely seems possible to one that he can glide along day after day and week after week, if need be, without encountering a single wave or a single ripple to disturb the motion of the vessel, and yet, at the same time, be all the time on the ocean and have the benefits of an ocean trip.

Those who have made the journey over Alaskan waters say the only realizing sense they have of the character of the voyage is the voracious appetite engendered, without the accompaniment of the much dreaded monster—sea sickness.

The islands, too, by which the vessel glides, will be a constant source of wonder. One will marvel how, and when, and why, these islands past which he rides were formed—islands, some of them no larger than a good sized house, and others large enough to be empires in themselves.

Channels a Menace.

Not infrequently the traveler has to pass through narrow and serpentine passages, which can only be navigated at slack and high tide on account of the teriffic current which rushes through the straits at other times. These channels are often hundreds of miles in length and as straight as an arrow. Many of them are almost unfathomable in depth and are banked on either side by perpendicular and gigantic mountains, whose untrod summits are clothed in ice and clouds.

The impression given the traveler is very much the same as that afforded by the somewhat similar scenery of upper Norway. In a general sense there is the same bleakness observable on the mountains, a somewhat similar stunted vegetation and an almost identical invasion of the mainland by the sea. But what the traveler will not find in Norway or in any other part of the world are the matchless glaciers that, in common acceptance, are one of the most remarkable features of Alaska scenery.

The traveler will see a number of them on his way to Juneau, glittering in the distance and apparently bleak and inaccessible. As he gets farther into the country, these glaciers become greater in size and more numerous. It has been said that the largest glacier in Switzerland would not make more than a respectable sized nose if it could be transferred bodily, to the face of one of these sleeping giants in the fastnesses of Alaska.

The Great Muir Glacier.

Here, again, a comparison will be of service to enable one to appreciate the wonders of Alaska scenery. Of the Norwegian glaciers, which may be most fairly used for comparison with the Muir, the Jodtesalbrae, the largest glacier in Europe, lies three degrees north of the Muir, at an elevation of 3000 feet above the sea. It covers 470 square miles.

The Muir glacier drains an area of 800 square miles, and the actual ice surface covers about 350 square miles. The mass of it is thirty-five miles long and from ten to fifteen miles wide, and lies but a few hundred feet above the sea level. It is fed by twenty-six tributary streams, seven of which are over a mile in width.

If all their affluents were named and counted, as in Switzerland, the Muir might boast two hundred branches or tributary glaciers in its system. The mountain gateway, two and a half miles wide, through which it pours to the sea, is formed by spurs of Mt. Case, 5510 feet high, and Mt. Wright, 4944 feet high. All the mountains in the immediate vicinity of the glacier average from 4000 to 6000 feet in height.

For further comparison it may be pointed out that the Svartisen, the snow glacier of the Norway coast, about eight degrees north of the Muir and on the line of the Arctic circle, is an ice mantle forty-four miles long and from twelve to twenty-five miles wide, occupying a plateau 4000 feet above the sea. The Swiss glaciers, all lying from 4000 to 6000 feet above the sea are like those of Mt. Ranier, and in no way to be

compared with the Muir, twenty of whose arms exceed the Mer
de Glace in size !

Apropos of the Muir glacier one cannot do better than to
quote a few words from the lamented Kate Field on Alaskan
glaciers in general and the Muir glacier in particular. Says
she:

"Soon after leaving Wrangel, the first Alaskan glacier is seen
in the distance, looking like a frozen river emerging from the
home of the clouds. The sea is glassy, and a procession of
small bergs, broken away from the glacier, float silently toward
the south. It is Nature's dead march to the sun, to melt in its
burning kisses, and to be transplanted into happy tears. Wild
ducks fly past, and from his eyrie a bald-headed eagle surveys
the scene—deeply, darkly, beautifully blue—apparently con-
scious that he is the symbol of the Republic.

"There are glaciers and glaciers. In Switzerland a glacier is
a vast bed of dirty air-holed ice that has fastened itself, like a
cold porous-plaster, to the side of an Alp. Distance alone
lends enchantment to the view. In Alaska a glacier is a wonderful torrent that seems to have been suddenly frozen when
about to plunge into the sea. Down and about mountains wind
these snow-clad serpents, extending miles inland, with as many
arms sometimes as an octopus.

A Frozen Niagara.

"Wonderfully picturesque is the Davidson glacier, but more
extended is the Muir glacier, which marks the extreme northerly
points of pleasure travel. Imagine a glacier three miles wide
and three hundred feet high at its mouth. Think of Niagara
Falls frozen stiff, add thirty-six feet to its height, and you have
a slight idea of the terminus of Muir Glacier, in front of which
your steamer anchors ; picture a background of mountains fifteen

thousand feet high, all snow clad, and then imagine a gorgeous sun lighting up the ice crystals with rainbow coloring.

"The face of the glacier takes on the hue of aquamarine, the hue of every bit of floating ice, big and little, that surrounds the steamer and makes navigation serious. These dazzling serpents move at the rate of sixty-four feet a day, tumbling headlong

MOUNTAIN SCENE IN ALASKA.

into the sea, and, as it falls, the ear is startled with submarine thunder, the echoes of which resound far and near. Down, down, down goes the berg, and woe to the boat in its way when it again rises to the surface."

If the tide is right, the traveler will hear the thundering crash caused by the icebergs breaking off from the glaciers and tumbling into the water. It is no unusual thing for a vessel on these inland arms of the ocean to be literally in a sea of ice.

A Picture of Beauty.

This is declared to be one of the most beautiful pictures man ever witnessed, and many of the thousands of people, who have left southern latitudes to wend their way into the fastnesses of Alaskan territory have written home in the most glowing terms of the wonders, witnessed, especially in the ocean part of their journey. Of these descriptions none, perhaps, is more striking or will convey a better idea of what travel in these solitudes really is than the words of Miss Skidmore, who threaded the wilderness and wrote a book on her experiences Says she :

" Life on the waveless arms of the ocean as a great fascination on one of these Alaskan trips, and, crowded with novelty, incidents and surprises as each day is, the cruise seems all too short when the end approaches. One dreads to get to land again and end the easy, idle wandering through the long archipelago.

" The voyage is but one protracted marine picnic, and an unbroken succession of memorable days. Where in all the list of them to place the red.letter or the white stone puzzles one. The passengers beg the captain to reverse the engines, or boldly turn back and keep up the cruise until the autumn gales make us willing to return to the region of earthly cares and responsibilities, daily mails and telegraph wires. The long nightless days never lose their spells, and in retrospect the wonders of the northland appear the greater.

" The weeks of continuous travel over deep, placid waters, in the midst of magnificent scenery, might be a journey of exploration on a new continent, so different is it from anything else in American travel. Seldom is anything but an Indian canoe met.

For days no sign of settlement is seen along the quiet fiords, and making nocturnal visits to small fisheries, only the unbroken wilderness is in sight during waking hours.

"The anchoring in strange places, the going to and fro in small boats, the queer people, the strange life, the peculiar fascination of the frontier and the novelty of the whole thing affects one strangely. Each arm of the sea, and the unknown, unexplored wilderness that lies back of every mile of shore, continually tempt the imagination."

No one can give so good an idea of the marvels and delights of this strange and virtually unknown country as those who have actually made an extended journey in it, and no apology, therefore, is made for the insertion of a passage written by another traveler, who, like Miss Skidmore, went where few readers of this book have been privileged to go. Speaking of the wonderful scenery of the country the writer says :

"It is, perhaps, a little remarkable that the marvelous panorama of fantastic peaks, rushing streams, huge glaciers and maddened cataracts in no way lessens the enjoyment or appreciation of the mountains by the-sea, that pass in review during the trip to Alaska.

Through Noisy Torrents.

"In one case the traveler is rushing onward, literally at railway speed, now passing through the shoulder of a mountain, and now round the base of another, sometimes through primeval forests, sometimes by the side of a noisy torrent or deep cañon, and sometimes through a secluded valley ; and in the other instance he is gliding along the deep but placid waters of the landlocked arms of the Pacific Ocean, on the undisturbed surface of immeasurable depths, while the snow capped heights are within pistol shot of where he sits, and the rugged precipices fall sheer into the depths almost at his side.

"The entire length of this inland passage of over 1000 miles is heavily timbered. Great avalanches of snow have swept down the mountains here and there, and in their devastating tracks long streaks of timber have been mowed down. At intervals, little Indian villages dot the shores, resting most picturesquely upon narrow shelves just at the edge of tide water, Throughout the whole stretch of country, travel by land is almost impossible owing to the dense timber and underbrush that cover its surface."

By Another Witness.

One who nas traveled far and wide (the Marquis of Dufferin and Ava) pithily describes the trip through these waters :

"Such a spectacle as its (British Columbia) coast line presents is not to be paralleled by any country in the world. Day after day, for a whole week, in a vessel of nearly 2000 tons, we threaded an interminable labyrinth of watery lanes and reaches, that wound endlessly in and out of a network of islands, promontories and peninsulas for thousands of miles, unruffled by the slightest swell from the adjoining ocean, and presenting at every turn an ever-shifting combination of rock, verdure, glacier and snow-capped mountains of unrivalled grandeur and beauty. "

H. Juneau, one of the founders of Juneau, Alaska, gives a similar account. Says he :

"Along the seacoast Alaska presents a grand and picturesque view for miles in extent, from an ocean steamer. It is a good idea to get acquainted with Alaska and enjoy its scenery. It is a grand country to visit, and its scenery surpasses any mountainous scenery in the world. Travel on water can be provided for in comfort, and be enjoyed without great risk of danger.

"Alaska is a country on edge. It is so mountainous. Basins are mainly filled with ice. The weather is always hard in great extremes. Where there is no ice there is moss and devil's club, the latter a vine that winds around everything it can clutch. Persons walking become entwined in a network of moss and devil's club, and passage is extremely difficult and ' torturous ' as well as tortuous."

Miss Skidmore has another interesting passage relative to the beauties to be seen on the trip north from Sitka. Speaking on the straits and narrows, she says:

" The tourist should not miss any part of this scenic passage ; the near shores, the forested heights and the magnificent range of peaks around the Stikines delta, composing one of the noblest landscapes he will see. The sunset effects in the broad channels at either end are renowned, and possessor of a Claude Lorraine glass is the most fortunate of tourists.

Marvelous Sunrise Effects.

" He who has seen the sunrise lights in the narrows has seen the best of the most marvelous atmospheric effects and color displays the matchless coast can offer. It is a place of resort for eagles, whose nests may be seen in many tree tops, and is a nursery for young gulls, who float like myriad tufts of down in the still regions.

" A hedge of living green rises from the water's edge, every spruce twig festooned with pale green mosses. At low tide broad bands of russet sea weed frame the islets and border the shores, and fronds, stems and orange heads of the giant kelp float in the intensely green waters.

" The tides, rushing in from either end, meet off Finger Point, whose two red spar buoys are prominent in the exciting navigation. The tide-fall varies from fourteen to twenty-three feet, and salmon, entering with the tide, turn aside at the red spar buoys, clear an islet, manoeuvre to the foot of the falls, leap its eight feet at high tide and swim to a mountain lake."

13

Nor is the element of the wonderful lost as one leaves these deep inlets of the sea and penetrates into the interior fastnesses. One leaves in a measure the stunted, yet luxuriant, forestation of the southern and coast districts for a bleaker and more repellant landscape. But the great water courses, such as the Yukon and the Klondike, with their numerous tributaries, in a sense take the place of the salt water inlets. The rivers alone would suffice to give a fair idea of the immensity of the country. And right here a word about the Yukon.

What the Amazon is to South America, the Mississippi to the central portion of the United States, the Yukon is to Alaska.

It is the great inland highway of the country. It makes it possible for the explorer to penetrate to the very heart of this unknown region.

This mighty stream rises in the Rocky Mountains of British Columbia, and the Coast Range Mountains in southeastern Alaska, about 135 miles from the city of Juneau, which is the present metropolis of Alaska. It is only known, however, as the Yukon River at the point where the Pelly River, the branch that heads in British Columbia, meets with the Lewis River, which heads in southeastern Alaska. This point of confluence is at Fort Selkirk, in the Northwest Territory, something like 125 miles southeast of Klondike.

Giant Among Rivers:

The Yukon River proper, therefore, is 2044 miles in length. From Fort Selkirk it flows northwest 400 miles and touches the Arctic Circle. Thence it bends in a southward course for a distance of 1000 miles and empties into Behring Sea. The mighty stream drains more than 600,000 miles of territory and discharges at least a third more water into Behring Sea than the Mississippi River discharges into the Gulf of Mexido.

At its mouth it is sixty miles wide. As far inland as 1500 miles it widens out from one to ten miles. Throughout its course it is dotted with inland islands, more than 1000 of these, it is said, sending the course of the stream in as many different directions. The stream thus merits being considered as a geographical wonder, and from mouth to head there is scarcely a point devoid of interest to the traveler.

SCENE ON THE YUKON RIVER.

Like most of the great streams of Alaska the navigation of the river is attended with danger, and the sense of constant peril affords one of the pleasures of the excursionist's trip to the interior. Only natives who are thoroughly familiar with the river are intrusted with the piloting of boats up the stream during the

season of low water. Even at the season of high water there
are places where the stream is so shallow that it is not navigable
by sea-going vessels ; but only by flat-bottom boats of a carrying capacity of from 400 to 500 tons.

Canon of Lewis River.

As an illustration of the danger incident to this river travel, a
few words may be quoted relative to the cañon of the Lewis River,
which were written by one who recently made a trip to the interior. Says he :

"Before reaching the cañon, a high cut bank of sand on the
right hand side will give warning that it is close at hand. Good
river men have run the cañon safely even with loaded rafts ; but
it is much surer to make a landing on the right side and portage
the outfit around the cañon three-quarters of a mile and run the
raft through empty.

" The sameness of the scenery on approaching the cañon is so
marked that many parties have gotten into the cañon before they
were aware of it. Below the cañon are the White Horse rapids—
a bad piece of water ; but the raft can be lined down the right
hand side until near the White Horse, three miles below. This
is a box cañon about a hundred yards long, and fifty in width, a
chute through which the water of the river, which is nearly 600
feet wide just above, rushes with maddening force.

" But few have ever a tempted to run it, and four of them
have been drowned. Of two men who made the attempt in
May, '88, nothing was found save a bundle of blankets."

Reference has been made to the intense cold of the northern
regions where gold abounds, and it must be borne in mind that
during the winter season, which is practically nine months of the
year, the Yukon is absolutely frozen solid and thus closed to
travel. The Frost King asserts his dominion and locks up all

approaches with impenetrable ice. Only for ten or twelve weeks, that is, from the middle of June to the early part of September, is the river for use in travel, except by way of sledges drawn by dogs.

When, however, in the early spring the bonds of ice are riven, a never-ending panorama of extraordinary picturesqueness is unfolded to the voyager. The banks of the stream are then fringed with flowers and carpeted with the all-pervading moss or tundra, as it is called. Then birds in countless number and of infinite variety in plumage, sing out a welcome to the traveler from every tree top. One may pitch his tent wherever he likes in midsummer, and a bed of roses, a clump of poppies or a bunch of blue bells will adorn his camp.

Above all the Glaciers.

One is never allowed to forget, however, that high above this brief paradise by the river side, which for a time is almost of tropical exuberance, the giant glaciers sleep in the summit of the mountains above the bed of roses. With the first days of September, and here the traveler will experience a deep sense of regret—everything is changed. The bed of roses has disappeared before the ice breath of the Winter King. This, as has been said before, often sends down the mercury to from eighty to ninety degrees below zero.

The birds, as might be expected, hie themselves southward. The white man has to take to his cabin and the Indian to his hut, and even the bears are early driven away from the field and begin their sleep of nine months. Throughout all northern Alaska, from September on, the rivers are but ribbons of ice, marking off the mountains, and the plains, and the forests, which are all alike covered with a coat of snow.

As might be expected from the general configuration of the

land, Alaska is a country of fine waterfalls. The most remarkable of these leap from the cliffs along Cook's Inlet, and the
alteration of snow peaks, volcanoes, forested slopes and fertile
prairies make a continually changing and charming picture to
the eye.

A Land of Volcanoes.

Go where you will you will find snow-clad peaks, glaciers,
cliffs, and ferreting their way through the country, innumerable
streams, the courses of which are often partially blocked, resulting
in waterfalls and rapids that would be regarded as sights worth
long trips were they anywhere else in the world than in the
distant and, as it is commonly supposed, forbidding territory of
Alaska. There is a whole line of volcanoes, curving down to
the southwest and joining those of Kurile Islands and of Japan,
which complete the Pacific's " ring of fire," as it is called.

Brilliant auroral displays are mostly to be witnessed in August,
and at such times mirages frequently appear. By refraction, the
ice floes are often magnified into ice cliffs 1000 or more feet high,
apparently barring a ship's advance or retreat. Many attempts
have been made by photographers to secure a sharp negative of
a mirage, but it is difficult to do so. The lines of glimmering
ice cliffs leave no definition or shadow, but waver and fade
quickly. The reflected light from these glaciers and snow fields
is thus often a bar to the most experienced photographer.

The world has been given, however, one great hoax in the
way of a picture of an Alaskan mirage. This was the so-called
Phantom or Silent City, which was issued in 1889 by Richard
Willoughby. Thousands of prints of a cloudy negative of
Bristol, England, were sold on his statement that he had seen
and photographed the city from Glacier Bay.

It is with the advent of the Winter King that the Alaskan
dogs come in play so conspicuously. And a word about these

dogs, which are really one of the wonders of Alaska, will be of interest. They really seem not dogs at all, but animals closely related to the wolf.

Strange as it may seem, they are all natural born thieves, or nothing. They are all prone to enjoy what is commonly called a "scrap," and they usually celebrate the arrival of newcomers by a general fight. Men who have spent years in the Alaskan wilds say that the dogs will steal anything from a pair of boots to a side of bacon, and in doing so will evince as great a degree of cunning and cleverness as the most expert thief who ever plied his calling in a metropolitan city.

To be on the safe side in the matter of their possessions, all the miners have adopted the plan of " caching " their harness, clothing, etc. This is done by erecting a strong house upon posts twelve or fifteen feet above ground for the safe keeping of all such articles.

Animals With Cracked Barks.

A peculiar thing connected with these dogs is the fact that they are all animals with cracked barks. In other words, their attempts at barking are simply a source of the most unheard discord. The howling of wolves, it is said, is pleasant music compared with the howling of these dogs at night.

What is more, on the slightest provocation, in the dead of night, some dog will raise an apology for a bark, and every animal within a radius of five miles will join in the general uproar. Alaska is not obliged to wait for the Fourth of July for discord. The dogs can make it on short notice at any time.

To the stranger in Alaska the sunlit nights and the moonlit days will for a long time be a source of constant wonder. Old Sol, when he is on duty, which, it must be remembered, is only part of the year, is no laggard in Alaska. He rises before three o'clock in the morning and keeps steadily at work until fully eleven o'clock at night. In the gold regions, therefore, during the mining months, there are a few short hours only when it is not sunshine.

Luna Takes Precedence.

During the long winter months, however, Sol takes a back seat and Luna takes precedence. Then there is an era of moon-lit days. Miss Anna Fulcomer, a plucky University of Chicago girl, adverts to this peculiarity—one may say wonder—in a letter written from Circle City, in the heart of the gold region. She says :

"While teaching at Circle City I went to school by the light of the setting moon—that was about nine o'clock in the morning—and went home at noon by the light of the rising moon. Literally I have lived in moonlight for the last year. Moonlight and cold. Still, the temperature last winter was not as intense as usual. The coldest we had it was only sixty-five degrees below zero, and that for Alaska in the northern latitudes was mild weather. It was quite cold enough, however, to make one feel the need of genuine Alaskan clothing, good shelter and good solid food.

"I pity the people who come here under the delusion that mining life in Alaska is anything comparable with what it was during the gold excitement in California. There they had mild weather, in which people could comfortably camp out. But people here must come with the expectation of meeting cold and hardship and possible suffering."

That many of the miners who penetrate into the wilderness in the hopes of amassing wealth do meet hardship and suffering is now an old story. The following words taken from the *Alaskan Searchlight* are in point at this time. Says the writer, who made the trip from Juneau to the Yukon in January :

"The miner of Alaska looks to the Yukon country for a repro-
duction of the scences of the Cassiar and Caribon districts. That
along that river and its numerous tributaries there are millions
of dollars hidden in the sands or locked within the mountains'
rock-bound walls there can be no doubt.

"For several years the more adventuresome of our placer
miners have been going to that Mecca of the North—Forty-Mile
Creek. Many of them have returned after one or two season's
sojourn none the richer, save in experience; others have struck
it rich and made for themselves snug little fortunes; and a thousand others are wintering there now hoping that next summer
may bring them that good luck for which they have so long
waited.

"Day after day, and season after season, the miners toil cheerfully at the bars and old water courses of the creeks and rivers
which form part of the Yukon system, and every year sees their
numbers increased, and every fall a larger quantity of gold finds
its way to the mints, and every spring the Alaskan steamers bring
several hundreds to join the fortune hunters of the interior,
Forty-Mile being the objective point of all going to the Yukon
gold fields."

Country Has Its Romance.

And this country so wild, so new, so unexplored, so lately
brought to the notice of the civilized world, virtually is not without its evidence of romance in the way of memorials that point
to former activities that now no longer exist, or mark the spot
of disaster or suffering. As far back as 1883 a forest of totem
poles rose in the great lodges of the Stikines village. In 1893
only a half dozen remained, and the "show pair" guarded a
cottage which replaced the ancestral lodge. One of these guards
relates the legends of the builder's family and the other that of
his wife.

Here and there on the route from southern Alaska to the gold fields the traveler will find similar relics, deserted hearths of a bygone day. This seems strange in a country so lately invaded by the white man. And this juxtaposition of the unknown, the unexplored and the relics of former peoples and former explorers will ever be a cause for wonder.

Speaking of totem poles, it may be said that this is one of the favorite occupations of the Indians. The traveler will be amused at the totem poles which are to be found wherever an Indian village dots the landscape. The natives make them by cutting down a good sized straight tree, dressing it to the desired size and then carving it in a very rude way with the figures of birds, Indian warriors and other fantastic shapes, which very much resemble Chinese carvings.

Totem Poles Come High.

After these poles have received a sufficient amount of labor and skill they are raised and planted on end before the owner's huts. Great value is attached to some of them, and the Indians who, strange to say, from their uncivilized condition, are the shrewdest of money makers, will not infrequently ask from $1000 to $2000 for a pole. This they consider a very reasonable price, and they are somewhat surprised when the traveler, who places no value on these rude works of art, smiles at what he deems exhorbitant figures which they place upon them. Mentioning the Stikines River naturally brings mention of the marvels of the fishing product of Alaska, owing to the fact that a large salmon cannery is located there. To one who has been accustomed to fish in southern waters, baiting a hook and pulling out an occasional fish, it would be nothing less than wonderful to sit down by the side of the Yukon or the Klondike or the Lewis or the Stikines rivers to fish for salmon. Fish not infrequently are so thick in these waters as virtually to impede navigation.

Salmon make their way up the Yukon in shoals 1000 or more miles, and are caught by the natives, or rather taken by the natives, by the ton. No Alaskan Indian would ever think of fishing with hook and line, or even spearing fish.

They will wait until the shoals come up the river. Then parties of Indians will get on either side of the stream with branches of trees, sticks and the like and beat the water, thus driving the fish to the shallow places. Here other Indians will be stationed with common pitchforks, and will stick and hand out the fish in quantities that would make them a drug in the southern market. These fish are often of an exceedingly large size, and when dried or otherwise cured make the staple of the native diet.

Greatest Salmon River.

It is worthy of note as one of the wonders of Alaska that the country has the greatest salmon stream in the world. This is the Karluk River. The stream rises on the west coast of Kadiak, and is sixteen miles long, from 100 to 600 feet wide and less than six feet deep. These figures, it is pointed out, give the dimensions of the solid mass of salmon that used to ascent the Karluk to a mountain lake before canners came with gill nets in 1884.

The largest cannery in the world is at Karluk. There used to be 1100 employes, and over 200,000 of forty-eight one-pound tins, containing 3,000,000 salmon, was the output. A single haul of the seine in this river has reached 17,000 salmon. Yet each ebb-tide then left thousands of stranded fish to die on the banks and bars.

In the palmy days of the canning industry the canners enjoyed a monopoly without tax, license or any government interference. The nearest United States commission was 700 miles away. Stores, employes and pack were conveyed to and

from San Francisco in the canners' own vessels, and the hun
dreds of Chinese, Greek, Italian, Portugese and American workmen constituted the most untrammelled community anywhere
to be found under one flag from May to September of each year.

Won't Cure Their Catch.

Often the supply of fish is so large that the natives will not
even take the trouble of caring for their catch! The fish are
simply piled up and allowed to rot for compost. It might be
mentioned right here that one of the favorite dishes of the native
Alaskans would be a marvel to southerners of a more refined
taste. They will cut off the heads of the salmon, put them in
a hole, bury them and leave them for weeks to rot. Then there
will be a general gathering of the clans, and the deposit of the
fish hole will be opened, and the unsavory mess will be parceled
out to be eaten by the natives as a delicacy. And nobody calls
stinking fish!

In this wilderness of mountains, with their snow-capped
peaks; plains, with their almost barren and desolate features;
and rivers, with their almost endless, tortuous courses, where,
until recently, and by recently one means the time of the purchase of the country from the Russians, few ever ventured, the
traveler will be surprised at the almost utter absence of game.
He would naturally suppose that where the white man has been
for so short a time, would be a sportsman's heaven. The contrary, however, is true. Here in this wilderness, there is almost
an utter absence of game for the reason that the miners, who
have been at work there, finding it impossible to get fresh meat
from the south, and wearying of canned goods, have literally
driven game from every locality into which they have set their
foot. The result is somewhat curious.

There are in Alaska districts comprising hundreds of square

SCENE NEAR JUNEAU CITY.

205

miles that are solitudes in the strictest and truest sense of the word. The white men have not been induced to settle there, natives have moved away, and all the animals have been driven away to such an extent that, barring insects, there is no indication of life in the territory. Solitude and silence reign supreme. If there is a sound, it is due to the wind sweeping down the gulleys, upturning trees or something of that sort.

It is worthy of notice that while Alaska may, in a certain sense, be said to be the home of the Aurora Borealis or Northern Lights, and displays are frequently seen covering the entire northern sky with a brilliancy of color that it would be worth going hundreds of miles to see, electrical storms are something of a rarity in Alaska. A cyclone is a thing unknown.

Still, in the summer season the rain-fall is marked, but it comes without the attending electrical disturbance that is so common a feature in southern latitudes. This may possibly be due to the comparative dryness of the northern air. The dryness by the way has the effect of tempering the air and mitigating the intense cold.

Cold Scarcely Noticed.

Even with the thermometer at eighty or ninety degrees below zero at Dawson City, Circle City or any of the other mining camps, the intense cold is really not noticed. It would seem very strange to a person used to southern weather to hear a native or a person who had lived for a series of years in Alaska, talking about its being a warm day or a mild day, with the thermometer at sixty-five below. Yet, this peculiar characteristic of the weather, extreme dryness with extreme cold, makes this a common saying among the people.

No chapter on the Land of Wonders, as we have called Alaska, would be complete without reference to the mosquitos, which are one of the greatest nuisances of the country. The Yukon mosquito is a giant among insects and is king of his tribe. It may seem like a yarn, but it is said to be an actual fact that the mosquito actually hunts and kills bears along the Yukon River.

Lieutenant Schwatka, the well-known explorer, who visited the Yukon some years ago, is authority for this statement. He assures us that the bears, under stress of hunger, sometimes come down to the river in mosquito time, and are attacked by the insects, who sting them about the eyes and cause them to go blind and die of starvation. A prominent Yukon miner, who has spent years in the country, has published the statement that he has known mosquitos to bite through a thick moose skin mitten.

The natives, who are born and bred to the nuisance, are forced to smear themselves with grease and soot to keep off the pests. Often miners are forced to resort to the same expedient or to work with helmets of gauze to protect themselves from the bites.

Natives of Great Interest.

Apart from any consideration of scenery, industries or resources, the natives themselves will ever be a source of interest if not of wonder to the voyagers. Shrewd and enterprising in their way, they are yet children of nature and have all sorts of notions that will strike the stranger as odd if not ludicrous.

Chatham Strait, for instance, is a playground of inferior whales, great totemic creatures, which the Indians believe were once bears, but going to sea wore off their fur on the rocks and had feet nibbled off by other fishes. The all-mischievous raven, they say, often creeps down the whale's throat, and causes such agony that the whale rushes to the shore and vomits the intruder on the beach. Paintings and carvings showing the demon in the

whale's body are often taken as proof that the Indians have a
Jonah legend, and are of direct Asiatic descent.

Another of these old Indian legends that is constantly told to
strangers concerns the all-present glacier.   They say that in
their fathers' time, which may be taken as an indefinite or inde-
terminate period anywhere from fifty to a thousand years, the

SCENE IN SOUTHERN ALASKA.

ice reached as far as Bartlett's Bay.   About 1860 it was in line with
Willoughby Island.   The Indians say that long, long ago the glacier advanced and swept away a city on the sands at the base of the
mountains, where the Beardsley Islands now rise.   They say it
came down in a day and did not go away in ten years, and tell how
the ice floods descended, ploughed up the fields, destroyed their
houses, as the Gorner glacier once devastated its valley.

Again they say, a great wave rushed in from the ocean, swept away the village near Bartlett's Bay, mowed down the trees with icebergs, and left no living thing. They say further that a glacier once crept down and damned up their best salmon stream. Two slaves were then offered up to the evil god that caused the mischief.

Tell Legends as Facts.

Legends like these, told as positive fact, coupled with odd ways of thought and dress and action, make the Indians an interesting study. They seem in a sense fitting denizens of the wilds of the territory. An ampler account of these Indians, however, will be given in the chapter on ethnology.

In conclusion, it may be said that one of the wonders of Alaska is the Treadwell mine, on Douglass Island, near Juneau. This is the largest quartz mill in the world, and one well worthy of a visit from anyone wishing to know the process of operation followed in that particular form of mining. It should also be remembered that it is only a short walk from Juneau to the placer mines, so that those who do not wish to penetrate into the barren wilderness of the North in search of adventure or wealth, but who wish to see placer mining and know how it is done without the hardships incident to the long overland journey on snow shoes or on sledges drawn by dogs, can have their curiosity gratified and can gain the information desired on a jaunt for pleasure.

14

CHAPTER VI.

Women at the Mines.

THE gold mines on the Klondike are not without their romance, and by this is meant, not the romance of speculation and adventure, but the romance of real life in which the gentle sex figure. The poet Compbell, years ago wrote the couplet:

> "The world was sad; the garden was a wild:
> And man, the hermit, sigh'd—till woman smiled."

Some Klondike Campbell sighed, and women all over the United States smiled. At least they were among the first to catch the gold fever and brave the dangers and the hardships of the Alaskan wilds.

What is more, they contracted the craze just as badly as the men, and many of their enterprises and their hobbies were no whit less out-of-the-way and outlandish than those of their brethren. From Maine to California women of enterprise and courage, many of them of education and gentle birth, flocked to the North in the wild rush to secure wealth by a lucky stroke.

Women who had never known hardship in any form, did not hesitate to leave comfortable homes and brave the unknown. From the very outset the officers of the great transportation companies received a numerous mail from the women of the

210

country, making inquiries as to the outfits necessary for them and the cost of transportation, and what they would likely have to undergo in carrying out their projects to penetrate to the interior of the gold region.

Women with Great Schemes.

Many of these women came with schemes by which they hoped to attain wealth, not by mining and prospecting, but by catering to some real or fancied needs of the miners. Others again expressed their determination to become prospectors and *bona fide* miners. Not a few did not hesitate to admit that they were going to the unknown country in hopes of meeting some miner who had made a happy hit and amassed a fortune, whom they might captivate by their charms and thus secure at once both husband and opulence.

Conspicuous among these women who lent the charm of their presence to camp life were several women of note, who, actuated by different motives than the great mass, made the long, perilous journey over the snow-clad plains and mountains, and up the dangerous rivers as far as Dawson City, Circle City and Klondike. Some of these had had previous experience of Alaskan summers and winters, and knew what it was to live in moonless nights and sunless days. Several of them left their homes with the avowed determination of wintering in the fastnesses of the North.

Among these women conspicuous for their social position may be mentioned Mrs. Eli Gage, wife of the son of Secretary Gage of the United States Treasury. Mrs. Schwatka, wife of Lieutenant Schwatka, the well-known explorer, and Miss Anna Fulcomer, who first went to Circle City, under the auspices of the United States Government, to teach the Indians and gather facts for the Smithsonian Institution. A word from such women will be deemed welcome to those members of the sex who may have it in mind to brave the perils of the North.

Mrs. Eli Gage came from St. Michael's on the ship which brought the Klondike argonauts back to civilization. Her husband, who is prominently connected with the North American Company, is the man who traveled 1500 miles overland last winter and brought out of the centre of Alaska the first reliable news of the wonderful strike in the Klondike region. He is a stockholder in many valuable claims in that vicinity.

Mrs. Gage returned in August to the far Nortwest to join her husband, with whom she will spend the winter at Dawson City. She was accompanied by W. W. Weare, second vice-president of the North American Transportation and Trading Company, and several friends of herself and her family were in the party.

They "went in" by way of Juneau and the Chilkoot Pass, the brave young wife making light of the perils incident to the 800-mile journey over the icy mountains and in an open boat in Arctic weather, to join her husband at the Klondike capital.

Voyage in a Yacht.

A specially constructed yacht was built for the party in Toronto, planned and fitted out expressly for the various exigencies of the voyage from Lake Linderman to Dawson. It was shipped in sections to Dyea, and thence was "carried" over Chilkoot Pass and put together on the shores of Lake Linderman, whence the long water voyage began. It was provided with many comforts and even luxuries to make the journey as little like the rough, hard experiences of rafting or canoeing as possible, and still was far from being suggestive of the winter luxury of the elegantly appointed home in Chicago which Mrs. Gage abandoned to share with her pioneer husband the rigors of a close season in the polar climate of Dawson City.

But her home on Evanston Avenue was, in her mind, at least, the most unimportant of the many things Mrs. Gage left behind her in Chicago when she started on the year-long trip into the northern wilderness. Her fifteen-months' old baby was thought too young to undertake the hard, hazardous journey, and was left with friends while the young mother hastened off to the Klondike to be once more with her husband. When she sees her darling again the baby lips will have learned the use of speech to welcome her, and the tiny feet will know how to fly to greet her coming.

Tells of the Gold.

In speaking of her trip down on the Portland in July, Mrs. Gage said:

"It is almost impossible to tell how much money the Portland brought into the States. The boat was filled with returning miners and prospectors, and the smallest deposit in the ship's safe was $15,000 in dust and nuggets. There were many others —so many that the captain's room was like the treasure store of a king. It was literally filled with gold in all forms, and while I sat in the midst of the wealth it occured to me that the old trade of buccaneering had missed a rare chance in not waiting by the sea road for this load of gold.

Mrs. Gage says there is a wonderful quantity of gold in the Yukon field and any man who has $500 for " grub-staking " a claim need have no fear in going to the Klondike region in hopes of a rich harvest, for he is sure of gaining it.

Even though a man go poorly equipped and supplied, he rarely receives poor treatment from the hands of his neighbors, and may find plenty of work to do which will enable him to earn from $15 to $17 per day.

Mrs. Gage speaks well of the people who make up the population, dwelling on the fact that they are a class who may be trusted, and that they form a desirable community. The valley of the Yukon is not populated with such men as constitute a large part of western mining camps.

One thing Mrs. Gage particularly emphasizes. It is that there is absolutely no truth in the report of famine. It has been said that starvation would overtake many who went to Alaska this fall, but Mrs. Gage is firm in her belief that enough supplies are being taken from Seattle and San Francisco by the two trading companies in Alaska.

" Those in charge of the business of these concerns," she said, " are making ample preparations for the coming winter. They fear no famine, and the individual miners are taking advice and are already supplying themselves with necessities. There is gold enough in Alaska for everyone."

Reverting again to the marvelous golden treasures of Alaska, Mrs. Gage said with enthusiasm :

" Four great Alaskan miners came down with us, and a more than interesting sight was to go down into the great safe on the ship and see the bags of gold dust. There have been many fortunes found in Alaska, yet there is gold enough to satisfy everyone.

" Mr. Gage is at Dawson and will not return until spring. He is constantly busy and likes the life. Since my arrival in Seattle I heard that a man whom he had ' grub-staked ' has dug up gold worth $35,000 in three months on a small claim. If a man goes out there without money he can very soon earn it, for wages paid, even for common labor, in all the region range from $15 to $17 per day."

No Fear of the Trip.

Just before leaving Chicago for Dawson City, Mrs. Gage said :

" My husband and I were separated over a year, and he spent the time in a log cabin at Circle City while I lived in Chicago.

This year I have decided to go to him. I am not afraid of the trip. I have been to Alaska and I know the stories of hardship are much exaggerated. If one is well prepared for the journey there is really no great danger. There is no use for doctors in Alaska."

Mrs. Gage is not a large women, but she said she never enjoyed better health than in Alaska, despite the cold.

"It is such a dry cold one hardly feels it," she said. "And I am not at all afraid. Women are always safe in the Yukon. Although beer and liquors are sold, the men are rarely disorderly and those who do become outrageous are quickly put in order by the majority. Dress, employment and other circumstances make the men of the Yukon often to look and seem uncouth and coarse, but at heart they are noblemen, and this is in no way more agreeably shown than by their courteous and gentle treatment of women. But women going to the Klondike must make up their minds to live in a primitive way, and be prepared to endure hardships incident to a new and Arctic country."

Mrs. Gage's Outfit.

Mrs. Gage's outfit in many things is like that of a man going in to "rough it" in the wilderness, and her brother, of course, looked out for the food supplies for the journey. Yet, it may be of interest to women who think of going to the Klondike overland to know that this dainty daughter of wealth carried for daily wear two short heavy skirts of waterproof cloth made a lá bicycle skirt, a heavy fur coat, warmly lined and with pockets enough for a man, besides a lined hood attachment to be drawn over the head and face in cold or stormy weather, several pairs of stout boots, warm leggings and overshoes, a mackintosh and a fleece-lined sleeping-bag. Then there was plenty of the softest, warmest underwear in the hamper, and at Juneau Mrs. Gage will supply herself with reindeer hide boots, made with the soft down inside, long, tight and loose, which will answer either to keep out the water in case of accident necessitating wading ashore or during a possible wet experience on a portage or going over the Chilkoot Pass, or to keep out the cold if at any time the more civilized boots and leggings fail to meet the demands of the Arctic temperature.

"Not a powder-box nor a curling-iron the outfit," Mrs. Gage said, with a merry laugh, as she enumerated the list of her baggage, or "luggage," as she preferred to term it, not inappropriately, because it would have to be "lugged" so far and often, "and only a small hand-mirror. Women don't have to 'dress up' to be appreciated on the Yukon, I assure you."

Mrs. Schwatka no Novice.

Mrs. Frederick Schwatka was no novice in Alaskan experiences. She had been there with her husband and had been over much of the ground that it is necessary for the prospectors to traverse on their way from the coast to the gold fields. She was fairly familiar with the various routes commonly followed by explorers and miners, and she expressed herself to the effect that the Taku Pass would prove to be a bonanza to the first trading company that established a system of pack trains from the Taku Inlet through to Juneau, which is the base of supplies for the mining region.

Besides being the easiest route for the miners themselves, it was, she thought, preferable, because a shallow draft steamer could be brought to run on the Taku river, which would leave only ninety miles of land to be crossed to get to Juneau.

Mrs. Schwatka, in discussing the difficulties of the journey from southern Alaska north, said that her husband had explored the Taku River and Pass a number of years ago and that he tried to get the people of Juneau to establish a pack train line through the pass to connect with a steamboat on the inlet. That, she said, was before there was much travel through Juneau. The people of the then thriving village did not believe that it would be a success financially.

Grounds for Her Belief.

Now she thought there was no doubt whatever that it would be a paying venture and would be a boon to the multitude of people who were pressing on to the gold fields. Said she:

" In fact, the pass contains an excellent railroad grade, and it would cost a comparatively small sum to build and equip a road through the ninety miles between Juneau and the inlet. The current of the river is strong and there are frequent floods, but a light draft steamer would have no difficulty in ascending it and making connections with the road to Juneau. It would be an easy matter to get supplies from Juneau then. The Canadian Pacific comes so near to that country it seems as if it could profitably build a line through the pass and connect the two branches by steamer.

" Lieutenant Schwatka made a map of the region, which I think I shall have published. He made the trip up the river by canoe and reported the current there very swift and strong. I am certain that the Taku route is the easiest for persons going from Juneau, however."

Mrs. Schwatka, like most people who have had any lengthy experience in Alaska, had much to say of the great territorial pest, the mosquitos. This nuisance—not nuisance, evil is a better word—cannot be overlooked by those who purpose to leave the States for the plains and mountains of Alaska.

" The pest," said she, " is not so observable, of course, very early in the spring or late in the fall, but during the mining months the mosquitos are simply intolerable. The Indians even, who are hardened to them, have to go about in summer with their hands and faces smeared with pitch and lampblack. The ordinary mosquito netting is no protection whatever, because the mosquitos force their way through it.

Mosquito Bites Fatal.

"Many of the miners, in addition to adopting the plan of the Indians and anointing themselves with pitch and lampblack, work in summer with their heads in a wire frame covered with close netting. I have even known persons to die merely from the bites of the mosquitos.

"This is something for the women who purpose to try their fortunes in the gold fields to take into consideration. They will find it is no place, either in summer or winter, for either the dress or manners to which they have been accustomed in their southern homes.

"Imagine, for instance, a society belle, or a woman who has had gentle rearing and been accustomed all her life to the ordinary convenience and comforts of civilized life, going into the wilderness of a country about which we know very little, donning the costume largely of the natives and subjecting themselves to all the hardships and privations necessarily incident to a residence in that country. Especially imagine such a woman smearing her face with soot and grease by way of cosmetic and wearing over her coiffure a helmet that would put to the blush in point of looks and inconvenience the shields commonly worn by the men who stand behind the bat in the game of base ball."

Speaking from personal experience, Mrs. Schwatka continued:

" In the summer it is so hot in the river regions that even the moose are driven away, and it is practically impossible to get game there, in spite of the reports that are sent out. It will not take a very great increase in the white population to kill off all the game there is. The Indians are pretty careful and don't kill any more than they need for food, but it will not be that way with the whites.

"The salmon do not ascend the Yukon as far as the Klondike, either, and fishing in that region is not nearly as good as it is made out to be. It would be taking a great risk to go there depending much on the natural resources of the country for food.

"Prospecting in Alaska is altogether different from what it was in California. There is as much difference between the mountains in Alaska and the most mountainous parts of California as there is between the latter and the Indiana avenue pavement. California is a flat plain compared with it. All of the Indians up there die of consumption, partly brought on by the climate and partly by the hardships they have to endure.

Steps in the Ice.

"Why, I have seen these Indians, who are used to the country, come in with packs from the very same passes which the miners are now crossing with welts across their backs from the pack straps almost as thick as my wrist. Their hands would be torn and lacerated horribly. The only way they can get through at all in the winter is by cutting steps in the ice."

Mrs. Schwatka gave many interesting recollections of what she had experienced and witnessed in Alaska. Adverting to the climate she continued :

"About the middle of August heavy frosts kill all vegetation, and the country begins this early to take on an Arctic aspect. Furious gales begin to blow from the north, which continue with little cessation all winter. In September or October, at the latest, the river is frozen hard, and sledging, as in the Arctic, is the only mode of travel in the country until the great spring freshets in May set the rivers free. As you can readily see, the journey to the gold fields by this route is not only a very long one, but a very expensive one, and wholly impracticable for numbers in winter. The average miner and prospector must enter Alaskan fields by a shorter and more accessible route, even though the hardships encountered are greater.

For a number of years past, miners going to and from the placer gold fields at Forty-Mile Creek, and Dawson City, and Circle City, have used the Chilkoot Pass, outfitting at Juneau, the principal town of the territory. The Chilkat and White Passes have never been as popular with the miners as the Chilkoot. Therefore, I shall speak of the Chilkoot, as they are all quite similar.

" From Juneau the Lynn Canal is entered at Chilkat Harbor. This is the most northerly channel in the inland passage route. This Lynn Canal is divided by a long peninsula. The southern side is Chilkat and the northern Chilkoot. It is up Chilkoot Inlet miners ascend, and thence canoe up a rapid, glacier-fed mountain stream known as the Dayay. They are then at the foot, or near the foot, of the great pass. This so-called pass is really no pass at all, but a precipitous climb of over 3500 feet up bare, rugged rocks, and over great snow peaks, and across treacherous glacier ice.

Must Climb by Hand.

"So steep is the ascent that the hands of the climbers must be used to help pull themselves up. No white man can carry unaided the necessary amount of provisions and material required even to keep him from starvation until he can reach the mines. For this reason they rarely make the journey alone, but always in parties.

" It is necessary to bargain with the Chilkat Indians to act as porters and carriers over the trail. They have in the past carried loads of 100 pounds at from $10 to $15 a load. These they take over the dangerous and difficult trail to the top of the mountains or down to the first lake, which forms the source of the great Yukon River. Here again obstacles are met with, and it now becomes necessary to build a whipsawed boat, and the little timber to be found is unsatisfactory and stunted."

Mrs. Schwatka had also much to say of the prospects of the people who went there and what they would have to expect. She was satisfied that there were great hopes for the man of pluck, energy and perseverance ; but she was also convinced that it was policy even for people of this stamp to go expecting worse than had commonly been represented at the time when she was interviewed. Said she :

"I believe that a great deal of gold is going to be found along White River also. That is in Alaska, and not much prospecting has been done there yet, I understand. When I was last in Alaska, five years ago, the so-called ' Klondike' was an unknown and untalked of region and almost unheard of. Lieutenant Schwatka explored the country, and brought back a good many photographs and maps of it which are very interesting. I believe the Klondike is nothing more than a little creek, which, as it was about the first place in that region where gold was found, gave its name to the whole region, and has assumed the importance of the Yukon River itself in the eyes of the people who read about it.

" I have already spoken of the lack of work during the long winter season. It must not be understood that no work can be done, then, for many miners spend the winter prospecting in places where it would be impracticable in summer. On some submerged bar they build a fire, and when it burns down they pick and shovel out the gravel as far as the warmth has penetrated. This is repeated until they sink a shaft to bedrock. In summer the water pouring through the loose gravel prevents deep shafting except by expensive works.

Mining Very Difficult.

"Again in summer the work of the miner is difficult. As I have said, the interior country is tundra land—that is, the earth is frozen to a great depth, never entirely thawing out. Wherever the sun strikes the surface great pools of muddy water are formed, and this prevents any sort of prospecting. These pools of stagnant water breed great swarms of mosquitos and gnats, which make it desirable to cover the head with mosquito netting, or, better still, adopt the Indian method, and smear the hands and face with a mixture of grease and soot, which prevents the pests from biting.

"At some seasons in this country they are in such dense swarms that at night they will practically cover a mosquito netting fairly touching each other, and crowding through any kind of mesh. I have heard it asserted by people of experience that they form co-operative societies and assist each other through the meshes by pushing behind and pulling in front. Others again say they are too mean for such generous action."

In Mrs. Schwatka's opinion, Juneau was bound to be the most important trading centre of Alaska for the mining district, and she thought that it was eminently desirable that capitalists with the means at their disposal should take steps without delay to make more sure and ample the food supply of the Yukon Valley. The main reason why she insisted on this was, that the game had largely been driven away from the mining districts and that it was a menace to the health of those who had courage to penetrate the wilds to have to live week after week and month after month on dried fish, as the Indians do, or on canned goods exclusively.

Speaking of the scarcity of game in the Yukon Valley, Mrs. Schwatka said :

The great Yukon Valley has but little game in it during the summer, for the mosquitos drive all game to higher altitudes. Formerly during the winter season a living could be made by experienced hunters in bringing moose and caribou meat to camp. I heard one miner say, who had spent four winters on the Yukon, that he had seen moose and caribou so numerous on the bald hills above timber limit, in the present gold-field district that they gave the snow a mottled gray appearance.

"Of course these have now disappeared with the advance of civilization, and fresh meat of any kind is now at a premium. To illustrate how abundant this game was but a few years ago, a hunter captured a couple of young moose and they were made great pets among the miners during the long winter.

"This scarcity of game of all kinds," continued Mrs. Schwatka, "coupled with the great number of people entering the country, will in the near future be productive of great suffering, unless positive and decisive steps are taken to make the food supply ample and sure, as I have said. Tin and canned goods are very high in price, and it seems a wrong to the miners that, for a lack of ample transportation facilities, which, in my opinion, might be easily provided, they are subjected to the dangers of the diet they have to put up with.

Scurvy a Terror.

"Scurvy is one of the greatest evils of camp life, and this is engendered and fostered by the diet the men and women in the Klondike region have thus far had to endure. It is only a hearty man who, in face of the hardships and privations to which the mining community is subjected, can survive the six or eight months of dim twilight of the winter season, with the thermometer ranging anywhere from forty to ninety degrees below zero."

Mrs. Schwatka thought that great care should be taken by those who tempted fortune in the wilds of Alaska in the matter of providing a suitable outfit. She was convinced that a great many had gone and would likely go, who were little fitted to the experiences they would have to face, but, said she :

" Those who are determined to go should not only take the necessary winter clothing, but be prepared to invest in Arctic furs—a reindeer coat, suitable boots and leggings, and a fur sleeping bag. Skins of the temperate zone do not make the best clothing for this purpose. A reindeer sleeping bag will keep one warm in the severest weather and is a necessity, especially if one is to try to pass the winter in a tent, as I have heard many will do.

" Even the Indians of the country take extra precautions in preparing their lodges in winter, building houses of brush and logs. With proper clothing and plenty of nutritious food the problem in this land is easily solved."

Warning to the Sex.

In conclusion Mrs. Schwatka wished earnestly to give warning to her sisters who were likely to seek their fortunes in the unknown country. She said she did not wish to discourage those who thought it to their interest to brave the perils, but considering all things, and speaking from her own hard experience, she thought that the average woman would find it more to her interest, and certainly more to her comfort, to leave the dangers incident to the extremes of climate, dangers of diet, and hardships of travel to the men, who are better able naturally to support what will have to be undergone. Said she :

" To keep from freezing it requires the same sort of clothing that the Arctic explorers wear—all furs and no woolens. The fur coats are made by the Esquimeaux from skins brought over from Siberia, and it is likely that they will cost a great deal more than they ever did before.

"Alaska is a poor place for women and no place at all for children. Of course, many women are able to endure hardships and fatigue just as well as the men, and it might not be so bad for them to go there in summer. It is a fearfully hard life there at best."

Miss Anna Fulcomer, like Mrs. Schwatka, has had a former experience in Alaska. She is of Norse descent, and is thoronghly imbued with all her race's traditional love of adventure. As said above, she went to Alaska on her second trip as a Government employe, receiving a good salary and being screened from many of the hardships to which other women who went to the Alaskan gold mines were subjected. But she, like the rest, became touched with the craze for gold, and determined to leave her school in Circle City, which, soon after the Klondike fever broke out, became virtually a deserted town, and try her fortunes with the rest of the prospectors.

Got a Man for Nothing.

So she hired a dog for $30, agreeing to pay $75 if anything happened to the animal, and had a man thrown in for nothing. A few days after her determination to quit Circle City, she was on the trail of the gold-seeking throng. It did not take her a great while to discover that it is not all gold that glitters, and before she had been many days on her enterprise her hopes were a good deal like Alaskan weather, so far below zero, that she could scarcely read the thermometer. Some of her experiences

15

can best be told in her own language.   Said she relative to the
difficulties of beginning her enterprise :

"A dog, a dog, my kingdom for a dog," is the general cry
here.   Horses have practically proved a failure here as a means
of transportation.   They have to be housed in tents in which a
fire is kept.   The dogs, however, live on next to nothing, and
often make quite astonishing time.   We had a visitor at the
house I am living in, some time ago, who came on a dog sledge
eighty miles in nineteen hours, without once stopping.   Another
man came here 240 miles in five days.

" The relative value placed on men and dogs is shown by the
fact that I could get an experienced man for my trip to Klondike
for nothing, but had to pay $30 rental for a dog, and had to
make a contract to pay $75 if anything happened to the animal.
The hopes of hundreds here rest on their ability to get a bob-
tailed dog.   When I set out on my gold-finding enterprise I
found that my case was not an exception."

Good Word for Morals.

Miss Fulcomer has a very good word to say for the morals of
the Klondike mining camps.   During her year of residence at
Circle City she knew of no murder being committed, and of very
little lawlessness of any sort.   The miners, she said, practically
make a law unto themselves, and woe betide the man rash
enough or dishonest enough to violate the unwritten code.
Continuing, she said :

"One of the peculiar features of the new camp is the lack of
shooting, due to the fact that the Canadian government does not
permit men to carry firearms.   Police disarm miners when they
enter the district, so that there is not any of the lawlessness and
crime which marked early placer mining in California.   There is
much gambling and play is high.

" ' Lawyers and other disturbers of the peace ' are kept out, and this is the reason assigned for the quiet and order that prevail.

" The camps are in no sense to be compared with the camps in California during the gold fever there," says Miss Fulcomer. " Their inaccessibility in a large measure protects them from desperate characters. It is a 900-mile trip over the snow from Juneau to the gold fields, and it is a hardy person who would enter upon a trip that none but Arctic explorers ordinarily would undertake. The climate, too, makes living out of doors impossible, and it costs money to live under shelter. These conditions, as you will readily understand, help to keep away mere adventurers.

Side-tracked in Desolation.

" But it is a dreary place to be side-tracked in. The average miner and prospector is buoyed up by the knowledge that there is gold in abundance on the Yukon, and the hope that he may make a fortune quickly. For the rest of one's personal experience, the less glowing accounts that are given the better."

Like Mrs. Schwatka, Miss Fulcomer wished to emphasize the fact that Alaska in the mining regions is anything but a paradise. She said she pitied the people who came there under the delusion that mining life there was anything to be compared with that which obtained in California in the days of the gold excitement in that State. There were only four months in the year, she said—May, June, July and August—when mining was possible, and even then the ground thawed no more than two or three inches. The rest of the time the soil was virtually like a solid rock, and to make matters worse the thermometer was likely to be from ninety to ninety-five degrees below zero.

" One of the great causes of suffering here," she said, " is that Americans put on their heaviest clothing almost as soon as

they get here.   The result is that when regular winter weather
sets in and the thermometer gets down to eighty or ninety degrees
below zero, they nearly perish.   This, with the difficulty of getting good, fresh, wholesome living, makes the Yukon gold region
anything but an Eldorado.

" This," Miss Fulcomer explains, " is not because there is not
gold at Klondike—there is gold in abundance, dirt rich enough
on some claims to yield from $100 to $500 per pan ; but it is
mined with difficulty, mined in a small way, mined slowly, so
that for the average experienced digger the profits are swallowed
up in the expenses.   Men who have been mining at other points
in Alaska and the British Dominion virtually abandoned their
old claims, owing to the craze over Klondike, hurried there and
staked off their claims, and are holding or working them.   This
was early in the movement, and consequently newcomers have
to be content with the leavings of the old men in the work."

First to Cross the Divide.

Dawson City at the time the Klondike fever broke out in its
full intensity, had a population of 2500 souls, and of these only
thirty-three were women.   To Mrs. Tom Lippy belongs the
unique distinction of being the first to cross the divide and go
into the new Klondike camp.   She is described as a little, lithe,
brown-haired, brown-eyed woman, to whom fear is practically
unknown.   Unlike many of the women in the camp, she, for a
long time clung to her costume of civilization, dressed neatly
and even stylishly.   She followed her husband and her husband's
fortunes, and did not think she was doing anything out of the
way in braving the same perils he was obliged to face.   Said she,
when asked about her trip and her life in the gold region :

" I was the first white woman on the creek and the only one
in our camp.   There was another one mile from us, Mrs. Berry.

She was the only white woman I had to speak to while we were at camp. When we got to Eldorado Creek we lived in a tent until Mr Lippy got our log cabin built. It is twelve feet by eight, eight logs high, with mud and moss roof and moss between the chinks, and has a door and window. Mr. Lippy made the furniture—a rough bed, table, and some stools. We had a stove—there are plenty of stoves in that country—and that was all we needed. The cabin was cozy and warm. I looked after the housekeeping and Mr. Lippy after the mining.

"Everything we had to eat was canned. Things were canned that I never knew could be canned before. Of course, we missed fresh food dreadfully, but we kept well and strong. We had no fresh milk or meats or fruits or eggs.

Dearth of Amusements.

"Amusements? Well, nobody bothered much about amusements. Everyone was busy and kept busy all the time. I did my work. Mining is hard work—one doesn't pick gold off the ground. It is genuine toil, and when Mr. Lippy finished he wanted to rest. All men were about alike on that point.

"Fashion? Well, we were not entirely cut off from the fashionable world. People were coming in all the time. We got fashion papers, a few months old, to be sure, but still they kept us fairly up to time."

Most people who have taken interest in the report of the Klondike region will remember Joseph Ladue, who owns the site of Dawson City. On returning to Plattsburg, New York, early last August Mr. Ladue had some interesting gossip about women at the mining camps. Several of those who had faced the dangers of the journey to Klondike, he said, were doing well and would likely be large gainers by their enterprise. Said he:

"There are women there who own property. Susie Lamar is

one. She is a single woman who came from Germany. She
has been cooking for me and my partner. I guess she has done
pretty well. I pay her $40 a month right along.

"Lottie Barnes also owns property there. She came over the
divide two years ago and settled on Second avenue. She was
formerly in Circle City.

"There is also a Mrs. Willis, who has quite a history. She
went in with my party two years ago. In the party were Ellis
Turner, from Schuyler Falls ; William Lamay, George Mulligan
and myself. She joined our party at Juneau, where she had been
working in the laundry. She is about forty-five years old, a
blonde, stout and rugged. She pulled her own sled weighing
250 pounds from Lake Linderman through to Lake Labarge,
about 700 miles.

Women of Enterprise.

"Before she came there she was stewardess on the steamer
Willipaw, when I first met her. She went first to Circle City,
where she started a laundry and bake shop. She did pretty
well. I think she got fifty cents a loaf for bread—pound loaves
made from wheat flour. She went out two years ago as a nurse
for the steamship company. I think she went as far as San
Francisco. She returned the next spring. That time she
brought in herself, with the aid of two dogs, about 750 pounds,
including a sewing machine.

"That was not the first sewing machine brought in. Mrs.
Behan, wife of a banana trader, brought in the first machine about
twenty years before. Two years ago I suppose there were probably forty or fifty sewing machines in the country.

"There were pianos there. The pianos and organs were
principally in the dance-houses and theatres at Circle City."

Klondike is not much of a place, as the reader will readily

understand, for style, but once in awhile there is a "boiled shirt" to be seen there, and to Mrs. J. P. Wills, of Tacoma, is due the honor of introducing the first one. She is described as a women of iron will, whose husband is a gun or locksmith and virtually a cripple from rheumatism. His illness made it impossible for him to undergo the dangers of the journey and penetrate to the frozen North, but his wife said she would go for him, and go she did.

She settled at one of the mining camps and for two years made so little money that she was practically disheartened. Then the Klondike mines were discovered and Mrs. Wills was among the first to join a party of cattle men and hurry to the new region. She began her career in Alaska as a washerwoman ; then she went to work as a cook for the Alaska Commercial Company, at Dawson City, and received fifteen dollars a day for her services.

Her Experience a Romance.

When she joined the throng heading for Klondike she asserted her determination to abandon the work she had been doing and take a claim. She did so, and in a few weeks struck it so rich that instead of being a poor washerwoman she was worth a quarter of a million dollars.

While doing washing Mrs. Wills introduced the first "boiled shirt" into the Yukon gold camp and paid $2.50 for the box of starch with which she starched it. Her first assistant in the laundry was a squaw, to whom Mrs. Wills paid four dollars a day and board. Her little log cabin cost her thirty-five dollars a month and her supply of wood for the winter cost $225. A twenty-five-cent washboard cost her six times that amount, and, while she made a small fortune washing and baking bread, Mrs. Wills complains that the trading company got most of it. Mrs. Wills parts her hair on the side like a man and is stout and jolly.

She is fifty years of age and is industrious and a good business woman.

The Catholic Church has long had a representation in the frozen wilds of the North, but almost immediately when the Klondike gold fields were discovered, two Sisters of Mercy, young women from Lachine, in the Province of Quebec, headed their way for San Francisco on their errand of mercy, braving all the severities of an Arctic winter, that they might render such service in the camp life that might be demanded of them. The two young women belonged to the Sisterhood of St. Anne. When they started they did not expect to be able to go any further than St. Michael's, completing the journey at the earliest possible moment.

Mercy Their Motive.

When the girls started there were already thirteen sisters of the Order of St. Anne in Alaska, some at St. Michael's, others at Holy Cross and St. Joseph, and the remainder at Circle City. At this latter town the sisters run a hospital, and it was to work in the hospital for a time and then push on farther into the wilderness that these two brave young women undertook their hazardous journey.

Importation of young women into Northern Alaska as wives for the miners is the project one elderly dame laid before the officials of the North American Transportation and Trading Company. She figured that at least 2000 of the 10,000 hardy prospectors in the Klondike would like to get married right away and would be willing to pay a good price for the proper kind of helpmeets.

"I am organizing a company," she said, "and want your indorsement. You can make money off the transportation and board of the women, and the commissions from the miners will insure my company a big profit. Now, I want you to take some of the stock in pay for the passage of myself and two or three

agents while we run up there to make arrangements and——."
But Mr. Weare shut her off and made his escape.

Charlotte Smith, the Eastern sociologist, wants to transplant
4000 or more working women from sweatshops and factories to
Klondike camps.   Hers is not a money-making scheme—she is
laboring solely in what she thinks the best interests of humanity.
Transposition from a life of drudgery, with a bare pittance in the
way of wages, to homes in Alaska would, in Miss Smith's
opinion, be a blessing which thousands of women would be
glad to embrace.   To carry out her plans funds are needed, but no
big subscriptions thus far have been reported.   In the meantime
an enrollment is going on of those women who are willing to
take their chances in the frozen North.

Went for Business.

Another woman wanted to get $2000 to use in organizing a
company to locate gold placer claims.   She was endowed with
powers of clairvoyance and could unerringly point out hidden
deposits of precious metals.   She had done so with great success in California and Colorado, and would now like to try her
hand in Alaska.   Suggestion that clairvoyance should enable
her to pick out a backer was taken as a personal insult, and she
departed in high dudgeon.

There is a touch of romance and good fortune in the story of
Mrs. Capt. Healy.   She went to the Klondike region a poor
woman and soon became a mine owner.   Opposite the Klondike
River on the rocky cliffs that project into the Yukon is the
pioneer quartz mine of the country.   It was at this point that
what is known as the great copper belt crosses the river.   Captain Healy of the North American Transportation and Trading
Company, a couple of years ago, located on a ledge after a very
superficial examination of it.

Quartz mines were at that time practically ignored, and after a while the captain forgot the circumstance of his owning a claim, and made a trip on the company's business to Sixty-Mile. It was on this trip that he recalled the circumstance of his owning the claim, and, while passing it, made the remark:

"It's good-looking rock," said the captain, "but I don't think I will bother with it. There will be plenty of time for considering quartz."

"Aren't you going to claim it?" asked Mrs. Healy.

"No; I don't care to bother with it—not now."

"If you don't want it, I do. I will locate it and pay for the assessment work."

"Well, it's your mine, then."

Mrs. Healy Begins Work.

And so Mrs. Healy re-located it, and they set a man to work out the first assessment and took samples of the ore. Mrs. Healy named it the Four-Leaf Clover, so if anyone sees it quoted in the mining exchanges, away up pretty high, he may know it is her mine.

They gave the samples to the assayer, and they show from $8 to $16 to the ton in gold, in addition to a good percentage in copper. The vein is eighty feet wide.

Early in August, Miss Georgia Osborne, of Jacksonville, Ill., a miss of twenty-two summers, accompanied by Mrs. M. L. Keiser, of the same place, set out for the Klondike diggings. Mrs. Keiser said she had scaled the Alps and knew how to rough it, but Miss Osborne had had no experience of that sort, but was brave enough to face the dangers without question.

Miss Mary Elizabeth Mellor, Superintendent of the United States Indian Training School at Unalaska, Alaska, took a trip to the Klondike regions, and for a time experienced all the dangers and hardships of camp life. She returned to Seattle on the steamer Portland, early in July of the present year, and in speaking of the short summers and long winters of the northern wilds, of the scarcity of food and inadequacy of the clothing supply, touched upon the hardships of the miners and said their sufferings were often something terrible. She said:

"When I left flour was selling at the rate of $50 a sack, and if the luxury of eggs was indulged in the consumers paid $4 per dozen. Then it must be remembered that each egg of the twelve was not what a Pennsylvania farmer would consider freshly laid. Clothing is also hard to obtain and is high in price, the majority of the gold-seekers wearing clothes made of coarse woolen blankets."

Romance of Courtship.

Clarence J. Berry is commonly called the Barney Barnato of the Klondike, and his bride the belle of the mining district. The couple made one of the most fortunate strikes at the diggings. He took out $130,000 from the top dirt of one of his claims in five months, all of which was clear profit, barring $22,000 which he paid to his miners. His wife, the bride of but a short time, was equally as energetic and fortunate. She had her own claim and is reported to have lifted out $10,000 or more in her spare moments.

Berry and his wife went to the Klondike on their honeymoon. They were gone but fifteen months, came back wealthy to San Francisco, the happy possessors of claims that are supposed to be worth millions of dollars. And behind these millions of dollars there is a pretty romance which is worth relating:

Berry was a fruit raiser in the southern part of California. He did not have any money. There was no particular prospect that he would ever have any. He saw a life of hard plodding

for a bare living. There was no opportunity at home for get-
tiug ahead, and, like other men of the far West, he only dreamed
of the day when he would make a strike and get his million.

This was three years ago. There had then come down from
the frozen lands of Alaska wonderful stories of rewards for men
brave enough to run a fierce ride with death from starvation and
cold. He had nothing to lose and all to gain. He concluded
to face the dangers.

His capital was $40. He proposed to risk it all—not very
much to him now, but a mighty sight three years ago. It took
all but five dollars to get him to Juneau. He had two big arms,
the physique of a giant and the courage of an explorer. Presenting all these as his only collaterals, he managed to squeeze
a loan of $60 from a man who was afraid to go with him, but
was willing to risk a little in return for a promise to pay back
the advance at a fabulous rate of interest.

Pluck Carried Him Through.

Juneau at that time was alive with men who had heard from
the Indians of rich finds of gold, and had seen samples of the
rock and sand which they had brought. A party of forty men
was formed and Berry was one of the forty. Each took a complete outfit and a year's mess of frozen meat and sufficient furs,
packed the stuff to the top of the Chilkoot Pass and pushed on
toward the interior. Thirty-seven of the forty turned back in
despair, but Berry was one of the three who had pluck enough
to hold out, he being obliged to borrow bacon and other supplies to get through, and landing at the diggings without a cent
in his pocket.

He reached Forty-Mile within a month and began work at
$100 a month. He soon secured a claim and on finding him-
self on the highway to wealth sent word to Miss Ethel D. Bush,

to whom he was engaged, telling her of his good fortune and holding her to her pledge. Berry then went for his bride, and soon the couple were on their way back to the diggings.

They both decided it was worth the try—success at a bound rather than years of common toil. Berry declared he knew exactly where he could find a fortune. Mrs. Berry convinced him that she would be worth more to him in his venture than any man that ever lived. Furthermore, the trip would be a bridal tour which would certainly be new and far from the beaten tracks of sighing lovers.

A Remarkable Bridal Trip.

Mr. and Mrs. Berry reached Juneau fifteen months ago. They had but little capital, but they had two hearts that were full of determination. They took the boat to Dyea, the head of navigation. The rest of the distance—and distances in Alaska are long—was made behind a team of dogs. They slept under a tent on beds of boughs.

Mrs. Berry wore garments which resembled very much those of her husband. They came over her feet like old-fashioned sandals, and did not stop at her knees. They were made of seal fur, with the fur inside. She pulled gum boots over these. Her skirts were very short. Her feet were in moccasins, and over her shoulders was a fur robe. The hood was of bearskin. This all made a very heavy garment, but she heroically trudged along with her husband, averaging about fifteen miles each day. They reached Forty-Mile Creek a year ago in June, three months after they were married. They called it their wedding trip.

Berry built for his bride a log house, leaving simply holes for doors and windows. The thermometer was then getting to from forty to fifty below zero.

Mrs. Berry trudged through the nineteen miles of hard snow

and took her place in the hut with her husband. There was no floor, but the snow bank. It cost the couple $300 a thousand feet to get firewood hauled, and there was but little chance to use fuel save to thaw out the moose and caribou which the Indians peddled.

The bride and groom kept warm by cuddling—a thing somewhat unknown in civilized communities, but absolutely necessary with the mercury disappearing in the bulb, and wood worth its weight in gold. They endured all the hardships without complaining, since by this time they knew they had reached the golden pot at the tip of the rainbow.

All Credit to His Bride.

Berry gives all the credit of his fortune to his young wife. It was possible for her to have kept him at home after the first trip. She told him to return—and she returned with him. It was an exhibition of rare courage, but rare courage fails. The wedding trip lasted about fifteen months. Berry says it was worth $1,000,000 a month. This estimate is one measured in cold cash—not sentiment.

The new gold king and queen made the first strike of a year ago in November. They were working along Eldorado Creek, a branch of the Bonanza, which empties into the Klondike about two miles above Dawson City. Their site was the fifth one above where the first discovery had been made in this particular region. It took nearly a month to get into paying dirt, but when the vein was opened it was simply awful.

The first prospect panned two and three dollars to the pan. It grew suddenly to twenty-five and fifty dollars to the pan, and kept increasing. It seemed they had tapped a mint, and one day Mr. and Mrs. Berry gathered no less than $595 from a single pan of earth. This they saved in a sack by itself, and the people who have listened to the strange stories of the young man and his young wife have no fear that they have been mistaken.

Many Catch the Fever.

Thirteen women left Seattle for Alaska very soon after the Klondike fever broke out, and with them went the Rev. Father Stippick, who had for years been stationed at Circle City. Among the women were Mrs. Holmer Chase, Miss Pauline Kellogg, Mrs. C. W. Romley, all of Chicago. They all declared they were going to the new Eldorado, not for pleasure, but to seek their fortunes, the same as the men who had undertaken the journey.

One of the most striking instances of good luck at the diggings in which the woman is in any way concerned, was that of Ulry Gaisford, a Tacoma barber. Heartbroken, it is reported, over a wayward wife, he fled from his Tacoma home and sought to bury himself in the Klondike camps. He arrived there penniless, and within eighteen months found himself the sole owner of a Klondike placer, which is conservatively estimated as being worth $1,000,000. Within a few days after beginning to work on his claim the barber had taken out $50,000.

Ulry, it is said, brooked the conduct of his wife as long as he could, and then furnished her the money, on her request, with which to secure a legal separation. This formality completed, Ulry hied him to the wilds of Alaska, where he and his companions were shipwrecked while navigating the Pelly River, and provisions and clothing were lost. With absolutely nothing left but the clothing on their backs, almost all became disheartened and returned to civilization.

He pressed on, for a time working in a saw mill and later running a little barber shop in Circle City. It is with the trifle he saved from his barber shop and some money he saved in a logging enterprise on the Yukon that he filed a claim on the Klondike.

Mention was made above of Joseph Ladue, and there is a pretty romance connected with his marriage and good fortune. Many years ago, it is reported, he became enamored of a Miss Anna Mason, of Schuyler Falls, and they soon became engaged to be married.

The parents of the young woman objected on account of Ladue's lack of financial resources, and he went out to the Black Hills during the mining craze in that region. He was lucky and struck it rich. He corresponded with his sweetheart, and at last he thought he had enough money to return and claim the bride.

Lost a Fortune.

Leaving the mines, he tarried at Deadwood, was enticed into a gambling game, and his fortune passed into the pockets of sharpers. He wrote his affianced and told her the facts, adding that he was going to Alaska to make another fortune and hoped she would wait for him. Correspondence was kept up and the young woman remained constant to her faithful and adventurous lover.

When he visited his old home two years ago he was already prosperous, but he was not satisfied with his accumulations, and it was decided to postpone the marriage awhile longer.

He returned to his sawmill and trading post on the Yukon, and when the rich gold discoveries there brought him wealth beyond what he had dreamed of, he shaped matters as soon as possible to return and fulfill his long engagement. The parents are satisfied with his worldly prospects at last, and the wedding was celebrated at Schuyler Falls lately. That quiet hamlet was in a fever of excitement over the nuptails which crowned this romance in real life.

These cases are but a few of the many which might be cited as illustrations of the interest women have taken in the gold craze, and the earnestness and determination with which they have entered upon their life of hardship, toil and often privation. It is these women who are largely responsible for the high morals observable in the mining camps in the Klondike region. As said by Miss Fulcomer in the interview given above with her, the morals of the Alaskan camps are in no sense to be compared with those of the mining camps of California in the days of the excitement there.   This in a measure is due to the fact that the diggings are so remote and the journey to them is attended with such hardship and danger, that the looser class are deterred from threading the wilderness to the camps.   Thus, only women of nerve and enterprise, who have some legitimate purpose to subserve, have thus far made the trip to the diggings.

If the gold excitement continues nobody contends that this state of affairs will last, as it never has in former periods of mining excitement.   But thus far, on the Klondike, the women adventurers have brought only romance, good morals, and comforts to districts where they have been needed.

Women as Promoters.

Scores of women, some of them good-looking and of seeming refinement, have announced their willingness to marry anybody in the shape of a miner who has made a lucky strike, and in evidence of good faith have put their names and house addresses on record.   Others want to visit the Klondike as cooks, as nurses, as domestics, in any capacity so long as they can get there without outlay for fare, and with prospect of big wages at the end of the trip.

Women appear also as promoters of mining and development projects.   Some of them can talk intelligently about the country

16

and its prospects, and have a convincing way of setting out their propositions. One, a little keener than her competitors in the hunt for the dollars of the public, has sprung a plan by which stock may be paid for on the installment basis at the rate of twenty-five cents a week a share. In the spring—most of these good things are going to come off in the spring—experienced prospectors will be grub-staked and sent into the Klondike to look for a paying claim. The company has nothing as yet in the way of assets save expectations, but these are very big and strong.

A midwife advertises for a partner to furnish money to open a hospital in Dawson City. "On an investment of $5000," she says, "I will guarantee a yearly income of $50,000 sure, with the chance of making double this."

Fictitious Klondike stocks, with the quotations regulated by clock mechanism, have made their appearance in some of the bucket shops frequented by women. It is simply the substitution of Klondike for the old names on the tape, but the gamesters stake their money on the turns with as much eagerness as if the figures were wired from a genuine stock exchange in Alaska, and there is an observable spurt in the business. "If I can win $1000 here I'm going to the real Klondike just as quick as I can," said one woman customer in a La Salle street shop. While she was speaking a whirl of the wheel wiped out her margin, and she hustled around to borrow car fare to pay her way home.

Mrs. John A. Logan Interested.

Early in August, 1897, Mrs. John A. Logan was asked to become the president of an association of New York women organized to send a business expedition to the Klondike. The promoters of the enterprise were Mrs. Eliza P. Connor and Mrs. S. W. McDonald, both newspaper workers. The aim of the association was to send women to the Yukon.   Mrs. Logan was to attend to the work at the New York end of the line.

A Women's Klondike Syndicate was also organized about the same time in New York.   Miss Helen Varick Boswell was president, and among the patronesses were Mrs. Jennie June Croly, Laura Weare Walters, Des Moines, Ia. ; Mrs Sarah E. Bierce, Cleveland ; Mrs. William Creighead, Dayton, O. ; and Mrs. Sarah Thompson, Delaware, O.

"We expect to leave New York on March 1, 1898," said Mrs. McDonald, one of the officers, "and a Pullman sleeping car, or two cars, if forty people join us, will be chartered from New York to Seattle, and will be occupied exclusively by the members of the expedition.   Three meals a day will be furnished on the cars and all fees and tips will be defrayed by the party. The distance is 3310 miles, and we will make it in seven days.

Details of the Journey.

" From Seattle to Sitka, another thousand miles, we go by steamer, and it will take us four days.   From Sitka to Klondike is an overland route of 700 miles.   We will make a short stay at Sitka in order to complete the outfit of the expedition, which will be ordered by telegraph on leaving New York.

" We may decide not to go over the Chilkat Pass, but to take the Schwatka route instead ; we will decide that question at Sitka.   We will travel by caravans when we leave Sitka, where the vans will have to be taken to pieces and carried on horseback over the pass ; so will the tools and provisions.

" On the other side of the pass the vans will be refitted and the journey continued as when leaving Sitka.   When we reach the lakes rafts will be built from timber on the banks and the rafts will float people, horses and vans across.   For twenty persons there will be five vans, each with four horses, and three of the vans will be fitted with portable sleepers to accommodate seven persons each. The two other vans will be used for provisions, with sleeping bunks in front. For those wishing to sleep alone tents and army cots will be provided."

Romance of a Seamstress.

Mrs. Chester Adams, of Winlock, a small sawmill town in Western Washington, has written a letter from Dawson City in which she says that the steamer leaving there early in July for St. Michael's carried $2,000,000 in gold. She promised her friends to write the truth about the Klondike stories that have been printed telling of the great wealth of the Alaska gold fields. Her letter confirms all that has been said, and Mrs. Adams says half has not been told.

She went to Dawson City with a view to making a few hundred dollars at dressmaking. In the first three days she cleared up $90 with her needle. She says she was the first woman in the diggings that could fit a dress, and, while there are no "bones" or "waist binding or canvas" or other articles about which women know everything and which go into a dress, Mrs. Adams says prices are kept up, ranging about as follows: Five to ten dollars for a plain Mother Hubbard, six dollars to twelve for an empress, eight dollars for a plain wool skirt, ten dollars to an "ounce" for a waist. These prices are simply for making the goods up, and Mrs. Adams says she and her partner have more work than they can do.

CHAPTER VII.

Poet of the Sierras' Vision.

Rushes off to the Diggings at the First Report—Mining in '49—Goes in to Rough It—Carries His Own Pack, Pick and Pan—Will Hunt for a Good Job—Coming Back With Bed-rock Facts—Contradicts Some Horse Stories—Schemes of the Pioneers—Not a Pistol in the Crowd—One Way to Get Bear Meat—Recalls Other Big Strikes—On Mary Island—With Father Duncan's Flock—No Jail Nor Police at Metlakahtia—Hay on the Klondike—None Coming From Yukon—Frolic with Indian Children.

JOAQUIN MILLER, "the Poet of the Sierras," known so long and well to admirers on two continents by his *nom de plume* that his real name, Cincinnatus Heine, has become more obscure than another man's "alias," was one of the first of the old California argonauts to catch the Klondike fever.

As a youth he was a miner in the rich placer beds and along the gold-laden lodes of the Sierras, and again in 1862 he was in the rush to Salmon River, when Idaho and Montana were found to be gold fields.

The news of the marvelous finds on the upper Yukon was more than he could stand, and July 26th, little more than a week after the arrival of the Portland with its golden store, found him on board the steamer City of Mexico, upward bound for Juneau, Dyea and the Klondike.

Goes In to Rough It.

In a letter to the *Chicago Tribune*, dated enroute in the Gulf of Georgia, the poet wrote of his Arctic quest and its object in these words :

" I have been asked, as I have asked so many of our party, what equipment I have for the route over to the mines, and you may also want to know.

" Briefly, then, I have twenty pounds of bacon, twelve pounds of hardtack, half a pound of tea.   I have a heavy pair of blankets, the heaviest socks, underclothing, boots, a rubber blanket, a mackintosh, a pound of assorted nails, 100 feet of small rope, a sail, and an ax.   My pack is forty pounds all told.   I have a pocketknife and an iron cup, a thermometer, and about $100.

" I hope to build a raft, carry my own pack over all the places, and travel hastily on ahead and alone.   You see I have spent years alone in the mountains and have been in almost all the ' stampedes ' for the last forty years, and I know what I am about.

"Of course, I am not doing this for fun, but for the information of poor men who mean to go to the mines next spring. This is what those who pay me to take this trip want and what I have promised to do if it can be done without too much risk of life or limb.   I shall report exactly all the desired details as I go along.   I am to apply for work at the first mines I reach and report exactly, work or not work, wages, hours of work— everything, in fact, that a man of small means needs to know.

Will Hunt for a Job.

" If I make this trip thus equipped, find work and good wages and all that sort of thing, why, any other man who wants to can do it.   For I am about fifty-five years old and a bit lame of the leg.   Of course I may have to change some of my plans, may join a party and go down in a boat instead of on a raft, and so on ; but I am going to ask for work at all events, get it if I can, and do it, for I am an old miner and can do almost twice the work of a new man.   Certainly I can do more good just now in that way than by describing clouds, snow peaks and Polar bears, although, of course, I shall not all the time keep my face to the earth, even though my feet do cleave solidly to it.

"After having got right down to the bed-rock of the cold, frozen facts, I shall take the steamer at Dawson and return straight to San Francisco. So you see my forty pounds will be about all I absolutely need. But the 'stayer' will not follow my example in this. Still, I am bound to say right here that it does not at this distance look like practical common sense to waste so much time and strength in getting in supplies by this land route when they are bringing thousands of tons by the water route. However, I am sent out to tell of things as I find them, and shall give plain facts, neither opinions nor advice.

"More than all this, if I find the mines limited, either in area or thickness, my first duty is to let the world know. I shall write again when we get to Alaska, also again from the other side, or base, of the so-called 'terrible pass.' But once launched on the swift river and link of lakes flowing the other way, there will be only a monthly mail. Yet, if we find anything of great importance in the way of facts we will find some means of sending it back. If we do not find plenty of faint-hearted fellows coming back, even after crossing the mountains, it will not be in line with other excitements from '49 up to this hour."

Refutes Some Horse Tales.

According to Miller, the stories that horses were not available in crossing the mountains were not founded on fact, for he wrote of there being many horses on the steamer, all intended for use in going over Chilkoot Pass.

The poet was reminded, by some of the stories he heard at Seattle and Victoria, of the men who discovered the Salmon River mines in Idaho in 1862, and who sent out runners and posted notices to keep the people from rushing in and sharing the treasure with the discoverers. "Starvation and intolerable hardship" was the awsome argument used then, but history recorded that nobody really starved, though a number perished in the snow.

He writes in this vein :

" It would seem that those on the outside, as well as those on the ' inside ' have been most willing if not eager to keep all new-comers in the dark.   The men who have horses and all sorts of comfortable equipment are those who live along here—Seattle, Port Townsend, and so on—and are more nearly in touch with the inside.   Frankly and truly, each day I come upon some sort of evidence that those who know the most are playing the same old game that we of the Idaho and Montana mines played a third of a century ago."

Not a Pistol in the Crowd.

The poet was struck by the wide difference, in bearing and dress, between the gold-hunters of '49 and those of '97.   When he wrote he had not seen a pistol among the scores of men aboard bound for the mines, though there were rifles and shot guns in plenty, and he argued well from this for the figure the prospectors would cut when they got into the diggings.   "A miner of to-day looks more like a bicyclist than a booted and crimson shirted argonaut " was his happy way of expressing the eminently peaceful appearance of his companions.

One passenger on the City of Mexico, a Californian, had an outfit whose extremes were a frying pan and a gilt-edged copy of Shakespeare.

The poet pricks the starvation bubble thus neatly :

" One man returning from the mines told me this morning that he always had to keep the bacon up on a high pole, and had to grease the pole, for the bears were so bad that they would tear the cabin down, and even climb the pole if they could.   Now, it seems to me that while the bear up the pole was eating the bacon a man of reasonable wisdom could get a little of the bear if starving."

Though he disclaims any direct knowledge of the reputed strikes, the poet cannot forbear some characteristic observations, thus :

" You have no doubt read daily of great strikes. I will not add to the fever by uttering what I have heard all along the line. I am almost certain, however, that the mines are immensely rich. At the same time, let it be borne in mind that only a few millions have been brought to light. True, only a few men have a hand in the work as yet, but when I hear it said on all sides that these are the richest mines ever found it sets me to looking back. At first in the Idaho mines about a dozen men in Baboon's Gulch took out more gold and in less time than any dozen or so in the Klondike. The Klondike has given up only $2,000,000 or $3,000,000, but Alden Gulch yielded more than $100,000,000 from 1863 to 1873. The McGregor Company took out $2,000,000 in ninety days from Mount Gulch. They built a boat and took it down the river to St. Louis armed with Winchesters.

" At the same time, the mines are so different and the means of working the mines so difficult that they never could be worked at all if not marvelously rich. No one ever heard before of $500, $800, or $1000 to the pan."

These notions of a veteran gold-seeker are at least worth contrasting with some of the awed ideas of "tenderfeet."

On Mary Island.

From Fort Wrangel, Alaska, Miller writes again to the *Tribune* as follows :

" Mary Island, the place of customs and the postoffice, lies to the left of this mighty river, so like the Columbia, so like the Hudson, only ten times its size and impressiveness, and right befo.e us lies what the prospectors who come and go with us call a mountain of gold.   Men, especially an ex-Federal Judge who is with us, say it is the richest piece of ground in the world, and that the famous Treadwell Mine, with all its millions, is but a babe in arms in comparison with this mountain of quartz and gold that lies right in our path as we push on from the Custom House toward the gold fields of the Klondike.

" But it is an Indian reservation, and the Indians, a community under the leadership of a wise and good old Scotchman, known as Father Duncan, are reputed to be by far the best and most wise on the continent, and so the Government is loath to disturb them.   More than that, it is a point of honor to keep strict faith with them, for they are guests of ours."

With Father Duncan's Flock.

Then he draws a pretty pen picture of this peaceful Indian settlement, thus :

" You see, Father Duncan had a difference with the Canadian authorities about his converts, and begged the United States for an island where his people could live apart from miners and travelers with rum, tobacco and bad ways of other sorts, and as he had a great and good name as a civilizer, we gave him the island.   This was in the early eighties.   In the early nineties gold was found all along the steep, starry new home of the Indians from the tide wash to the snow that caps the peaks.

" Many efforts and appeals to dislodge the Indians have been made, but the Indians are so humble, and virtuous, and kindly disposed that they are pretty safe unless a very incompetent man comes to be at the head of this department at Washington.   A decision was rendered only quite recently entirely favorable to these simple savages.

" Their little city, Metlakahtia, is fairer to see from afar off as well as close at hand, than almost any city of the white man's side ; clean streets, a church that is almost a cathedral in stateliness, sidewalks, three or four fire companies, little houses for hose and hook and ladder companies at several points ; in fact, everything that the white man has except a jail, policemen and politicians.

" ' No,' said good Father Duncan with a smile, ' we have no need of either jail or police.   As for politicians, we have no need of them, and they, perhaps, have no need of us.' "

" The place is built and maintained on the co-operative plan, and is certainly prosperous, for the people are pefectly content and happy, and not one of the several hundred has any notion of going to the mines.   Let us take note of their condition here."

Raising Hay on Klondike.

A miner who had spent several winters in Alaska took the raw edge off the climate stories to the poet by telling him the climate at the Klondike was the exact counterpart of that at Metlakahtia.   The old Alaskan added:

" They raise the best hay there I ever saw.   I have seen grass as high as my head there in June, and cattle driven in from Juneau to Dawson are in better condition when they arrive than when they are started from the trail."

Miller said he followed up the cattle story and found it true.

He found out something about the Chilkoot Pass, also, and this is the way he puts it :

"And now for news, the newest news about the dread mountain pass, which, according to all received accounts, was to be undertaken only at the peril of life and limb.   Well, men all along here at the Indian villages and postoffices where we find men to talk to, tell me that the true news was not one-quarter as bad as published ; that last winter two mails were brought this way by English mail-carriers and three by American mail carriers, making the monthly mail trips over the sky-scraping glaciers and impassable pass as regularly then in the midwinter as they make it now in the midsummer.

"More than this, Mr. White went, almost a month ago, to cut a trail below and around the so-called death trap, and now it is comfortable. It is three or four miles longer, but it is of easy grade and a good, safe pack trail four feet wide.

"The first five miles is already a wagon road, so you see, as I prophecied on leaving Seattle, there was a whole lot of big stories told for the benefit of the far-off poor man who was trying to get to the mines.

"The nearer we approach the less formidable are all the obstacles before us. The walls of Jericho are already down and we have not once trumpeted.

"Why, if this keeps on, in thirty days more we will enter the Klondike country at Dawson in palace cars."

None Coming from Yukon.

Then, almost as he had penned the cheerful words, there came a shock to him and all the other 497 souls on board the City of Mexico. Let him tell it in his own graphic way as he wrote it to the *San Francisco Examiner :*

"A strange, a pathetic scene took place a little time ago. In the mildness of all this stillness, solitude, might and majesty of nature, we met a steamer, the Alki, San Francisco, coming right down upon us out of the clouds and snow. She had come from Dyea, the nearest possible point for ships to the Mecca of all good gold-hunting pilgrims. She came straight on as if to take us in her arms. Seeing that there was news and good news for all, she lay right alongside. The great ships ground their sides together. Our eager gold-hunters came on the decks by hundreds.

"News? News? What is the news from Klondike?

"Not the ghost of news from there, good or bad, thousands had gone forward and down the great river Yukon, but not a single one returned. A good sign, perhaps, but it was as if questioning the dead. And they were so few and so reserved and faint of speech and action, compared to our own great big-hearted and open-handed men, begging for news from the gold fields, that it was as if we had landed Charon's ship and demanded the secrets of his dead.

No Bacon nor Bread.

"Only one bit of news did they have to tell, and that was doleful enough; not a bit of bacon or bread at the trading posts ahead of us; and the Klondike, where there are plenty of supplies at some price, away over Juneau, on and on, hundreds of miles beyond the glittering mountains of snow before us. Men looked each other in the face, for many of the miners in their haste to get forward had brought no supplies at all, but expected to outfit at the posts at the base of the mountains, and that is why some will not sleep to-night. They will have to turn back or wait for the traders' ships to come from far away.

"It would seem that more men have gone into the mines by this mountain route than had been believed. Yet think how many are coming. We hear that ships by the score had been chartered and every berth taken in them by the time we were setting out. They will be along here the next week or the next, and likely enough lots of them, like some of our own boys, will have no supplies at all. But then, of course, there can be no suffering. There is plenty in the loads of the more provident, and these waters are always open and ships go up and down all the year. It is not like finding this state of things on the other side of the mountain, but it may make delays for a number of bold, good men, who have neither patience nor money to spare."

The poet had a charming experience with some Indian children on Mary Island.   He wrote to the *Examiner:*

" I was walking out of the edge of town, trying to get a knowledge of the place, when some children who saw me almost up to my knees trying to get some jack-in-the-pulpit plumes came to look and help if need be, perhaps.   Seeing at last what I wanted, they nimbly came into the brush and nettles and elder bushes and got all I could hold in my two hands ; great heaps of yellow, fragrant wild plumes, set off by red elderberries.   Now, when I got my wild flowers well in hand I said : ' Thank you, my little lady ; now, what is your name ? '

Five Cents and Ten Cents.

" She was about seven or eight.   She put her fat little hands behind her, and, turning about a great deal, her eyes down to the plank walk, where we now stood, she was silent.   Then I said again :

" ' What is your name, my good little girl ? '

" She turned about a great deal more, with her eyes held to the levels, and then said :

" ' Ten cents, ten cents.'

" I offered her ten cents, but she would not take it.   Then I offered her a quarter, but the little brown hands were in hiding and would not come out, coax as I might.   Then I turned to another little girl, her sister, perhaps, and said :

" ' What is your name, little girl ? '

" She was not so shy, but, lifting her tiny black eyes to mine said :

" ' Five cents, five cents.'

" I offered her the quarter, but she tried to dig her little big toe into a crack in the plank, turning her bushy black head to me, smiled, and tried to laugh a little, but she would not put out her hand. When the whistle blew I hastened aboard the steamer, they following at a little distance. Then, having a moment to spare, I turned and said again :

" ' Now, pretty, what is your name? I like you and would like to tell my friends about such a good little girl. Please, now, what is your name ? '

" ' Ten cents, ten cents,' she answered.

" 'And her name ; what is your little sister's name ? '

" ' Five cents, five cents.'

" I laid some little bits of coin on a stump and ran away for the steamer, and I reckon I never will know whether they wanted money or not, but am inclined to believe their names were Ten Cents Ten Cents and Five Cents Five Cents."

CHAPTER VIII.

History and Purchase of Alaska.

One of the Happiest Deals Ever Made by American Statesmen—Seward's Glory—His Prophecy on Retiring to Private Life Verified—Comparatively Few People in the Territory—Story of the Early Days of Russian Occupation—The First Massacre—Country Onee Offered to the United States for Nothing—Appropriation for Money to Pay for the Tract Opposed by Congress Bitterly—Efforts to Provide Country with a Government—Interior containing Gold Fields onee thóught Worthless was Parceled Out in Thirds between as many Nations—Recent History.

L ITTLE as is known of Alaska among the sisterhood of countries having a place in history, its records go back early into the Eighteenth Century and are more replete with interest and romance, than most people suppose.

Its discovery was due to Peter the Great's craze for exploration, and from the time Vitus Bering sailed by commission of the Czar to find the fabled land of Vasco da Gama, to the days when the Klondike fever broke out in its intensity and became the talk of the world, it has ever, in some form or other, had something of a conspicuous place in the public mind.

The purchase of this vast tract by the United States was one of the happiest deals our statesmen have ever negotiated. The country was bought from Russia in 1867 at the ridiculously low figure of less than half a cent an acre. From the very outset the investment has been a paying one, as is clearly shown by Dr. Dall's figures.

Alaska paid a net profit of eight per cent. on the purchase price during the first five years it was owned by the United States. The government leased two tiny seal islands, which alone paid four per cent. on the original cost of the entire territory, which was $7,200,000.

In addition to the profit returned by the fisheries and the seal islands and the mining of baser metals, the output of the gold mines before ever Klondike was thought of, yielded to the United States a sum far greater than the purchase price.   As an indication of the profit of the fisheries it may be pointed out that in six years, from 1884 to 1890, the salmon industry alone yielded $7,500,000.

Few There to Work.

In considering these figures relative to the profits of this great and virtually unknown country, it must be borne in mind, that it is one of the most sparsely settled regions in the world.   In 1893 there was but one inhabitant to each nineteen square miles. Thus far in the history of our country it has been a territory practically without a government, and only of late, since the gold fever broke out, has the general public given it much attention. A review of its history therefore will be acceptable to the reader.

It was in 1728 that Vitus Bering discovered the straits separating Asia and America, and it was in 1741 that he started out to find the fabled land.   He had two vessels on this journey which were separated in a storm about the latitude of 46 degrees north.   Bering sailed northeast and reached Kayak Island on St. Elias Day, July 17, 1741.

There he saw and named the great mountain, touched at the Shumagins, and was wrecked on the Comandorski Islands. There, too, the commander died.   But the scurvy-stricken crew survived and reached Kamschatka, with the pelts of the sea otter on whose flesh they had lived.   The sight of these furs stimulated traders, and from that day on Alaska had something of an interest for the Russians.

Tschirikow reached the coast near Sitka and sent a boat's crew to explore the bay.   The party spent six days in recon-

17

noitering and at the end of that time a search party was sent after them. The natives at this time were defiant and paddled out to the ship, and raised such a din on shore as probably was never equaled in the region.

Gregory Shelikoff, a rich Siberian merchant, was practically the first to establish a regular post in the country. This was done in 1783, on Kadiak Island. A regular trade was then established with the Russians in Siberia. Baranof pushed his enterprise also when he started it in May, 1799, in every possible way. He reached Sitka Sound and built a stockaded post three miles north of where the present city of Sitka stands. An imperial charter, with monopoly of the American possessions for twenty years, was also obtained by Resanol, the son-in-law of Shelikoff, and Baranof now became the virtual head of the Russian-American Fur Company, in which eventually nine rival Siberian firms were consolidated. In this great concern several members of the Imperial family were stockholders.

The First Massacre.

Such was the discovery of Alaska, and such the founding of its capital, Sitka. The old fort at Sitka was destroyed in 1802, and all, save a few Russians, who found refuge on a British trading ship, were murdered. At the time of the calamity Baranof was absent, but he returned two years later, in the month of August, with 800 Aleut and Chugach hunters. At the sight of Baranof and his band the Indians, who had murdered the Russians, fled, and, retreating through the country, destroyed villages wherever they came upon them.

Soon afterward, Baranof contemplated building a fort on the Columbia, but, through Resanof, he opened trade with the Spanish colonies in California. Resanof, whose wife had died, paid court to Donna Concepcion Argeuello, daughter of the Alcade of San Francisco Bay. They were betrothed, and it was while on his way to St. Petersburg to obtain the Czar's consent to their marriage that Resanof died in Siberia.

It was about this time in the history of Alaska that the Fur King, John Jacob Astor, began to figure. Baranof was suspicious of him and his many ships, and distrusted the New York trader's offer of a permanent alliance of interest.

It is worthy of note here that Baranof was the first man to attempt agriculture in this barren region. He established a regular agricultural colony. He was popular among the natives, who uniformiy called him " Master," and apparently none of the Russian governors of the country after him were quite so acceptable to the Indians.

Emperor Nicholas' Offer.

American interest in Alaska, of course, dates from the negotiations which terminated in the purchase of the country. The Emperor Nicholas always had a warm spot in his heart for the American nation, and in 1844 he offered to the United States the entire Alaskan territory for the mere cost of transfer, if President Polk would maintain the United States line at 54 degrees and 40 minutes and thus shut out England entirely from frontage on the Pacific. This generous offer, however, was not accepted, owing to diplomatic considerations.

Again, in 1854, the country was offered to the United States, and still again in 1859, when $5,000,000 was refused. From 1861 to 1866 surveying parties traversed a good portion of Southern Alaska, choosing a route for a telegraph line to Europe, via Behring Strait. The success, however, of the Atlantic cable in 1866, after the failure in 1859, ended this project, and the cable line to the west was abandoned.

Then, seeing that the government evinced so little interest in the great country to the north, about whose resources there was a great difference of opinion, a California commercial syndicate proposed to lease and then purchase the entire country in 1864, and still again in 1866. This project went so far as to receive serious consideration at St. Petersburg.

It was at this time that Secretary Seward took up the matter of the purchase of Alaska. Seward always deeply appreciated Russia's tacit alliance in sending its fleets to the harbors of San Francisco and New York in 1863, and keeping them there at that critical time, when France and England were on the point of recognizing the Richmond government. This sense of gratitude on the part of Seward is, in a sense, responsible for our possession of Alaska and its priceless gold fields to-day.

When the Czar intimated that he wished to sell Russian America to any nation, excepting England, Secretary Seward entered into negotiations with Baron Stoeckl in February, 1867. The following March a treaty of purchase was sent to the Senate. This was reported on April 9th, was ratified on May 28th by 30 yeas to 2 nays and was proclaimed by President Johnson on June 20, 1867.

To Senator Charles Sumner is due the honor of giving the permanent name to Alaska. This, as was shown in a previous chapter, is simply the corruption of the Indian word meaning "great country." But the natives gave the name to Captain Cook, and Sumner apparently chose the name from its connection with the explorer, whom he admired.

Honor for Garfield.

It is also an interesting fact that the intention was to make General Garfield, one of the martyr Presidents, the first governor of the territory. It was further proposed to divide the great tract purchased into six territories. All these schemes, however, fell through.

Immediately upon the purchase of the country military occupation was decided upon. General Lovell H. Rousseau, as commissioner on the part of the United States, and Captains Pestschouroff and Koskul, on the part of Russia, met at Sitka on October 18, 1867.   Three men-of-war, the Ossipee, the Jamestown and the Resaca, and General Jefferson C. Davis and 250 regular troops were in waiting.

At half-past three o'clock that afternoon, Maksoukoff and vice-governor Gardisoff and the commissioners met the United States officers at the foot of the governor's flag-staff. The formality of transfer was short and simple. The men-of-war fired a double national salute, as did also the land battery. The Russian flag was lowered and the American flag was raised, and the country which has proved thus far such a source of wealth, and which promises to be the most prolific gold bearing region in the world, was American property. The only speech recorded as having been made at the time was that of Captain Pestschouroff, who said, as he advanced and the Russian flag fell :

"General Rousseau, by authority of His Majesty, the Emperor of all the Russians, I transfer to you, the agent of the United States, all the territory and dominion now possessed by His Majesty on the continent of America and in the adjacent islands, according to a treaty made between those two powers."

Territory is Accepted.

General Rousseau, metaphorically speaking, accepted the gigantic territory, and his little son slowly raised the new flag. Following this formal tender and acceptance, Prince Makasoukofl gave a dinner and ball. The ships were dressed in bunting, and there was a display of pyrotechnics.

That day ended all Russian dominion in the western continent, and there was an immediate exodus of all Russians who were able to leave the country. The Russian Government soon offered its subjects free transportation across the Pacific to the Amoor settlements, and within a comparatively short time there was scarcely a Russian to be seen on Alaskan territory.

This transfer of the country resulted almost immediately in an important change. The Russians used the Julian calendar, and this gave way to the Gregorian calendar, and a day was dropped from the Sitkan records, to correct the difference of twenty-four hours between the Russian day, coming eastward from Moscow, and our day, going westward from Greenwich.

Soon after the American occupation of the land scientists began to evince an interest in the country and, during the summer of 1867, Prof. George Davidson and eight other eminent specialists made a tour of investigation of southeastern Alaska. It is an interesting fact that their report and Senator Sumner's speech were the two strong arguments Secretary Seward offered for the purchase of Alaska in " Russian America."

Appropriation was Opposed.

Despite the fact that this valuable tract of land was purchased for half a cent an acre, there was the bitterest opposition to the appropriation of $7,200,000 in gold, equal to about $10,000,000 in paper at that time, to pay for the territory. It was not till July 14, 1868, that the House agreed by vote of 98 to 49, and the draft was handed to Baron Stoeckl.

As in most great government deals, the cry of corruption was raised, and it was alleged that there had been misappropriations and private gain in the negotiations. As it has been put relative to this alleged corruption, there was a "winter of investigation following a winter of contest and ridicule."

Connected with the purchase and early occupation of the country some pleasant reminiscences are recorded. Mr. Seward, returning to the United States by way of Kootznahoo, visited the country and addressed the citizens in the Lutheran Church at Sitka. He made a trip to the Taku glacier, visited the mining camps of the Stikine River and Fort Wrangel, and, as he afterwards expressed himself, was convinced of the wisdom of his course in purchasing the country from Russia.

Lady Franklin, too, visited Sitka in 1870, going there on the troop-ship Newbern, and, with her niece, Miss Cracroft, was the guest of the Commandant on the Kekoor. The following year the discovery of gold caused excitement to the garrison life, and the army pay vouchers were sunk in mining experiments at Sitka. The efforts then made, however, were as profitless as were those made at Juneau ten years later.

Garrison is Withdrawn.

On June 14, 1877, the last garrison of United States troops left Sitka, and the control of the military department over Alaskan affairs came to an end. It was but a few months thereafter that the Indians had destroyed all the government property outside the stockade. They threatened a general massacre, and appeals were sent to Washington for protection. This cry for help, however, was unheeded.

The residents at the stockade were besieged in the old fur warehouse. A last desperate appeal came from Victoria, and finally Captain Holmes A'Court hurried to their relief, without orders or instructions. But for this act of bravery and assumption of responsibility, it is probable there would have been a general massacre of all the Americans then living in Sitka.

From that time a man-of-war has constantly been stationed in southeastern Alaska, and the commanding officers have virtually been naval governors of the place.

Between the time of the transfer of the country from Russia

to the United States and of the passage of Senator Harrison's bill, May 13, 1884, which gave the nondescript tract a skeleton of civil government, thirty bills aiming to provide some form of government for Alaska were introduced. The Harrison bill finally passed, and gave to the country a governor, a district judge, a marshall, a clerk and a board of commissioners, with right to enter mineral claims, but distinctly withholding the general land laws.

In 1867 the Russian archives, manuscript journals, records, logs and account books were transferred from Sitka to the State Department at Washington. These, with Tikhmenieff's history of the colony, are among the most interesting relics of the country in our possession.

Some Account of Sitka.

A word may here be said about Sitka, the capital and seat of government of the territory of Alaska. It is situated on the west coast of Baranof Island. It is described as the merest apology for a town, but it, of course, has a certain importance, owing to the fact that it is the official residence of the governor and other officers appointed by the United States. Ten years ago it had a population of about 1000, of whom only 295 were whites.

The town is built on a level stretch of land at the mouth of the Indian River. Its main street is named after Lincoln, and extends from the government fort to the old Russian sawmill and the Governor's Walk, which is a beach road built by the Russians. Fronting the harbor is a large parade ground. Conspicuous among the buildings is the so-called castle, which was mentioned in Chapter V. Here, as everywhere in Alaska, the traveler will find an interesting display of Alaskan totem poles.

One interesting building in Sitka should not be passed by without mention.   This is the old log structure next to the Custom House, occupied by the Sitka Trading Company.   It was at one time the old fur warehouse, and many a time in its history it has held pelts to the value of $1,000,000.

Following the transfer of Alaska to the United States several grave international questions arose.   Among these was that of the international boundary line.   This matter really runs back to quite an early period.   Succeeding the Nookta Convention of 1790, the Northwest Coast became what is termed virgin soil, open to free settlement and trade by any people.   As a result three nations claimed it.

The Russians asserted ownership as far down as the Columbia. Then they withdrew to the fifty-first degree, or approximately to the north end of Vancouver Island.   The British Government laid claim to the coast from the Columbia River to the fifty-second degree ; and the United States to everything west of the Rocky Mountains, between forty two degrees and fifty-four degrees forty minutes.

Treaty of Occupation.

Then the United States and Great Britain, in order to avoid complications, agreed in 1818 to a joint occupancy of the region.   In 1819 the United States bought Florida from Spain, and with it acquired all the Spanish rights and claims on the coast north of the forty-second degree.   As a matter of fact, the United States was now virtually in possession of the region.   Still the British fur traders were pushing westward from the interior and there was likelihood at any time of trouble.

Two years later, in 1821, the Emperor of Russia took a hand in the matter, and by his ukase forbade all foreign vessels from approaching within 100 Italian miles of his possessions in the Pacific Ocean.   This brought about the conventions of 1874 and 1875 to adjust the rival claims to North American territory and to regulate the trade relations.   A treaty was formed with the United States in 1824, and in 1825 a somewhat similar treaty with Great Britain.   Russia then agreed to 54 degrees and 40 minutes as the southern limit of her possessions, and allowed the vessels of the other two nations to trade freely, without let or hinderance, for the period of ten years.

Interior Thought Worthless.

At that time the interior, which, of late, has given such promise as a gold producing country, was uninhabited, and indeed wholly unknown, except to the fur trader.   Its resources were not suspected, and it was deemed practically worthless.   It was parceled out in even thirds.   Russia took that part to the northwest, or what is commonly called the Yukon region. England took the Mackenzie region, and all the country between Hudson Bay and the Rocky Mountains.   The Oregon territory, that is, all west of the Rockies and north of 42 degrees fell to the United States.

Four years later an agreement was made between the United States and Great Britain, by which the occupancy of the Northwest coast was indefinitely extended.

President Tyler, in his annual message to Congress in 1843, declared that the United States' rights appertained to all between 42 degrees and 54 degrees 40 minutes.   At that time slave interests were being negotiated relative to Texas.   To gain the State without interference, Calhoun was discussing a settlement with the British Minister, with the forty-ninth parallel as the Oregon boundary.

The British Minister, however, rejected the proposition as his predecessor had done in 1807, when Jefferson had made proposals on practically the same lines.

Then arose the so-called " Fifty-four Forty " fight. These words became a political slogan, and Polk was elected as the champion of the cause. Polk took occasion in his inaugural message to say : " Our title to the country of Oregon is clear and unquestionable." and in his first message he reiterated the statement : "All of Oregon or none."

" The boundary question has been fought over time and again and it may be well in this connection to give the exact words of the treaties of 1884 and 1885, by which the Russian possessions are defined :

" Commencing from the southernmost point of the island, called Prince of Wales Island, which point lies in a parallel of 54 degrees 40 minutes north latitude, and between 131 and 133 degrees of west longitude (meridian of Greenwich), the said line shall ascend the channel called Portland Channel, as far as the point of the continent where it strikes 56 degrees of north latitude ; from this last mentioned point the line of demarkation shall follow the summit of the mountains situated parallel to the coast as far as the point of intersection of 141 degrees of west longitude (of the same meridian); and finally from the said point of intersection the said meridian line of 141 degrees in its prolongation as far as the Frozen Ocean.

The Boundary Line.

" With reference to the line of demarkation laid down in the preceding article it is understood (1) that the island called Prince of Wales Island shall belong wholly to Russia (now by this session to the United States). (2) That whenever the summit of the mountains, which extend in a direction parallel to the coats from 56 degrees of north latitude to the point of intersection of 141 degrees of west longitude, shall prove to be of the distance of more than three marine leagues from the ocean, the limit between the British possessions and the line of coast which is to belong to Russia, as above mentioned (that is to say, the limit of the possessions ceded by this convention), shall be formed by a line parallel to the winding of the coast, and which shall never exceed the distance of ten marine leagues therefrom.

It is an item of historical interest that, for the last twenty-eight

KILLING SEALS ON ST. PAUL ISLAND.

years of Russian ownership of Alaska, the thirty mile strip, as it was called, was leased to the Hudson Bay Company, which paid an annual rental for the territory which Canada now claims as her own.

Dr. G. M. Dawson, of the Dominion Geological Survey, in 1887 and 1888 invented a new map showing the boundary line claimed by his government, as drawn by Major-General R. D. Cameron. This narrows the thirty-mile strip to five miles in some places, and absorbs it entirely as part of British Columbia in others.

This Cameron line includes all of Glacier Bay, Lynn Canal, and Taku Inlet. It also incorporates all of the Stikine River, and, ignoring the channel known as Portland Channel, it strikes to tide water at the head of Burroughs Bay, and follows Behm Canal and Clarence Strait to Dixon Entrance.

By this map Canada lays claim to a large strip of territory about which there has been the bitterest contention, among other spots, the island which the United States used for a military post and then for a custom house for twenty years, and even Mary Island, where the United States Custom House now stands.

Claiming all the Alaska coast up to 56 degree by this arrangement, the late Sir John Robson, premier of British Columbia, even suggested that the United States yield up the small remaining strip of mainland between 56 degree and St. Elias, for certain concessions in sealing matters.

It is to be noted that all Canadian maps are now drawn according to the Cameron line, and, that Canadians, realizing the advantages of possessing this territory, are loud in their assertion of claims about which apparently the United States is apathetic.

Russians Find Gold.

Apropos of the Klondike gold fields one recalls the fact that it was the discovery of gold that awakened the Russians' interest in 1862. The leasing of the thirty-mile strip to the Hudson Bay Company did away with the necessity of precisely marking a boundary line. The Russians showed very little interest in the matter until the gold discovery.

It was incorporated in the Russian-American Company's lease that all mineral land should belong to the Crown, and following the report of the discovery of gold, the Czar ordered Admiral Popoff to send a corvette from Jaoan to see if the British miners were on Russian soil. Possibly his Imperial Majesty had in mind some tax similar to that which Canada has recently imposed upon all the American miners in the Klondike regions.

Apropos of the boundary quarrel San Juan Island nearly caused a war between Great Britain and the United States. According to the Oregon Treaty of June 5, 1846, both countries claimed ownership. The treaty did not specify whether the boundary line should pass through Canal de Haro or Rosario Strait. As a result, James Douglass and Governor Isaac Stevens both claimed jurisdiction of the island.

The matter came to an issue in consequence of petty quarrels. An American citizen shot a British pig, the owner of which did not think that $100 was an equivalent. Sentiment waxed hot over the matter. The sheriff of Whatcom County sold Hudson Bay Company sheep for taxes. General Harney dispatched troops to the scene of trouble and established a military post on one end of the island in 1859. This was just about the time the British and American Boundary Commissioners had begun their work of peaceable settlement.

War Ship on Guard.

A British war ship was stationed guard. The garrison was increased and General Scott came from Washington and offered joint occupation until the boundary line should be definitely decided. For two years a company of United States soldiers held the southern end of the island and an equal number of British blue jackets the northern point. The two garrisons had as pleasant a time as the circumstances would permit, exchanging visits and entertaining each other as best they could.

Then came the treaty of Washington in 1871. The Emperor of Germany as arbitrator decided that de Haro was the main channel and the water boundary. In obedience to this decision, the British withdrew in November, 1872, carefully replanting gardens and leaving everything as nearly as possible as they found it.

San Juan, by the way, is an important point, commanding the straits, and its thousand-feet-high hill makes one of the most effective batteries in the world. As might be expected, the diplomats who had the settling of this controversy split hairs, the representatives of each country doing their best to secure permanent right to the important military point. The importance placed upon this island by the British may be gleaned from these words of Lord Russell:

"San Juan is a defensive position in the hands of Great Britain. It is an aggressive position in the hands of the United States. The United States may fairly be called upon to renounce aggression, but Great Britain can hardly be expected to abandon defense."

Mr. Seward's Glory.

The discovery of gold on the Yukon in 1897, and the exodus of people from the southern States into the wilderness to seek their fortunes, recalls the words of Secretary Seward, and confirms their wisdom. A public dinner was given him on retire‧ ment to private life, and in the course of the evening the question was asked him:

"Mr. Seward, what do you consider the most important act of your official life?"

"Sir," said the secretary, without a moment's hesitation, "I think the purchase of Alaska was by far the most important official act of my life. It will take two generations, however, for the public to appreciate the value of this purchase."

The old statesman was right. It has taken two generations and the world is now convinced of the truth of Seward's words. It may safely be said that it was Seward's crowning glory to add to his country's domain a new empire of such vast extent and of such untold wealth.   An empire whose very name signifies great country or continent, and whose mountains are supposed to hold the mother lode of the gold supply of America.

Early last August when the gold fever was at its height the boundary question naturally came up again, especially in Canadian circles.   R. W. Scott, Secretary of State, at Ottawa, Ontario, was then interviewed regarding the statement from Washington, which claimed that Great Britain, in its official maps, had drawn the boundary line on the Pacific coast so as to deprive the United States of hundreds of miles of territory adjoining the Klondike regions.

He said he had gone into the question when a member of the Mackenzie administration in 1878, and the point now raised was discussed then.

" The treaty of St. Petersburg of 1825," said he, " defines the line dividing Russian territory, now Alaska, from British by a line drawn north from the foot of Prince of Wales Island through Portland channel until it struck the mountains, when the method of delimitation was set forth.

" The map will show that a line running north from the foot of Prince of Wales Island must go through the Behm Canal, and that to reach Portland Canal the line would have to go east through the open sea a considerable distance before it could reach Portland channel or canal.

The British Contention.

" The British contention as shown by the dispatches of George Canning to Sir Charles Bagot, when British Ambassador at St. Petersburg, is that Portland Canal was to be in British territory and that the words ' Portland Canal ' in the convention was a mistake for ' Behm Canal,' or else that what is now called Portland Canal was not then so called.

This is supported by the physical impossibility of running a line due north through Portland Canal from the foot of Prince of Wales Island, so that Canadian maps show the boundary line as running north through the Behm Canal. The difference is great in view of the discoveries of gold, and it can only be settled by an international arbitration.

" The disputed territory with the ten marine leagues back from the coast added would not, however, embrace the present gold fields of the Klondike, which are clearly in British territory, because they are well east of the one hundred and forty-first meridian, which is the recognized boundary to the north."

Dispute Will Not Down.

The claim of Great Britain to a big share of Alaska promises, on account of the gold fields, to occupy a large amount of public attention for years to come, and it will be of interest to the reader to have the opinion of Secretary Scott, the Canadian representative in the matter, offset by that of an American who can speak as one having authority. The British claim is regarded by American officials in general as preposterous, and it will likely cause grave diplomatic complications between the United States and Great Britain.

The Senate, before which the boundary question was brought as the outcome of a treaty negotiated by Secretary Olney and Sir Julian Pauncefote, did not place itself on record in the matter. Before a vote was taken Congress adjourned, so that the location of the divisional line, which has been in dispute since 1884, is no nearer settlement than it has been at any period in the last thirteen years.

General Duffield, Superintendent of the Coast and Geodetic Survey, was a member of the boundary commission. The survey authorized by it has until of late been deemed official. The

18

following statement, therefore, from General Duffield is of value :

" Up to 1884 both countries were practically united as to the boundary line from Mount St. Elias to the southeast.   According to the terms of the treaty between Russia and Great Britain, the United States in purchasing Alaska in 1867 acquired all of Russia's rights.   In describing the southeastern boundary the Russian treaty read :

" The line of demarcation between the possessions of the high contracting parties upon the coast of the continent and the islands of America to the northwest shall be drawn in the following manner : Commencing from the southernmost point of the land called Prince of Wales Island, which point lies in the parallel of 54 degrees 40 minutes north latitude and between the 131st degree and the 133d degree of west longitude, the same line shall ascend north along the channel called Portland Channel, as far as the point of the continent, where it strikes 56 degrees of north latitude.

Fixing Landmarks.

" From this last mentioned point the line of demarcation shall follow the summit of the mountain situated parallel to the coast, as far as the point of intersection of 141 degrees of west longitude of the same meridian, and finally from the said point of intersection, the said meridian line of 141 degrees in its prolongation as far as the frozen ocean, shall form the limit between the Russian and British possessions on the continent of America to the northwest.

"Wherever the summit of the mountains, which extend in a direction parallel to the coast from 56 degrees north latitude to the point of intersection of 141 degrees of west longitude, shall prove to be a distance of more than ten marine leagues from the ocean, the limit between the British possessions and the line of coast which is to belong to Russia, as above mentioned, shall be formed by a line parallel to the widening of the coast and which shall never exceed the distance of ten marine leagues therefrom.

"On all maps from 1825 down to 1884 the boundary line has been shown as in general terms parallel to the winding of the coast and thirty-five miles from it. In 1884, however, an official Canadian map showed a marked deflection in this line at its south end. Instead of passing up Portland Channel this Canadian map showed the boundary as passing up Behm Canal, an arm of the sea some sixty or seventy miles west of Portland Channel, this change having been made on the bare assertion that the words ' Portland Canal,' as inserted, were erroneous.

By this change an area of American territory, about equal in size to the State of Connecticut, was transferred to British territory. There are three facts which go to show that this map was incorrect. In the first place, the British Admiralty, when surveying the northern limit of the British Columbian possessions in 1868, one year after the cession of Alaska, surveyed Portland canal, and not Behm Canal, and thus, by implication, admitted this canal to be the boundary line.

Second, the region now claimed by British Columbia was at that time occupied as a military post of the United States without objection or protest on the part of British Columbia. Third, Annete Island, in this region, was, by Act of Congress four years ago, set apart as a reservation for the use of the Metlektala Indians, who sought asylum under the American flag to escape annoyances experienced under the British flag.

Another Change Made.

"Another change was made at Lynn Canal, the northernmost extension of the Alexander Archipelago, which runs north of Juneau, and is the land outlet of the Yukon trade. If the official Canadian map of 1884 carried the boundary line around the head of this canal another Canadian map, three years later, carried the line across the head of the canal in such a manner as to throw its headwaters into British territory.   Still later Canadian maps carry the line, not across the head of the canal, but across near its mouth, some sixty or seventy miles south of the former line, in such a way as practically to take in Juneau, or at least all overland immediately back of it.   And the very latest Canadian map, published at Ottawa within a few days, while it runs no line at all southeast of Alaska, prints the legend ' British Columbia,' over portions of the Lynn Canal which are now administered by the United States."

A report was made early in 1897 by United States surveyors as to the boundary line in dispute.   It said :

Effect of Determinations.

" These determinations threw the diggings at the mouth of Forty-Mile Creek within the territory of the United States.   The whole valley of Birch Creek, another most valuable gold-producing part of the country, is also in the United States.   Most of the gold is to the west of the crossing of the 141st meridian at Forty-Mile Creek.   If we produce the 141st meridian on a chart the mouth of Miller's Creek, a tributary of Sixty-Mile Creek, and a valuable gold region, is five miles west in a direct line or seven miles, according to the winding of the stream—all within the territory of the United States.   In substance the only places in the Yukon region where gold in quantity has been found are, therefore, all to the west of the boundary line between Canada and the United States."

It can readily be seen that the claim of the United States is directly opposed to that of the Canadians.   It is true that the arbitration of the 141st meridian was favored by the United States surveyors, but some of them were angered at the claims of the English in regard to Lynn Creek and the whole south-eastern boundary, and expressed the belief that the United States would refuse to arbitrate the claims of this portion of the boundary.

An interesting chapter of Alaskan history is now making, and the prospect is that in the near future the name of Lincoln will be given to a territory or state in the great northwest, as that of Washington was some years ago.   There are enthusiastic advocates of the movement who think the proposed territory will eventually become a sovereign, if not the banner state of the Union.   Any account of the history of Alaska, therefore, should include this possibility by anticipation.

Long before the great gold discoveries in the Klondike region of the Northwest Territory became known a movement was quietly inaugurated to divide the great Territory of Alaska.   In May active work was begun and the project is now ready for public attention.

Petitions for division are now in circulation in the interior along the Yukon River, and in all the mining camps, and should reach Washington early in September.   The name of Lincoln for the new territory met with a quick response on the part of the hardy miners, who are delighted with the prospect of a territorial form of government that will give them direct governmental supervision, land laws and titles, and some incentive to good citizenship.

Recognition of Russia.

When the purchase was made it was construed by the administration papers as an act of courteous recognition of Russia's friendship in the civil war, it being remembered that a Russian fleet of three vessels appeared in New York harbor during the excitement over the Trent affair, when it looked as if war with Great Britain was certain to result.   It was said at the time—and is still maintained in diplomatic circles—that the Russian admiral had sealed orders, which directed him, in case of war between the United States and Great Britain, to announce Russia's alliance with America, and proceed to capture any British vessel possible.

How much the purchase of Alaska served as an expression of our gratitude for Russia's assistance at a critical period no one accurately knows. The "true inwardness" of the transaction was kept under cover for diplomatic reasons, but it pleased Great Britain as much then as the developments of the seal fishing controversy, and the uncertainty of the boundary line, at the present date.

In fact, the "national iceberg," as it was termed in 1867, has been from the beginning a torrid source of unpleasantness between the two great nations of the English speaking tongue.

Early Day Statistics.

When Alaska was annexed the population was stated by the Russian missionaries at 33,426, of whom but 430 were whites. The mixed race—termed creoles—counted 1756, and were the practical leaders, using the Indian tribes for hunting and fishing. Fur trade and the fisheries were at that time the only known resources. As early as 1880, however, the sea otters shipped represented a value of $600,000, the fur seals over $1,000,000, the land furs $80,000, and the fisheries from $12,000 to $15,000.

Mineral riches were hinted at by the early explorers. In 1885 the Director of the Mint credited Alaska with $300,000 in gold $2000 in silver, the chief contributor being the Alaska mill at Douglas City. In 1896 the gold product reached $1,948,900, showing a gain over 1895 equal to $386,100. For 1897 the gold output is placed by good judges at not less than $10,000,-000, which is nearly twice that of Colorado in 1892.

Small lots of smelting ore—from which some silver is recov-

ered—are shipped to Tacoma for treatment, but the main producers are the large mills on Douglass Island, equipped with
stamps, concentrators, and modern appliances for saving gold
values.   The grade of the quartz mined and worked, as early as
1892, showed an average value of $2.42 per ton.   This material
is taken from an immense quarry, which has none of the marks
of a glacial deposit.   The exposure of the quarry by glacial
action is entirely probable.

What will be Left to Alaska.

After the division there will be left to Alaska all of the territory along the Northern Pacific sea coast and the Aleutian
Islands.   This includes all the agricultural lands in Alaska and
that part of the territory which enjoys a comparatively mild and
equitable climate on account of the well-known influences of the
Japan current.   The proposed Territory of Lincoln will embrace
within its boundaries the valleys of the great Yukon River and
its tributaries and the coast along Behring Sea.

The city of Weare, at the mouth of the Tanana River, 800
miles from the sea, and on the Yukon River, as shown on the
map, will be named in the act as the seat of government of the
new territory.   Tributary to the capital on all sides will be the
great placer mining gold fields.

The influx of population into these gold fields is so great that
the residents of the interior of the present Alaska, and all who
have investments there, are unanimous in their demands for such
recognition from the Government as will give them protection to
life and property.   They are ready for the active development of
a rich, great country, too long kept closed.

There are mines of gold, copper, coal, iron, silver, and lead
within the proposed Territory of Lincoln, and to these must be
added the recently discovered rich oil fields.

Organization will immediately follow the territorial creation, and it is likely "the delegate from Lincoln" will soon be recognized in Congress. He will be on an equal footing with delegates from other territories, and will have a voice in argument, but no vote on roll-call.

There is political significance, too, to the movement that, in the eyes of many, is of great importance.

"The people of Sitka have little time and less inclination to attend to the affairs of the interior of Alaska," is the complaint that is most often heard.

The new division will give to Alaska the coast trade, the great quartz mines of Douglas Island, and all the land in the territory at present known to be adaptable to agricultural purposes—in round numbers 80,000 square miles. The Territory of Lincoln will comprise 500,000 square miles of the interior and northern coast country.

It is a reasonable supposition that a great deal of wealth will be taken out of these gold fields, and it should not be forgotten that the Canadians and their Government are vigorously extending their settlements and their sphere of influence north and west of British Columbia. A subsidy of $11,000 a mile is about to be given to a railway branching northward from the Canadian Pacific for over 200 miles, which is to be constructed with a view to open up that portion of British Columbia and drawing to it from the interior of Southeastern Alaska whatever trade may develop in that region. The American Government will at least be careful that its political rights and territorial jurisdiction are carefully guarded, in order that the enterprise of its people may have safe opportunity for achievement.

CHAPTER IX.
Topography.

AN account of Alaska naturally includes a description of its topographical features, somewhat more in detail than was given in the chapter on the Wonders of Alaska. As was there said, the very name signifies " great country " or continent. And it is a great country, great in every way, covering an area equal to the original thirteen States of the Union, with the great Northwest Territory added.

Put in other words, Alaska is as large as all of the United States east of the Mississippi and north of Alabama, Georgia and North Carolina, extending 1000 miles from north to south and 3500 miles from east to west. It is a remarkable fact that the shore line up and down the bays and around the islands, according to the United States coast survey, measures 25,000 miles or two and one-half times more than the Atlantic and Pacific coast lines of the remaining portions of the United States. The coast of Alaska alone, if extended in a straight line, would belt the globe.

Beginning at the north end of Dixon Inlet, in latitude 54 degrees, 40 minutes, the coast line sweeps in a long, regular curve north and west to the entrance of Prince William's Sound, a distance of 550 miles. From that point it extends 725 miles south and west to Unimak Pass, at the end of the Alaska peninsula. At this pass the chain of the Aleutian Islands begins and extends 1075 miles in a long curve almost across the Pacific Ocean to Asia.

The dividing line between Asia and Alaska, according to the treaty made with Russia, is the meridian of 193 degrees west longitude.   To the north of Unimak Pass the coast has a zig-zag line as far as Point Barrow, on the Arctic Ocean.   The general shape of Alaska is thus that of the head and horns of an ox inverted, the mainland forming the head and the chain of the Aleutian Islands the horns.

The surface of this immense tract falls naturally into three distinct districts.   The first is the Yukon, extending from the Alaskan range of mountains to the Arctic Ocean.   The second is the Aleutian, which includes the Alaska Peninsula and all the islands west of the 155th degree of longitude.   The last is the Sitkan, embracing Southeastern Alaska.

A Vast Moorland.

Of the Yukon district, in which most of the gold fields lie, we know comparatively little.   Until the hardy miners and prospectors were lured into the mountains and plains and along the river beds in the hope of securing fortune, few ever ventured into the region.   As might be expected, little or nothing of scientific value comes from people of this stamp.   The prospectors and miners in a large measure have but a single purpose and have been dependent upon the natives, who are familiar with the passes, to conduct them into the interior.   No body of scientific men has thus far undertaken a thorough exploration of the region.   Only in its greater outlines or details do we know it.

The "Coast Pilot," a publication of the United States Coast Survey, gives a passage which is worth transcribing, descriptive of the country between Norton's Sound and the Arctic Ocean. It says:

BREAKING UP OF THE ICE-PACK ON BEHRING SEA.

283

" It is a vast moorland whose level is only interrupted by promontories and isolated mountains, with numerous lakes, bogs and peat beds. Wherever drainage exists, the ground is covered with a luxuriant herbage and produces the rarest as well as the most beautiful plants. The aspect of some of these spots is very gay. Many flowers are large, their colors bright and, though white and yellow predominate, other tints are not un-common. Summer sets in most rapidly in May and the landscape is quickly overspread with lively green."

The Aleutian district is for the most part of mountainous and volcanic formation. There are, however, many natural prairies between the mountains and the sea, with a rich soil of vegetable mould and clay, and covered with perennial wild grasses. Speaking of grasses recalls the statement of Dr. Kellogg, botanist of the United State Exploring Expedition. Says he : " Unalaska abounds in grasses, with a climate better adapted for haying than the coast of Oregon."

The Rev. Sheldon Jackson says that in 1879 at Fort Wrangel he cut wild timothy that would average five feet in height, and blue grass that would average six feet. He measured one stem that reached seven feet three inches. Prof. Muir, State Geologist of California at one time, also declares that he never saw such rank vegetation outside the tropics.

Some Characteristic Features.

Alaska is remarkable for the boldness of its shores, and its deep water, numerous channels and innumerable bays and harbors, the great mountainous islands of Vancouver, Queen Charlotte, Prince of Wales, Wrangel, Baranoff, Chichagoff, and many others forming a complete breakwater, so that it is possible for the traveler to have an ocean voyage of 1000 miles or more without once getting out to sea. Says the Rev. Sheldon Jackson :

" The labyrinth of channels around and between the islands, that are in some places less than a quarter of a mile wide, and yet too deep to drop anchor ; the mountains rising from the water's edge from 1000 to 8000 feet, and covered with dense forests of evergreen far up into the snow that crowns their summits ; the frequent track of the avalanche cutting a broad road from mountain top to water's edge ; the beautiful cascades, or the glaciers, or the overflow of high inland lakes, falling over mountain precipices or gliding like a silver ribbon down their sides ; the deep gloomy sea fiords, cleaving the mountains into the interior ; the beautiful kaleidoscopic vistas opening up among the innumerable islands ; mountain tops, domed, peaked and sculptured by glaciers ; the glaciers themselves, sparkling and glistening in the sunlight dropping down from the mountain heights like some great swollen river, filled with drift wood and ice, and suddenly arrested in its flow, all go to make up a scene of grandeur and beauty that cannot be placed upon canvass or adequately described in words."

Archipelago is Divided.

This great archipelago of Alaska is naturally divided into three portions, the southern portion being in Washington Territory, the central in British Columbia and the northern in Alaska proper. This last was named, in honor of the Czar of Russia, the Alexander Archipelago. It is seventy-five miles from east to west and 300 miles from north to south. The aggregate area of these islands is 14,142 square miles.

To the westward is Kadiak, 600 miles distant, with an area of 5676 square miles ; then comes the Schumigan group, containing 1031 square miles ; and then the Aleutian chain which has an area of 6391 square miles. Then, to the northward, are the Seal Islands, containing, with the other islands in Behring Sea, about 3963 square miles. Thus, it will be seen that the total area of the island of Alaska alone is 31,205 square miles, an extent of territory equal to that of the State of Maine.

Alaska is also the home of great mountain peaks. It has the highest peaks in the United States. The coast range of California and the rocky range of Colorado and Montana trend together in Alaska and form the Alaskan mountains. Here, we may notice the fact that the old atlases misrepresent the range of mountains that is thus formed. It does not continue northward to the Arctic Ocean, as was supposed, but turns to the southwest, extending through and forming the Alaskan peninsula and then gradually sinking into the Pacific Ocean. Only a few of the highest peaks are here visible above the water. It is these peaks that form the Aleutian chain of islands, which are only the mountain tops.

Island Mountain Peaks.

The islands of the Alaskan archipelago naturally decrease in size and frequency as the mountain range sinks deeper and deeper into the sea. Unimak, the most eastern of the chain, is noted for that most magnificent of volcanoes, Shishaldia, 9000 feet high; then comes Unalaska, 5691 feet; after this Atka, 4852 feet; then Kyska, 3700 feet; and finally Attu, which is the most western of the group, and has an altitude of only 3084 feet.

Alaska has the highest mountain peaks in the United States, and some of them are worthy of special mention. Mount St. Elias towers aloft 19,500 feet; Mount Cook, 16,000 feet; Mount Crillon, 15,900 feet; Mount Fairweather, 15,500 feet. There are many others, whose altitudes are no less striking.

In Alaska, too, is to be found the great volcanic system of the United States. Grewingk enumerates sixty-one volcanoes. These are mainly on the Alaskan peninsula and the Aleutian

Islands. It is said that the violence of the volcanic forces is decreasing, and that only ten of these volcanoes are now active. Mount Edgecombe, near Sitka, is one of the extinct volcanoes. On the Naas River, just across from southern Alaska, there is still to be seen a remarkable lava overflow from a volcano in the neighborhood.

Interesting Indian Legends.

About these volcanoes the fancy of the Indians has linked any number of curious legends. To these children of the wilderness the volcanoes are little less than living entities and, naturally, reasons for their activity have been sought by the savages and have been expressed in some terms of ordinary life.

Again, it is in Alaska that we find the great glacial system of the United States, chief of which is the great Muir glacier, which has been described in Chapter V. One can hardly go anywhere along the coast of Alaska without finding these great sleeping giants, as they have been called, debouching slowly into the ocean. Their number is literally legion. Prof. John Muir describes one of these monsters and his description is worth transcribing, partly from Prof. Muir's reputation as a scientist and the accuracy of the facts he marshals, and partly from the picturesque language he uses. The glacier he visited and described particularly was one near Cape Fanshaw. Said he :

"The whole front and brow of this majestic glacier is dashed and sculptured in a maze of yawning crevasses, and a bewildering variety of strange architectural forms, appalling the strongest nerves, but novel and beautiful beyond measure—clusters of glittering, lance-tipped spires, gables and obelisks, bold outstanding bastions and plain mural cliffs, adorned along the top with fretted cornice battlements, while every gorge and crevasse, chasm and hollow, was filled with light, shimmering and fulsome in pale blue tones of ineffable tenderness.

"The day was warm, and back on the broad, waving bosom
of the glacier water streams were outspread in a complicated
network.   Each, in its own frictionless channel, cut down
through the porous, ice-decaying surface into the quick and
living blue, and flowed with the grace of motion and with a ring
and gurgle and flashing of light to be found only on the crystal
hills and dales of a glacier.

Reflecting God's Plan.

"Along the sides we could see the mighty flood grinding
against the granite with tremendous pressure, rounding the out-
swelling bosses, deepening and smoothing the retreating hollows,
and shading every portion of the mountain walls into the forms
they were meant to have when, in the fullness of appointed time,
the ice-tool should be lifted and set aside by the sun.   Every
feature glowed with intention, reflecting the earth plans of God.

"Back two or three miles from the front the current is now
probably about 1200 feet deep, but when we examined the walls,
the grooved and rounded features so surely glacial showed that
in the earlier days of the ice age they were all over-swept, this
glacier having flowed at a height of from 3000 to 4000 feet
above its present level."

The rate of recession of glaciers is one of the unsettled questions of Alaska.   It seems, however, that rain withers and breaks
away the ice most rapidly.   A close watch was kept in July and
August of 1891 by Miss Skidmore, who concluded from her
observations that the tide had little or nothing to do with the
fall of the ice.   On many warm, clear days she noticed, when a
hot sun fell upon the ice front for sixteen and eighteen hours
continuously, there was no sound.   After days of silence, on
the contrary, came tremendous displays, one-quarter or one-third
of the long wall falling away apparently without cause.   As a

general rule, these falls occurred in the middle of the night or at early daybreak.

Attempts have been made by photographic evidence to determine the recession of the glacier, but with limited success. In this way it has been shown with reasonable sureness that one glacier, at least, retreated 1000 yards between 1886, when Professor Wright visited it and 1890, when Professor Reid visited it. Photographs were again taken in 1891, which showed a recession of 300 yards in a year. Professor Muir noted a retreat of a mile between his visits to a glacier in 1880 and in 1890.

The effect of this irregular coast line, with its setting of mountain peaks and glaciers, is striking. The surroundings are fascinating. The shores are sentineled by gigantic mountains, on whose broad sides recline a dozen or more huge glaciers—amongst them the Davidson. But to reach the greatest of these "frozen Niagaras," Lynn Canal must be retraced to appropriately-named Icy Straits, north of which is Glacier Bay, into whose pellucid waters descend Titantic glaciers, king among which is the Muir.

In matchless beauty and colossal structure it is overpowering to the senses. Here, right in front, a wall of ice nearly two miles long and several hundred feet high, and rising in a glittering cliff out of the waves, marks the end of the Muir Glacier, which is formed by the union of twenty-six tributary glaciers, and the united mass of ice covers 1000 square miles.

A Giant Among Peaks.

A little further to the north is the Melaspina Glacier, lying beneath a grand circle of snowy peaks, the loftiest of which, Mount St. Elias, is 18,360 feet above the sea. The Melaspina Glacier is a great sea of ice, formed by the junction of many glaciers descending from the mountains.

19

These rivers of ice, at their confluence, spread out in one vast united ice-sheet, and from this great congealed, constantly moving mass, as it debouches into the sea, huge pieces break from the forefoot and with terrific force, lashing the waters into great waves, drop into the sea, accompanied by loud reports which reverberate like the booming of heavy artillery. From the summit of the Muir Glacier, the eye beholds a frozen world.

In Alaska also, are to be found numerous boiling springs, veritable geysers, from which the water bubbles up with a temperature that is really surprising. There are some large ones south of Sitka, and several more on Perenosna Bay, on Magat Island and at Fort Moller. Boiling springs are also to be found in numbers on many of the islands, and so hot is the water that gushes from them that for ages the natives have been accustomed to boil their food in them. In the crater of Goreloi there is a vast boiling spring eighteen miles in circumference. On Beaver Island there is a lake very strongly impregnate with nitre. Some of the springs are likewise touched with sulphur.

Like the glaciers and the volcanoes, these boiling springs have been subject for marvel on the part of the Indians. Noises proceed from them similar to the roaring of cannon, and it is natural that the unlettered savages, being unable to explain these mysterious phenomena, should surround them with a tissue of their own imagination and resort to legend for an explanation.

Fine Auroral Displays.

As part of the natural phenomena of the country, mention must be made of the magnificent auroral displays. Of these, Bancroft gives a pretty description. He describes them " as flashing out in prismatic corruscations, throwing a brilliant arch from cast to west—now in variegated oscillations, graduating through all the various tints of blue and green and violet and crimson, darting, flashing or streaming in yellow columns, upward, downward, now blazing steadily, now in wavy undulations, sometimes up to the very zenith, momentarily lighting up the surrounding scenery, but only to fall back into darkness."

It is recorded that on the occasion of one of these beautiful auroral displays the air was so thickly charged with electricity that sparks flashed from the points of the soldiers' bayonets.

In a previous chapter mention was made of the great Yukon River, and it remains here to be said chiefly that the Yukon, while it is the greatest, is only one of many mighty streams. Indeed, in Alaska are to be found some of the largest rivers, not only of the United States, but of the world. The Yukon is the great artery leading from the coast into the interior.

Its course throughout its 2500 miles of length is marked by features which make it one of the most remarkable water courses on the globe. For the first 1000 miles it varies in width from one to five miles and often, owing to the islands in its course, it is twenty-five miles in width. It is navigable for 1500 miles. Its upper waters are within the Arctic Circle and along its banks live thousands of people who know nothing of its mouth or of its head. To them it is simply an unexplored immensity.

Climate Extremely Varied.

Among the other principal rivers of the territory are the Stikine River, 250 miles long ; the Chilkat, the Copper, the Fire, the Nushergak and the Kuskokuim. This last is next in size to the Yukon, and is from 500 to 600 miles in length. The Tananeh is 250 miles in length, and half a mile wide at its mouth, and has a very strong current. Two of Yukon's principal tributaries are the Nowikakat, 112 miles, and the Porcupine.

The climate of Alaska, owing to the vast extent of the country, is as varied as in the United States. In Southern Alaska the

temperature is so mild as to give no suggestion of the extreme
rigor of the north.   The greatest cold recorded on the Island of
Unalaska during a period of five years was zero.   The average
for five years at seven o'clock in the morning was thirty-seven
degrees above.   The average of weather for seven years shows
53 clear days, 1263 half-clear days and 1255 cloudy days.   This
indicates a climate very similar to that of northwestern Scotland.
At Sitka the record is not very dissimilar.   During a period of
forty-three years there was an average of 200 rainy or snowy
days per year.   During the winter of 1877 the coldest night at
Sitka only formed ice about the thickness of a knife blade.   At
Fort Wrangel, which is at a distance from the ocean and near
snow-covered mountains, the climate is colder than at Sitka.
And when one reaches the regions of the North, where the gold
mines are located, it is no uncommon thing to find the temperature falling from eighty to ninety below zero.

Testimony of Travelers.

The mild climate of Southern Alaska is due to the Japan Gulf
Stream, which first strikes the North American continent at the
Queen Charlotte Islands, in latitude 50 degrees north.   At this
point the stream divides, one portion going northward and westward, along the coast of Alaska, and the other southward along
the coast of British Columbia, Washington Territory, Oregon
and California.   Thus the climate of the States just named is
made mild and pleasant in precisely the same way that the shores
of Spain, Portugal, France and England are made mild by the
ocean currents of the Atlantic.

As the climate is one of the terrors of the country in the
popular estimate, the testimony of people who have been in the
gold region in recent years will be acceptable to the reader.   The
prospector is willing to scale mountains, traverse plains, cross

rivers, shoot rapids, and brave a thousand perils, but the thought of living in a country whose temperature is often represented as being comparable with that of a vast refrigerator is appalling.

Owing to the popular association of the idea of extreme frigidity with the word Alaska, many people will doubtless be surprised to learn that the average temperature in the Klondike region during the four coldest months of the year is not ordinarily much lower than 20 degrees below zero.

The average winter's snowfall in that part of Alaska is only about two feet, whereas on the coast it is ten times that much.

Facts from Mr. Weare.

" The snowfall in the vicinity of Fort Cudahy is only about two feet during the winter, although it is as much as twenty feet along the coast where the influence of the Japan current is felt.

" It is bitterly cold in Arctic Alaska. There is no denying this. Forty degrees below zero for days at a stretch is not uncommon. But they have the same kind of weather in Northern Russia, and one does not hear any plaints of hardship from there. Peary and other Arctic explorers have spent whole winters hundreds of miles nearer to the pole without actual suffering.

" In Russia and other cold countries the people prepare for the long eight months' winter by building tight log houses in which they keep comfortable over their queer-looking tile stoves which give an immense amount of heat from a small bunch of wood. The same thing will have to be done in the Yukon country. Frail tents are not suitable shelter in winter.

" It's too much like a man trying to get along with a linen duster for a topcoat. If the prospectors are well housed, well clothed, and well fed, they can bid defiance to the cold, and those who are not able to secure these three important items should not tempt fate by making the trip."

The following is important as being exact figures direct from the gold region :

Table Showing Highest and Lowest Temperature at Fort Constantine, Yukon, Jan. 1st to May 31st, 1896.

Day of Month	January High	January Low	February High	February Low	March High	March Low	April High	April Low	May High	May Low
1	—24	—38	—20	—32	— 7	—26	11	—24	30	5
2	—29	—46	— 5	—22	— 1.5	—16	9	—13	19.5	5
3	—45	—55	—11.5	—43	12.5	—11	19	— 3	22	5
4	—46	—56.5	— 4	—40	17	8	23	—23	32	11
5	—54	—61.5	— 5	—21	18	0	8	—38	50	30
6	—50	—62.5	0	—15	13.5	— 1	6	—34	51	30
7	—40	—61	4	—20	13.5	—30	13	—38	46	31.5
8	—26	—54	7	—20	11	—24	8.5	—34	58	35
9	—17.5	—28	—17	—47	8	—23	12	—31	65	28
10	—12.5	—25	—27	—45	12.5	—20	15	—31	61	30
11	— 8	—23	—45	—61	23	— 1	21	—21	60	30
12	— 9	—25	—40	—62	34	2	20	—23	53	35
13	—14.5	—32	—46	—56	23	0	16	·26	56	30
14	—27	—41	—33	—56	35	7	16	26	55	29
15	—31	—42	—35.5	—55	39	6	21	1	56	38
16	—26.5	—36	—34	—50	31	10	39	20	55	33
17	—22	—42	—32	—47	39	19	45	31	54	30
18	—20	—39	—26.5	—56	34	2	48	30	59	28
19	—15	—26	—16	—53	34	10	38	14	62.5	40.5
20	—16.5	—42	4	—20	33	14	33	20	55	37
21	—21	—54	17.5	1	15	—35	40	17	47	33.5
22	—45	—58	24.5	10	13	—20	16	— 5	54.5	24
23	—45	—61	21	—15	20	— 5	28.5	5	59.5	32
24	—48	—60	25	—22	21	3	34	19	65	33
25	—48	—56	— 3	—15	28.5	11	43	29	58	35.5
26	—49	—64	1.5	—35	27	10	42	22	58	39
27	—57	—65	—10	—41	24	—29	32.5	6	61.5	35
28	—44	—59	—18.5	—41	21	—10	29	12	58.5	33
29	—18	—55	—10	—33	20	5	22	— 8	55	26
30	—13	—42	. . .	. . .	9	— 5	39	19	63	28
31	— 8	—27	. . .	. . .	7	—17	. . .	. . .	60	30
Means	—30	—46	—12	—35	20	— 5	25	— 4	53	28
Mn. tem. Mth.	—38		23.5		7.5		10.5		40.5	

CHAPTER X.
Flora, Fauna and Climate.

Agricultural Industries in Alaska—Vegetables and Small Fruits in the Southeastern Portion—Grasses and Fodder—Panorama of Blossoms in the Short Summer—Seasons in the Yukon Basin—Sea Otters and Fur Seals—Food Animals and Carnivoræ—Moose and Caribou—Value of Pelts—Fish of the Territory—Salmon Canning and Salting—A Dog Fish Story—Birds of Alaska—Among the Cetaceans—Mosquitos and Gnats—Weather Bureau Report—Temperature at Klondike—Animals and Vegetation in British Columbia.

ALASKA, bisected by the Arctic Circle, bounded by a vast coast line and culminating in the loftiest peak of the Rocky Mountain system, possesses a climate of remarkable variations and possibilities. From pleasant Sitka to ice-locked Barrow, from sea-girt Baranoff to the Alpine crest of St. Elias, from the Torrid summers to the hyperborean winters of the great Yukon basin, almost every extra-tropical range of temperature may be noted and almost every kind of meteorological condition experienced.

The effect of these wide climatic ranges is manifest in the fauna and flora of the territory. The former corresponds quite closely to the sub-arctic type; the latter presents a variety of brilliance and sobriety at once delightful and astonishing. The animals belong largely to the fur-bearing species, though natives of more temperate regions survive and even thrive with proper care, but vegetation ranges with charming prodigality from the luscious fruits and vegetables of the Southland to the frost-defying firs and spruces of the extreme north. Agriculture may never be a leading industry of the territory for the season is too short and crops are too uncertain of maturity. Yet below the

Arctic Circle it is easy to grow enough for food, and even farther north herbs and vegetables of quick growth make a rapid and even rank response during the short, hot summer.

Alaska, superficially, is either mountain, plain or archipelago. The country between Norton Sound and the Arctic Ocean is a vast moorland with numerous bogs and peat beds. The Yukon basin is a broad, alluvial plain with a rich soil of unknown depth. The islands and the adjacent coasts are generally rocky, but not sterile. Magnificent timber abounds in the uplands and along the lower coasts and summer from the Arctic Circle south is a jubilee of luxuriant herbage and beautiful plants and flowers.

In the Southeast.

In the southeastern portion of the territory nearly all the vegetables, herbs, grasses and smaller fruits of the middle temperate zone flourish without stint or extra care. Potatoes, carrots, beets, parsnips, radishes, lettuce and turnips grow large and sweet. Cabbages weighing seven pounds are on record and all " garden truck," in fact, except cucumbers and beans, does well. The best arable land in the territory is in this region, and in several districts agriculture is carried on with considerable success on a fairly extensive scale. Timothy, blue-joint, wood-meadow, marsh and the Kentucky blue-grass raise rank crops, and clover has done well wherever tried. These afford unexcelled grazing in summer and the best of fodder in the winter for stock. Cattle thrive in this climate, but sheep, despite the excellent feed, suffer from the extreme moisture which rots their hoofs. There is a poultry ranch at Fort Wrangel.

This region is noted for its bountiful berry crops. Red and black currants, raspberries, strawberries, huckleberries, Killikinick berries, bearberries, dewberries, heathberries, mossberries, roseberries, salmonberries and cranberries grow abundantly.

The Indians gather the salmonberries for local trade, and large quantities of cranberries are annually picked and sent down the coast.

The timber of the southeast is remarkable for its size and general excellence.   The spruce, hemlock, red and yellow cedar, poplar, alder, willow, birch, larch and pine abound of great size and general excellence.   Nearly all the barrels for the salmon canneries and salteries are manufactured from the Alaskan spruce and an excellent quality of shingles is also made from the same wood.   The yellow cedar, because of its peculiar hardness and lightness is highly prized by the Indians for their paddles, which in the peculiarly dangerous navigation of the intricate and swift waterways, need to be of the best material to insure immunity from serious and often fatal mishaps afloat.

This yellow cedar is also a very beautiful wood when polished, easy to work, of a bright canary and delightful odor, and is esteemed in the manufacture of furniture and all sorts of fancy articles.   It possesses also another point of excellence which, being strictly utilitarian, bids fair some day to largely deprive the arts of its use.   It is one of the few known woods which the destructive teredo refuses to attack and hence is invaluable for piling.   Except for its expensiveness it would long ago have run the Oregon pine out of the market for this purpose.

Making a Canoe.

Out of these great cedars the Southern Alaskan natives also hew their huge canoes.   The task is long and laborious, but the finished vessel has been rightly deemed a work of boat builders' art, and, for the waters where it is used, is unequalled.   No journey in these canoes seems long or hazardous enough to appal the Indian voyageur ; in fact, the natives have been known, on their forays, to paddle in them as far as Puget Sound and

back again.   To make a canoe a large and perfectly symmetrical
log is chosen and properly beached.   The outside is shaped with
a heavy axe and then the inside is roughly hollowed out with
fire and tools.   Then with a small. home-made hand adze the
boat carpenter goes over the entire vessel, inside and out, carefully chipping away until the smooth and perfect outline has
been produced throughout.   The boat is then steamed by filling
it with water into which heated stones are dropped, and the final
shaping or " spreading " is given by putting in the cross braces
while the wood is thus pliant.   Some of these single log canoes
are forty-two feet in length.

A peculiar feature of these southeastern forests, noticed by the
first white explorers, and for a time a scientific puzzle, was found
in the great number of yellow cedar trees standing outwardly
dead and yet not decaying, but sound to the core.   It was finally
ascertained that this was due to the thickly overshadowing
branches of the taller surrounding spruce and hemlock, slowly
smothering the cedars to death.

Reserve Lumber Region.

Alaska is the great reserve lumber region of the United States.
William H. Seward, returning from a trip to Alaska, said in a
public address :

" I venture to predict that the North Pacific coast will become
a common shipyard for the American continent and speedily for
the whole world.   Europe, Asia, Africa, and even the Atlantic
American States have either exhausted or are exhausting their
native supplies of timber and lumber.   Their last and only
resort must be to the North Pacific.   Then the country will
appreciate these thousands of square miles of cedar, spruce,
hemlock and balsam firs."

Although in the mountainous interior vegetation and fauna

partake of Arctic characteristics, near the sea in the southeast the summer is a season of delicious sounds, and sweet perfumes, the voices of birds, ripple of running water, and music of waving branches making it difficult for the traveler to believe that he is in the marches of the Empire of Ice. The flowers and orchids are almost tropical in the luxuriance and beauty.

In the Aleutian Islands the cereals will not mature, though numerous and persistent experiments to that end have been made. Vegetation of speedier growth flourishes in season, and the grasses are especially rank in growth. The state of the stock industry, however, is problematical. The timber of the islands is similar to that of the mainland, both as to variety and size.

On the Kadiak Islands are great forests and vast grassy plains where cattle thrive with little feeding and shelter. Sheep also do well here, except for a tendency to hoof rot.

Summer in the Yukon.

The brief summer in the Yukon Basin, enduring only from the middle of June to the first of September, presents an unending panorama of extraordinary picturesqueness and beauty. The banks are fringed with flowers, carpeted with the all pervading moss. Birds, countless in numbers, and of bewildering variety of plumage, pipe out a song from every treetop. Let the voyageur pitch his tent where he will in summer, a bunch of roses, a clump of poppies, and a bed of bluebells will adorn the camping.

High above this almost tropical floral exuberance, giant glaciers sleep in the summits of the mountain wall which rises from a bed of blossoms. In September they waken and everything is changed. The roses disappear before the frosty breath from the peaks, the birds fly to the southland, and mountain, and plain, hide for the long winter beneath a sheet of snow.

In the Yukon basin vegetables of the hardier sorts do fairly well. Turnips, radishes and salad plants and even potatoes have been successfully cultivated at St. Michael's and at Fort Yukon.

At Fort Selkirk, on the British side, gardening has become a science and the results are pleasing in size and variety. The whole Yukon basin raises fine berries and grass, but other crops are hard to mature, and though the fodder is plenty and good, the long winter precludes success in stock raising. It is believed the dairy industry would thrive, however.

The timber of the Yukon is principally willow, alder, cotton-wood, spruce, low fir, hemlock and birch. North of the basin the growths become stunted and finally disappear.

Dr. Jackson's View.

Dr. Sheldon Jackson, Commissioner of Education, has given to the Department of Agriculture, his views of the agricultural possibilities of Alaska as follows:

" The warmest friends of Alaska do not claim that it is rich in agricultural resources, or that it will agriculturally bear comparison with the rich valleys of the Mississippi River ; but they do claim that while there are large areas of mountains and unproductive land agriculturally, yet there are valleys and plains where, with suitable care, many of the earlier vegetables, fruits, and grains can be raised.

" On Kadiak, on adjacent islands, and on the shores of Cook's Inlet, where there are small Russian Creole settlements, they have for three-quarters of a century supplied themselves with vegetable food from their own gardens.

" Not only in the mild belt of Southern Alaska, but also in the arctic and subarctic belt of Northern Alaska, various wild berries grow and ripen in profusion (cranberries, currants, raspberries, huckleberries, blackberries, strawberries), and there is no

question that if the government places Alaska on an equal footing with the other States and Territories in the establishment of
one or more experimental stations it will be demonstrated that
sufficient vegetables can be raised for the consumption of its people. And if there is found a section so far north that the profitable raising of vegetables and grains becomes impossible, that
region can be utilized by the introduction of herds of domestic
reindeer.

"Taking Norway and Sweden, where complete statistics are
to be had, as a basis of calculation, and applying the same average to Alaska, it is found the country is capable of sustaining
9,200,000 head of reindeer, which will support a population of
287,500 living like the Laps of Lapland.

"The stocking of Alaska with tame reindeer means the opening up of the vast and almost inaccessible central region of Northern and Central Alaska to white settlers and civilization and the
opening up of a vast commercial industry. Lapland, with
400,000 reindeer, supplies the grocery stores of Northern Europe
with smoked reindeer hams, smoked tongues, dried and tanned
hides, and 23,000 carcasses per annum to the butcher shops.
On the same basis, Alaska, with its capacity of 9,200,000 head
of reindeer, can supply the markets of North America with
500,000 carcasses of venison annually, together with tons of delicious hams and tongues and finest leather."

Dall's Statement.

William H. Dall, of the Smithsonian Institution, wrote as follows :

"I am convinced, after careful inspection, that Alaska is a far
better country than much of Great Britain and Norway and even
part of Prussia. Excepting for the extreme cold in midwinter of
the interior, the Alaskan climate and productions are not unlike

those of the northwestern part of Scotland or the Shetlands and Orkneys."

As the Canadian territory contiguous to Alaska is at present the site of the gold craze and contains many of the avenues by which access is had to the British Klondike, the interest attaching to this alien region at the headwaters of the Yukon warrants a few words in notice of its flora and agricultural possibilities.

Surveyor Ogilvie's Report.

William Ogilvie, Dominion Land Surveyor, reported on this region to the Canadian Department of the Interior, as follows :

" The agricultural capabilities of the country along the river are not great, nor is the land that can be seen from the river of good quality. When we consider further the unsuitable climatic conditions that prevail in the region, it may be said that as an agricultural district this portion of the country will never be of any value.

" My meteorological records show over eight degrees of frost on August 1st, over ten on the 3d, and four times during the month the minimum temperature was below freezing.

"Along the east side of Lake Bennett, opposite the Chilkoot or western arm, there are some flats of dry gravelly soil, which would make a few farms of limited extent. On the west side, around the mouth of the Wheaton River, there is an extensive flat of sand and gravel, covered with small pine and spruce of stunted growth.

" Along the westerly shore of Tagish Lake there is a large extent of low, swampy flats, a part of which might be used for the production of such roots and cereals as the climate would permit. Along the west side of Marsh Lake there is also much flat surface of the same general character, on which I saw some coarse grass which would serve as food for cattle. Along the east side the surface appeared higher and terraced, and is probably less suited to the requirements of the agriculturist. Along the head of the river, for some miles below Marsh Lake, there are flats on both sides, which would, as far as surface conformation goes, serve as farms. The soil is of much better quality than any heretofore seen, as is proven by the larger and thicker growth of timber and underbrush which it supports. The soil bears less the character of detritus, and more that of alluvium, than that seen above.

"Some miles down the lake an extensive valley joins that of the lake on the west side. This valley contains a small stream. Around this place there is some land that might be useful, as the grass and vegetation is much better than any seen so far.

"On the lower end of the lake, on the west side, there is also a considerable plain which might be utilized; the soil in parts of it is good. I saw one part where the timber had been burned some time ago; here both the soil and vegetation were good, and two or three of the plants seen are common in this part of Ontario, but they had not the vigorous appearance which the same plants have East.

In Ogilvie Valley.

"Northward from the end of the lake there is a deep, wide valley, which Dr. Dawson has named 'Ogilvie Valley.' In this the mixed timber, poplar and spruce, is of a size which betokens a fair soil; the herbage, too, is more than usually rich for this region. This valley is extensive, and, if ever required as an aid in the sustenance of our people, will figure largely in the district's agricultural assets.

"Below the lake the valley of the river is not, as a rule, wide, and the banks are often steep and high. There are, however, many flats of modern extent along the river and at its confluence with other streams. The soil of many of these is fair.

"About forty miles above the mouth of the Pelly River there is an extensive flat on both sides of the Lewis. The soil here is poor and sandy, with small open timber. At Pelly River there is a flat of considerable extent on which the ruins of Fort Selkirk stand. It is covered with a small growth of poplar and a few spruce. The soil is a gravelly loam of about eight inches in depth, the subsoil being gravel, evidently detritus. This flat extends up the river for some miles, but is all covered thickly with timber, except a small piece around the site of the fort.

Vegetables for Miners.

"I think ten townships, or 360 square miles, would be a very liberal estimate of all the places mentioned along the river. This gives us 230,400 acres, or, say, 1000 farms. The available lands on the affluents of the rivers would probably double this, or give 2000 farms in that part of our territory, but on most of these farms the returns would be meager. Without the discovery and development of large mineral wealth it is not likely that the slender agricultural resources of the country will ever attract attention. In the event of such discovery, however, some of the land might be used for the production of vegetable food for the miners, but even in that case, with the transport facilities which the district commands, it is very doubtful if it could compete successfully with the South and East.

"The amount of timber fit for use in building and manufacuring in the district along the river is not at all important. There is a large extent of forest which would yield firewood and timber for use in mines, but for the manufacture of lumber there is very little.

"To give an idea of its scarceness, I may state that two of my party made a thorough search of all the timbered land around the head of Lake Bennett, and down the lake for over ten miles, and in all this search only one tree was found suitable for making such plank as we required for the construction of our large boat. This tree made four planks, fifteen inches wide at the butt, seven at the top, and thirty-one feet long.

" Such other planks as we wanted had to be cut out of short logs, of which some, ten to fourteen inches in diameter and ten to sixteen feet long, could be found at long intervals. The boat required only 450 feet of plank for its construction, yet some of the logs had to be carried nearly 200 yards, and two saw-pits had to be made before that quantity was procured, and this on ground that was all thickly wooded with spruce, pine and some balsam, the latter being generally the largest and cleanest-trunked.

" The great bulk of the timber in the district suitable for manufacture into lumber is to be found on the islands in the river. On them the soil is warmer and richer, the sun's rays striking the surface for a much longer time and more directly than on the banks.

Quantity of Timber.

" To estimate the quantity of timber in the vicinity of the river in our territory would be an impossible task, having only such data as I was able to collect on my way down. I would, however, say that one-fourth of the area I have given as agricultural land would be a fair conjecture. This would give us two and one-half townships, or ninety square miles, of fairly well-timbered ground ; but it must be borne in mind that there is not more than a square mile or so of that in any one place, and most of the timber would be small and poor compared with the timber of Manitoba and the easterly part of the Northwest Territories.

" It may be said that the country might furnish much timber, which, though not fit to be classed as merchantable, would meet

20

many of the requirements of the only industry the country is ever likely to have—viz. : mining."

The native animal life of Alaska, whether of land or sea, fish or fowl, is in general that of a northern country with its peculiar climatic conditions. The fur bearing land animals and amphibians are important, and the fisheries are not surpassed. The insect life partakes of a tropical nature and in summer time the pest of mosquitos and gnats is almost unbearable. There is some compensation in the absence of snakes from the territory.

Alaska's first value in the eyes of civilization was in its furs of land and sea, and for a century the fur industries were the chief occupation of the Russian colonists and their aboriginal allies. Only within a decade has gold been a rival to furs in the territory.

The fur producing amphibians are principally the valuable and comparatively rare sea otter and the fur seal, the ambition of every woman's heart on two continents and the cause of a hundred years of international complications. The fur of the sea otter is among the most beautiful and highly prized known, and until within a very few years has brought enormous prices in the London market. Of the fur of the seal it is unnecessary to speak further than to say that it is still the basis of the most extensive commerce, and it furnishes a livelihood ashore and afloat to many thousands of hands, and employment to many millions of capital.

The Sea Otter.

The sea otter was once abundant along the whole southeastern and southwestern coast of Alaska, how abundant may be gathered from the fact that the estimated total value of all the sea otter skins taken up to 1890 is $36,000,000. The Russians encouraged the natives to slaughter the valuable animal, and the

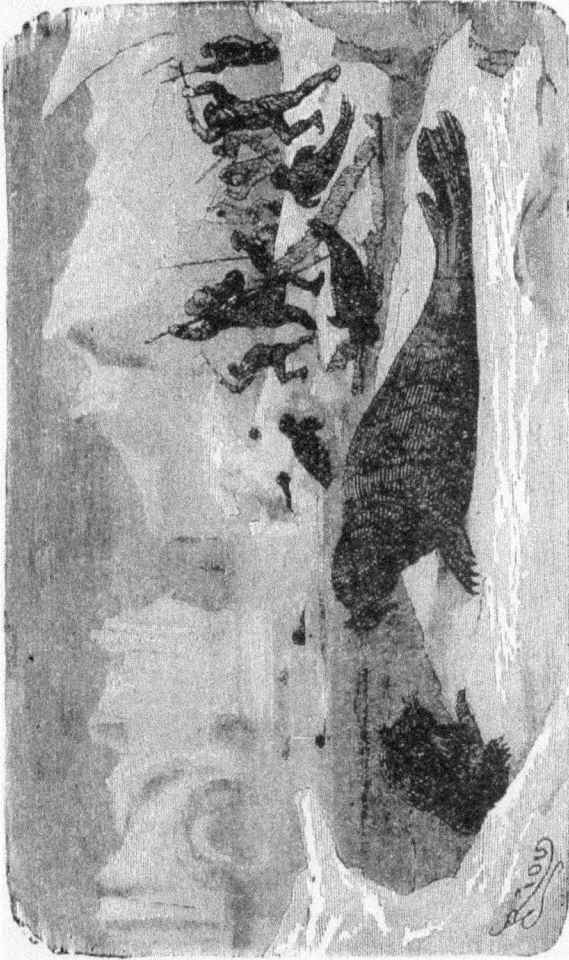

KILLING SEALS ON BEHRING SEA NEAR ALASKA.

307

Yankee fishers and their British brethren were no more inclined to mercy or thrift than the Muscovites. Gradually the furry amphibian was driven from the southeastern archipelago until to-day the chief and, in fact, almost the only grounds where it is successfully hunted are along the Aleutian chain and to the eastward in the neighborhood of Kadiak Island and the mouth of the Copper River.

La Perouse sent the first sea otter skins home to France in 1788. Their magnificent beauty soon made them the talk of the courts of Europe, and as they were easily approached by hunters in those early days their slaughter grew apace with the demand. The female otter is very tender of its young and, sailors say, often gathers the little one upon its breast between its fore legs and floating on its back on the water, croons a lullaby to the baby otter which the hunters aver is almost human in its tones.

Romance of the Otter.

A bit of romance which colored the lives of the native women in the early days of the Russian occupation of Alaska was due entirely to the sea otter. The right to hunt them was proscribed to all except natives or the husbands of native wives. As the pursuit was exceedingly profitable and the women not altogether bad looking, there came about a marrying epidemic among the white sailors, especially the Scandinavians, which gave the dark-skinned belles a chance to be courted into a home of their own, which it is safe to say they had never enjoyed before. From these unions grew up a race of hardy half-breed otter hunters whose prowess is still famous on the coast.

The fur seal, $47,000,000 worth of whose skins had been taken up to 1890, once had a habitat coextensive with that of the sea otter, but like the latter has been driven to the westward, and now only an occasional specimen is seen in the waters of the southeastern archipelago. Its principal Alaskan resting places are now the Islands of St. Paul and St. George and the adjacent rookeries.

Other seals which are native to Alaskan waters are the hair, leopard, saddle and big black seal or maklak. They are hunted by the natives for their skins, but the fur is of small commercial value..

Land Animals.

The land animals, native to Alaska, include several species of the fox, the land otter, beaver, brown, black, cinnamon, grizzly and polar bears, mink, marten or sable, lynx, wolverine, muskrat, marmat, ermine, squirrel, moose, caribou, deer, mountain sheep, mountain goat, barren ground caribou, musk-ox and wolf. The Esquimo dog, though comparatively domesticated, is also entitled to a place among the native animals of the territory. Some of the animals enumerated are of value for their skins or for food; otters are merely the brute Ishmaels of the wilderness.

The black, or silver fox (the same species with different markings), is easily the king of the vulpine Alaskans. Traffic in its skins makes up the bulk of the fur trade of the Yukon Basin. They are the highest priced of any of the native fox skins. The red fox is found all over the territory and has even been known to take a voyage over to the Aleutian Islands on an opportunely drifting ice cake. Its skin is as cheap as it is plenty. The cross fox, so named because it is a cross between the black and red, is likewise all over the country, and likewise cheap. The Arctic fox, both white and blue, is found on the mainland and in the Seal and Aleutian Islands. Its skin has little value. General characteristics of the Alaskan foxes are their perpetual famine, their absolutely omniverous taste and their lack of shyness which often leads to unpleasant experiences for "tenderfeet" when camping out. The Alaskan Commercial Company ten years ago established a "fox farm" on Semidi Island, bringing the black, blue and silver colonists from the mainland and leaving them to multiply.   The venture is said to have proved a financial success.

Otter and Bears.

The land otter, whose skin has considerable commercial value, both for itself and because of the ease with which it can be made into an imitation of seal skin, is found along the whole coast, among the islands, especially around Kadiak, and in the Yukon Basin.

The habitat of the beaver is within the timber limit.   The demand and supply in this fur are growing less together and the skins are cheap.   The old currency of the territory was beaver skins and the denominations are worth recalling as a matter of curiosity.   One beaver was worth four mink, two marten or two white fox skins ; a beaver and a half was equal to one red fox and three beaver skins were fair exchange for a land otter.

The brown bear is found all over the territory, and his pelts are plentiful and cheap.   Like all the Alaskan carnivoræ, he is a good fisher and can be found hanging around the salmon and trout streams in season.   He is the great road maker of the country and his broad trails over plains and through swamps are of no little use to travelers.   The black bear is widely at home on the mainland, generally in the timber, and his skin brings high prices.   The grizzly bear is found in the southeast.

The mink, which is common on the mainland, and the marten, which sticks close to standing timber, both supply cheap furs.

The animals of Alaska are all diligently hunted by the Indians and Esquimo for the flesh and for the skins, which form the natural clothing of the aborigines.   Those whose flesh is edible, as well as the more valuable fur-bearers, are also the white hunter's quarries, and the double chase is beginning to tell on the numbers of some of the species.

The moose and caribou are found in the Yukon basin and now and then furnish a dainty variety to the post trader or the miner for his menu. Deer are found mainly in the southeast, where the mountain sheep and goat are also comparatively plentiful. All are hunted for their flesh and skins.

Mrs. Frederick Schwatka says of the game in the Yukon basin :

" The great Yukon Valley has but little game in it during the summer, for the mosquitos drive all game to higher altitudes. Formerly during the winter season a living could be made by experienced hunters in bringing moose and caribou meat to camp. I heard one miner say, who had spent four winters on the Yukon, that he had seen moose and caribou so numerous on the bald hills above timber limit, in the present gold field district, that they gave the snow a mottled, gray appearance. Of course these have now disappeared with the advance of civilization, and fresh meat of any kind is now at a premium."

Canadian Fauna.

Dominion Land Surveyor Ogilvie's official report on the fauna of the Canadian territory adjacent to Alaska is as follows :

" The principal furs procured in the district are the silver-gray and black fox, the number of which bears a greater ratio to the number of red foxes than in any other part of the country. The red fox is very common, and a species called the blue is very abundant near the coast. Marten, or sable, are also numerous, as are lynx ; but otter are scarce, and beaver almost unknown.

" It is probable that the value of gray and black fox skins taken out of the country more than equals in value all the other furs. I could get no statistics concerning this trade for obvious reasons.

"Game is not now as abundant as before mining began, and it is difficult, in fact impossible, to get any close to the river.

"A boom in mining would soon exterminate the game in the district along the river.

"There are two species of caribou in the country; one, the ordinary kind, found in most parts of the Northwest, and said to much resemble the reindeer; the other, called the 'wood caribou,' a much larger and more beautiful animal. Except that the antlers are much smaller, it appears to me to resemble the elk or wapiti.

"The ordinary caribou runs in herds, often numbering hundreds.

Bear in Abundance.

"There are four species of bear found in the district—the grizzly, brown, black and a small kind, locally known as the 'silver-tip,' the latter being gray in color, with a white throat and beard, whence its name. It is said to be fierce, and not to wait to be attacked, but to attack on sight. I had not the pleasure of seeing any, but heard many 'yarns' about them, some of which, I think, were 'hunters' tales.' It appears, however, that miners and Indians, unless traveling in numbers, or especially well armed, give them as wide a berth as they conveniently can.

"Wolves are not plentiful. A few of the common gray species only are killed, the black being very scarce.

"The Arctic rabbit or hare is sometimes found, but they are not numerous. There is a curious fact in connection with the ordinary hare or rabbit which I have observed but of which I have never yet seen any satisfactory explanation. Their numbers vary from a very few to myriads, in periods of seven years."

The Alaskan birds include the grouse, ptarmigan, snipe, mallard and teal duck, goose, loon, gray and bald eagle, sea parrot, gulls, auks and many other sea fowls. One of the ornithological

TRAPPING FUR-BEARING ANIMALS IN ALASKA.

313

wonders of the territory last year was a pair of humming birds which nested in Sitka. The sea birds supply the Indians with a profitable pursuit gathering their eggs from the rocks. The eggs are a staple article of diet with the natives.

The piscatorial wealth of Alaska ranks next to the furs. The food fishes are numerous, but the salmon easily leads them all in importance, and the canning and drying of this dainty fish make the third industry of the territory, gold being now the first, of course, and furs the second.

The first salmon cannery was established at Old Sitka in 1878, but another was started in 1883 at Kadiak Island, and since that time the canneries and salteries (though the salmon was never accused of singing like the catfish it still has salteries) have spread all along the coast.

Species of Salmon.

The king or " tyee " salmon has the highest standing in the market. Less highly esteemed are the silver or red, cohoe, dog and humpback salmon. The cod, which is found all along the south shore, comes next in commercial importance. It much resembles the cod of the North Atlantic. Halibut are found all along the coast, in the channels and to the western extremity of the Aleutian Islands at Attu. No great quantity of this fish is shipped, but the natives catch it in great numbers, smoke or dry the flesh, and esteem it highly for food. Herring are found in immense shoals in the bays and estuaries and throughout the island chains. They supply material to a large oil and fertilizer factory at Killisnoo, the product of which is shipped to the Sandwich Islands. The salmon trout is a fish of magnificent size and fine flavor and mountain trout are caught freely in the southeast. There are also many other edible fish in the waters of the southeast.

The uliken, or candle fish is found in the southeastern waters, and is highly prized by the Indians for food and medicinal purposes. It is so oily that it cooks to a turn in its own oil and is said to be then a delicious morsel. The oil has a flavor not unlike that of olive, and the natives esteem it highly as a remedy for lung troubles and for dyspepsia.

It would not be fair to the dog-fish to pass him by without at least a mention. He is useless for food, even to the strong-stomached native, who deems blubber a delicacy and whale oil a libation to pour to his heathen gods ; but the dog-fish can stand more abuse and make less fuss about it than any other known member of the animal kingdom. When by any ill luck a tourist, fishing off the wharf at Sitka or Juneau, pulls up a dog-fish on his line, some stolid native is sure to beg the prize. The Indian rips the squirming dog-fish, takes out his liver to try out for oil, and flings him back into the water, where he swims off apparently as lively as if he was in the habit of having such things happen every day. It is said that the only dog-fish that was ever killed at Sitka was one which, having been originally delivered by an Indian, insisted on being caught by a white man and hauled up and thrown out to a native, as if in mockery of the latter's desire for liver. The Indian thought the joke had been played on him once too often, and smashed the dog-fish's head with a stone. A valuable lubricating oil is obtained from the dog-fish, and the natives use the skin of its belly for sandpaper.

Finds Vast Fishing Banks.

The United States steamer Albatross, in making soundings for the Coast Survey, developed vast and thitherto unknown fishing banks all along the Aleutian Chain. It is on these banks the best cod fishing is had.

Of the cetaceans the whale, beluga or white grampus, and porpoise are found all along the Alaskan coast.

The regular whale fishing gounds are on the Arctic shore, where Herschel Island, at the mouth of the Mackenzie River is a common station for all whalers. A large American fleet is constantly on the grounds. Black whales often appear in the channels around the Southeastern Archipelago in such numbers as to terrify the Indians who are out in their canoes.

The obese walrus, once the principal food supply of the region of its habital, has been hunted nearly to extermination.

Many beluga are taken each season by the Esquimeaux south of Norton Sound, with whom it is a food staple. The porpoise is also a constant object of the watery chase.

Crabs and clams are plentiful on the southern coasts, but no oysters are found.

Insect Pests.

It would be a vital defect in the story of the animal life of Alaska if no mention was made of the insects which make life a burden in the short, hot summer of the interior. Horseflies, gnats and mosquitos nearly drive men and beasts wild. The horsefly is larger and more "pointed" than the insect of the same name in the States. In dressing or undressing it has the pleasant habit of detecting any bare spot in the body and biting out a piece of flesh, leaving a wound which in a few days later looks like an incipient boil. Schwatka reports that one of his party so bitten was completely disabled for a week. "At the moment of infliction," he adds, " it was hard to believe that one was not disabled for life."

The mosquitos, according to the same authority, are equally distressing. They are especially fond of cattle, but without any reciprocity of affection. "According to the general terms of the survival of the fittest and the growth of muscles most used to the detriment of others," says the lieutenant in an unusual burst of humor, "a band of cattle inhabiting this district in the far future would be all tail and no body, unless the mosquitos should experience a change of numbers."

Mrs. Schwatka, in speaking of the trials of the miner's life, touches on his sufferings from these insect pests in these words:

"Again in summer the work of the miner is difficult. As I have said the interior country is tundra land—that is, the earth is frozen to a great depth, never entirely thawing out. Wherever the sun strikes the surface, great pools of muddy water are formed, and this prevents any sort of prospecting. These pools of stagnant water breed great swarms of mosquitos and gnats, which make it desirable to cover the head with mosquito netting, or better still, adopt the Indian method, and smear the hands and face with a mixture of grease and soot, which prevents the pests from biting. At some seasons in this country they are in such dense swarms that at night they will practically cover a mosquito netting, fairly touching each other and crowding through any kind of mesh. I have heard it asserted by people of experience that they form co-operative societies and assist each other through the meshes by pushing behind and pulling in front. Others again say they are too mean for such generous action."

Climate of Alaska.

The climate of the Alaskan coast regions is much milder, even in the higher latitudes, than it is in the interior or in corresponding latitudes on the Atlantic coast. This is easily explained and understood when the natural forces of production of this milder temperature are contemplated.

The most important among them is the thermal current resembling the Atlantic Gulf Stream, and known as the Japanese or Kuro Siwo, or Black Water. It has its origin under the equator near the Molucca and Philippine Islands, passes northward along the coast of Japan, and crosses the Pacific to the southward of the Aleutian Islands, after sending a branch through Behring Sea. On the coast of British Columbia it divides again, one branch turning north toward Sitka, and thence westward to the Kadiak and Shumagin Islands.

The comparatively warm waters of these currents affect the temperature of the superjacent atmosphere, which, absorbing the latent heat, carries it to the coast with all its mollifying effects. Thus the oceanic and atmospheric currents combine in mitigating the coast climate of Alaska, while the almost impenetrable barrier of lofty mountains deflects the ice-laden northern gales from the interior.

The mean winter temperature of Sitka is slightly above 30 degrees, while that of Portland, Maine, is about 27 degrees. The lowest in winter in 1889, in Sitka, was 3 degrees; in Halifax, Nova Scotia, 7 degrees; and in Portland, Maine, 15 degrees.

Weather Bureau Report.

Under the direction of Secretary of Agriculture James Wilson, Willis L. Moore, Chief of the Weather Bureau, makes public the following :

" The general conception of Alaskan climate is largely due to those who follow the sea, and this is not strange when we consider the vast extent of short line (over 26,000 miles) possessed by that territory.

" The climate of the coast and the interior is unlike in many respects, and the differences are intensified in this, as, perhaps, in few other countries, by exceptional physical conditions.

" The natural contrast between land and sea is here tremendously increased by the current of warm water that impinges on the coast of British Columbia, one branch flowing northward toward Sitka, and thence westward to the Kadiak and Shumagin

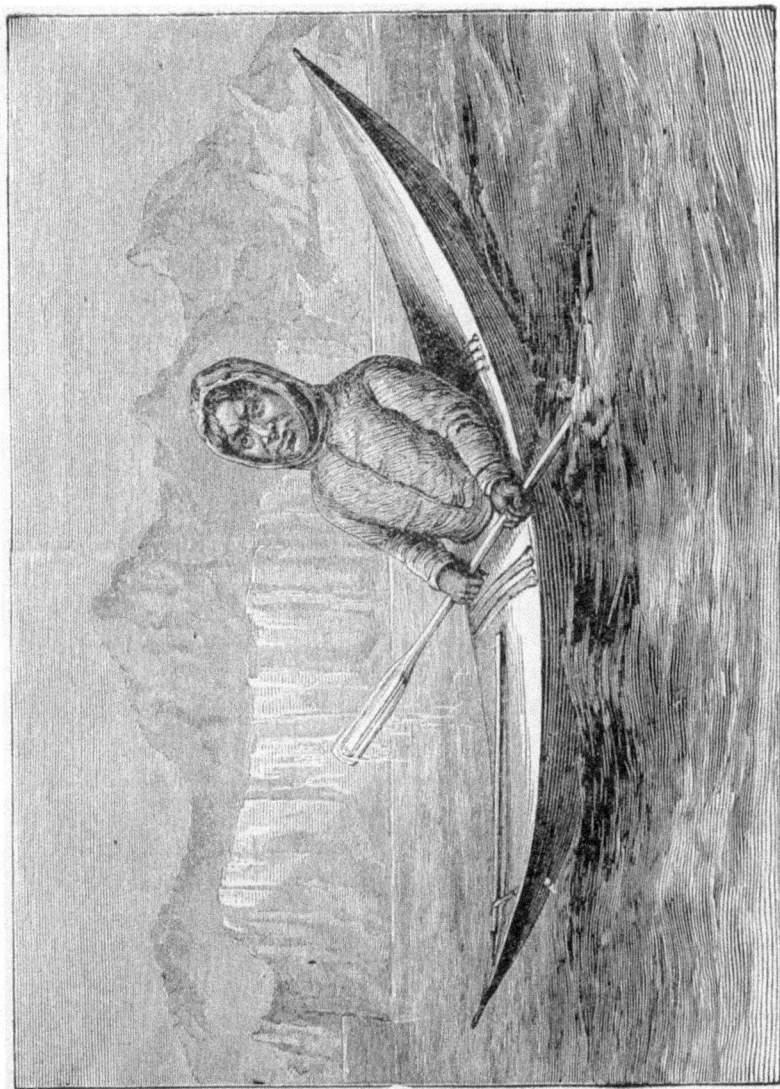

ESQUIMO IN HIS WATER-PROOF CANOE.

Islands.   The fringe of islands that separates the mainland from
the Pacific Ocean, from Dixon Sound northward, and also a strip
of the mainland for possibly twenty miles back from the sea,
following the sweep of the coast as it curves to the northwestward, to the western extremity of Alaska, form a distinct
climatic division which may be termed temperate Alaska.

" The temperature rarely falls to zero.   Winter does not set
in until December 1st, and by the last of May the snow has disappeared, except on the mountains.   The mean winter temperature of Sitka is 32.5 degrees, but little less than that of Washington, D. C.   While Sitka is fully exposed to the sea influences,
places farther inland, but not over the coast range of mountains,
as Killisnoo and Juneau have also a mild temperature throughout
the winter months.

Small Changes of Temperature.

" The temperature changes from month to month in Alaska
are small, not exceeding 25 degrees from midwinter to midsummer.   The average temperature of July, the warmest month
of summer, rarely reaches 55 degrees, and the highest temperature for a single day seldom reaches 75 degrees.

" The rainfall of temperate Alaska is notorious the world over,
and not only as regards the quantity but also as to the manner
of its falling—viz. : in long and incessant rains and drizzles.
Cloud and fog naturally abound, there being on an average but
sixty-six clear days in the year.

" Alaska is a country of striking contrasts, both in climate as
well as topography.   When the sun shines the atmosphere is
remarkably clear, the scenic effects are magnificent; all nature
seems to be in holiday attire.   But the scene may change very
quickly.   The sky becomes overcast, the winds increase in
force, rain begins to fall, the evergreens sigh ominously, and
utter desolation and loneliness prevail.

" North of the Aleutian Islands the coast climate becomes
more rigorous in winter, but in summer the difference is much
less marked.   Thus, at St. Michael's, a short distance above the
mouth of the Yukon, the mean summer temperature is 50
degrees, but four degrees cooler than Sitka.   The mean summer
temperature of Point Barrow, the most northerly point in the
United States, is 36.8 degrees, but four-tenths of a degree less than
the temperature of the air flowing across the summit of Pike's
Peak, Colorado.   The rainfall of the coast region north of the
Yukon delta is small, diminishing to less than ten inches within
the Arctic Circle.

" The climate of the interior, including in that designatior
practically all of the country except a narrow fringe of coasta'
margin and the territory before referred to as temperate Alaska,
is one of extreme rigor in winter, with a brief but relatively hot
summer, especially when the sky is free from clouds.

" In the Klondike region in midwinter the sun rises from 9:30
to 10 A. M. and sets from 2 to 3 P. M., the total length of daylight being about four hours.   Remembering that the sun rises
but a few degrees above the horizon, and that it is wholly
obscured on a great many days, the character of the winter
months may easily be imagined.

Temperature or Yukon.

" We are indebted to the United States Coast and Geodetic
Survey for a series of six months' observations on the Yukon,
not far from the site of the present gold discoveries.   The observations were made with standard instruments, and are wholly
reliable.   The mean temperature of the months October, 1889, to
April, 1890, both inclusive, are as follows: October, 33 degrees;
November, 8 degrees; December, 11 degrees, below zero; January, 17 below zero; February, 15 below zero; March, 6 above;

21

April, 20 above   The daily mean temperature fell and remained below the freezing point (32 degrees) from November 4, 1889, to April 21, 1890, thus giving 168 days as the length of the closed season of 1889–'90, assuming that outdoor operations are controlled by temperature only.   The lowest temperatures registered during the winter were : Thirty-two degrees below zero in November, 47 below in December, 59 below in January, 55 below in February, 45 below in March, and 26 below in April.

"The greatest continuous cold occurred in February, 1890, when the daily mean for five consecutive days was 47 below zero.

"Greater cold than that here noted has been experienced in the United States for a very short time, but never has it continued so very cold for so long a time in the interior of Alaska.   The winter sets in as early as September, when snow-storms may be expected in the mountains and passes.   Headway during one of these storms is impossible, and the traveler who is overtaken by one of them is indeed fortunate if he escapes with his life. Snow-storms of great severity may occur in any month from September to May, inclusive.

"The changes of temperature from winter to summer are rapid, owing to the great increase in the length of the day.   In May the sun rises at about 3 A. M. and sets about 9 P. M.   In June it rises about half-past 1 in the morning and sets about half-past 10, giving about twenty hours of daylight and diffuse twilight the remainder of the time.

"The mean summer temperature in the interior doubtless ranges between 60 and 70 degrees, according to elevation, being highest in the middle and lower Yukon valleys."

Dominion Climate.

Describing the country in the coast range mountains near Taiya Inlet, Dominion Surveyor Ogilvie writes :

" It is said by those familiar with the locality that the storms which rage in the upper altitudes of the coast range during the greater part of the time from October to March are terrific. A man caught in one of them runs the risk of losing his life, unless he can reach shelter in a short time.

" During the summer there is nearly always a wind blowing from the sea up Chatham Strait and Lynn Canal, which lie in almost a straight line with each other, and at the head of Lynn Canal are Chilkat and Chilkoot Inlets. The distance from the coast down these channels to the open sea is about 380 miles.

The mountains on each side of the water confine the currents of air, and deflect inclined currents in the direction of the axis of the channel, so that there is nearly always a strong wind blowing up the channel. Coming from the sea, this wind is heavily charged with moisture, which is precipitated when the air current strikes the mountains, and the fall of rain and snow is consequently very heavy.

" In Chilkat Inlet there is not much shelter from the south wind, which renders it unsafe for ships calling here. Captain Hunter told me he would rather visit any other part of the coast than Chilkat."

Mounted Police Report.

The report of the Canadian Mounted Police shows that on twenty-four days during the winter of 1896-97, the thermometer registered 50 degrees or more below zero. The report continues :

" Apparently the temperature first touched zero on November 10th, and the last zero recorded in the spring was on April 29th.

Between December 19th and February 6th it never rose above zero. The lowest actual point, 65 degrees, occurred on January 27th and on twenty-four days during the winter the temperature was below 50 degrees."

CHAPTER XI.

Industries and Industrial Development.

Chief Occupations of the Natives and the Settlers—The Four Remarkable Seal Islands—How the Animals Have Been Ruthlessly Slaughtered—When the Fur is at Its Best—The Great Fishing Plants of the Country—Alaska the Home of the Salmon—Cod and Other Fish Abound—Trapping and Hunting on the Decline—Current Belief that the Outlook for Lumbering is Not Good—Probability that this Opinion may be Reversed by Later Discovery—Trees on the Islands—Agricultural Development one of the Great Needs at the Present Time—Land Simply Needs Tilling—Vegetables and Berries Grown in Quantities—Reports of Travelers.

THE resources of Alaska are, as has been shown in another chapter, as diversified and remarkable as the surface of its vast district. With a few noteworthy exceptions, however, these resources are largely undeveloped.

The country is so remote, its fastnesses have been so inaccessible, the lack of transit facilities has imposed such a barrier on imigration, that few are the hardy souls who have traversed its boundless plains, its mighty rivers and its snow-capped mountains; and still fewer are the capitalists who have had the hardihood to seek the country for investment.

The result is that in most lines of industry the possibilities of the country are largely a matter of conjecture.

Two or three occupations received early attention and have been followed systematically. The Russians recognized the value of the fur-bearing animals and were pioneers in the enterprise that John Jacob Astor made so memorable throughout the Northwest. The wealth of fish in the vast rivers of the country also appealed to the commercial sense of the Russians. The same is true of the seal islands, from which such revenue was derived in the days of Russian occupation.

The mineral wealth of the Territory has only in a limited measure tempted the capital of more civilized communities, with the exception of gold mining. The story, therefore, of the industries of the country would be a meager one were it not in a large measure told in the language of opinion and prophecy. A resumé is here given of the interests that have claimed attention outside the gold fields, and a forecast of the future on other lines.

The Four Seal Islands.

The much-talked-of seal islands are one of the features of Alaska. These are four volcanic islands, which lie 220 miles northwest of Unalaska. They are veiled in perpetual mists and fogs in the summer season and are closely hedged round with drift ice in winter. They are absolutely treeless, but are covered with moss and grass, and in the proper season are brilliant with wild flowers.

Hundreds of thousands of seals gather annually on these islands, and the slaughter grounds, where millions of seals have been killed in the last century, are rarely visited except by those engaged in the business and by a few hardy tourists. The odors of these rookeries, as they are called, can be perceived far out at sea, and not infrequently the barking of the animals is the mariner's only guide in the dense fogs that settle over the waters.

No vessels other than those belonging to the government are allowed to enter or even to approach the harbors. The largest of the islands is called St. Paul, and is twelve miles long and from six to eight miles wide. St. George Island, thirty miles north, is a little smaller; and between these two lie Otter and Walrus Islands.

Practically the only inhabitants of the islands are the Aleuts, who have rather tidy villages, Greek churches and school houses. The islands are the government reserve, and are leased by the

United States Treasury Department for the term of twenty years. It has been said, and that with truth, that for over a century these four islands have yielded more wealth than any gold mine in the world.

With the settlement of the northwest coast, however, the prosperity of the islands has somewhat diminished, for the reason that the seals have been exterminated ruthlessly.

A word here about the discovery of these islands. For forty years Siberian traders hunted for the fabled island of Amik, where, it was believed, the sea bears lived. In 1786 Gerassim Pribylov heard the barking of the animals through the fog and found the summer home of the fur seals. It is said that 2,000,000 seals were killed that year, and the wholesale destruction of the animals has practically kept up ever since, barring a short interim when steps were taken of a preventive character to allow the rookeries time to recuperate .

In 1835 the islands were ringed with ice so that the seals could not land and their offspring died in the surf with their mothers. Some years later the herd was nearly extinct again. In 1844 Sir George Simpson found the company having control over the islands taking from 200,000 to 300,000 skins annually. The market at that time was so overstocked that the skins did not pay for carrying.

In cases of a glut of the market there have been times when from 700,000 to 1,000,000 skins were thrown into the sea to keep prices up. It was not until about the time of the transfer of the country to the United States that the vast importance of these four little islands was realized.

Seven Companies at Work.

No protection was afforded them in 1868, and at that time seven companies had the privilege of devastating the islands and slaughtering the animals. The next year, however, the islands were declared a Government reserve, and a guard of soldiers was stationed there. In 1870 the islands of St. Paul and St. George and the seal fisheries we're leased for a period of twenty years to the Alaska Commercial Company, of San Francisco. This company had previously purchased all the buildings and the good will of the Russian-American Fur Company throughout Alaska.

The company was permitted to kill 100,000 seals each year, 80,000 on St. Paul and 20,000 on St. George, for an annual rental of $55,000. It is believed that the company divided from $900,000 to $1,000,000 profits each year between twelve original stockholders. In 1890 a twenty-year lease was awarded to the North American Commercial Company, of San Francisco, at an annual rental of $100,000, a tax of $9.62 on each 100,000 skins taken, the islands then to return over a million a year to the Government, or 14 per cent. on Secretary Seward's investment.

Miss Skidmore points out the fact that pelagic sealing and rookery raiding by the Victoria fleet had so diminished the herd that the lessees were only permitted to take 20,000 animals the first season, and for three seasons, while the seal question was a matter of diplomatic discussion, only the few seals necessary for the food supply of the natives were killed.

Fur at Its Best.

The seal fur, she also states, is in its best condition immediately on the arrival of the animals at the islands, but they assume new coats in August, so that they are in fine condition when they leave at the end of September. Only male seals from two to four years of age are killed.

The bachelors herd alone, and the natives run in between them and the water, in the early morning, and drive them slowly to the killing grounds, where they are dispatched by a blow on the head. They are quickly bled and the skins taken to the salting house.

It may be mentioned as a matter of interest that Miss Anna Fulcomer, with whom an interview was given in another chapter, had the privilege of visiting the seal islands and the killing grounds. She crept up behind a herd of animals as they were sleeping, and softly stroked the ears of a big male. Her caress awakened the animal, and, with hissing and barking, he roused the rest of the herd, and the whole lot scampered off as fast as they could.

The Fishing Industry.

The fisheries of the country have been one of the leading sources of wealth to the time of the discovery of gold. It is to be remembered that productive as sealing has been, a limit has been reached in that industry which makes it, and will for some time make it, comparatively unproductive. The vast rivers of Alaska, however, annually teem with a wealth of fish, and the wholesale netting of them seems in nowise to diminish the number.

These fish vary in kind and are excellent in quality, and will, therefore, remain a constant source of wealth to the populace. In Southern Alaska and along the coast line many very large canneries have long been in operation, and their output has been something remarkable. There is no reason to believe that there will be any falling off in this line of occupation. Thousands of people of every nationality are engaged in the fisheries, the product of which is sent all over the world.

Unlike the great mineral wealth of the country, which lies hidden from view, and has to await some chance discovery, the fish that abound in the waters are open to view, and hence, there was no delay in the early development of this industry. Besides this, the canneries are for the most part located near the coast line, and hence those engaged in the business were not compelled to go hundreds or perhaps thousands of miles over snow-clad plains and mountains.

It was not necessary, further, to import into the country expensive machinery, and it was not difficult to get natives and other laborers from all over the world to engage in the work of catching the fish. As a consequence, Alaska soon built up a trade in the line of fisheries that placed it on a rank with the greatest fishing centres of the world.

Home of the Salmon.

Alaska is the home, practically, of the salmon, of which there are five distinct varieties. It has been pointed out that the Pacific salmon and the Pacific trout differ so from the Atlantic species that the question has been raised whether there are true salmon or trout on that coast, and whether any game laws can be enforced under such names.

The king salmon is generally called the tyee, which means chief. It averages from sixty to eighty pounds in the Stikine River, and often exceeds one hundred pounds in the Yukon. The fish commonly come in pairs and not in great schools, and hence it is not the whole pack of any cannery.

The red salmon is the blueback or Oregon Salmon, and is the canners' favorite. It averages from six to ten pounds in weight, comes in schools of vast size, and has flesh of a deep red color. The silver salmon is the gamiest of the lot, and is the most beautiful. Its flesh is pale, but has to be cared for almost immediately. Otherwise it is unfit for canning purposes. The fish always chooses clear water and shows a remarkable agility in leaping waterfalls.

The humpbacked species is the most abundant. It averages

from five to ten pounds in weight, and has flesh of a pale color which cooks soft, and hence is not very desirable for packing purposes. This fish has been known to jump falls sixteen feet high. In addition to these salmon there are the Dolly Varden trout, which follow the salmon in from the sea to devour their eggs, and the cut-throat trout, which are often used at the canneries.

Cod in Numbers.

The cod, which abound in Chatham Creek, are among the more important fish of the territory. The natives used to receive two cents apiece for the 8000 or 10,000 fish of five pound average, which they brought in daily from their trawls. The cod are dried artificially, and an excellent quality of cod liver oil is made.

Herring, too, which have been said to decide the destiny of nations, also abound in these waters. They come in great shoals or schools, and it is a matter of record that once in August the mail steamer passed through one school for four hours, the water being silvered as far as could be seen with the fish.

The natives do not take the trouble to fish for them in the usual way with the line and hook or even with nets. They simply rake them out with a lath set with nails, and an Indian or two can usually fill a canoe in an hour or so. The factory crew at Killis-snoo often gets from 300 to 600 barrels of herring at a single haul. Often 1000 barrels are seined at once, and it is not a great while since 1500 barrels were taken by one cast of the seine in Sitka Harbor.

There is every reason to believe that the number of people engaged in the fisheries in 1898 will be greater than in any preceeding year. As is said, the fish come annually in shoals that are simply marvelous in point of extent, and are thus wholly unlike the animals that for a long time afforded a source of revenue to the hunters and trappers.

KILLING POLAR BEAR IN NORTHERN ALASKA.

It may be said here that hunting and trapping, while still pur-
sued in Alaska, is in a certain sense, a thing of the past.   It is
true, that the country abounds in foxes and bears that make
trapping for a limited number a remunerative source of employment.   But the work of the Russians in the early days of the
country's history and of the men employed by John Jacob Astor,
has largely reduced the number of animals which would make
hunting a profitable venture for a great number.   The great
companies of the olden time live now only in recollection, and
it is thought there is little prospect that their activities will be
renewed.

Hunting for sport will doubtless for a long time, claim attention, but, even this, in the districts invaded by the prospectors
and miners, is likely to lose its charms, for reason that the lack
of fresh meats in the mining camps has virtually made every
miner and prospector a foe to the animals whose flesh may be
used for food.   In whole sections of the country, where claims
are now being worked, it is almost impossible to find the first
sign of game.

Lumbering Prospects Not Good.

There also seems to be little prospect for a development of
the lumbering industry, since there is a marked unwillingness on
the part of capitalists to invest money in lumbering camps and
machinery unless the timber possibilities are such as to promise
good lumber in large amounts and under conditions that make
its handling not too expensive.   This Alaska does not promise.
William Ogilvie, who made a thorough investigation of what
may be termed the timber lands of Alaska, speaks discourag-
ingly of the development of the lumbering industry.   He says:

"The amount of timber fit for use in building and manufacturing in the district along the river is not at all important.

There is a large extent of forest which would yield firewood and timber for use in mines, but for the manufacture of lumber there is very little.

"To give an idea of its scarceness, I may state that two of my party made a thorough search of all the timbered land around the head of Lake Bennett, and down the lake for over ten miles, and in all this search only one tree was found suitable for making such plank as was required for the construction of our large boat. This tree made four planks, fifteen inches wide at the butt, seven at the top, and thirty-one feet long.

"Such other planks as we wanted had to be cut out of short logs, of which some, ten to fourteen inches in diameter and ten to sixteen feet long, could be found at long intervals. The boat required only 450 feet of plank for its construction, yet some of the logs had to be carried nearly 200 yards, and two saw-pits had to be made before that quantity was procured, and this on ground that was all thickly wooded with spruce, pine, and some balsam, the latter being generally the largest and cleanest-trunked.

Trees on the Islands.

"The great bulk of the timber in the district suitable for manufacture into lumber is to be found on the islands in the river. On them the soil is warmer and richer, the sun's rays striking the surface for a much longer time and more directly than on the banks.

"To estimate the quantity of timber in the vicinity of the river in our territory would be an impossible task, having only such data as I was able to collect on my way down. I would, however, say that one-fourth of the area I have given as agricultural land would be a fair conjecture. This would give us two and a half townships, or ninety square miles, of fairly well timbered ground; but it must be borne in mind that there is not more

than a square mile or so of that in any one place, and most of
the timber would be small and poor compared with the timber
of Manitoba and the easterly part of the northwest Territories.

"It may be said that the country might furnish much timber,
which, though not fit to be classed as merchantable, would meet
many of the requirements of the only industry the country is
ever likely to have, viz., mining."

Largely a Mining Region.

The general impression seems to be that, barring an enormous
fishing industry, and a possibly limited lumbering trade, the country
is destined to be largely a mining region.   Still, the necessity of
providing food for the miners has forced upon the attention alike
of prospectors and capitalists the desirability of developing as far
as possible in the frozen north some form of agriculture and gardening that will obviate the necessity of the mining community
living virtually the year round on canned goods that are imported
from the south.

It is true that a large portion of the Territory is covered a
good share of the year with fields of ice and snow, but, while
there is a marked difference of opinion, there is ground for the
belief that the country has a future in an agricultural way quite
comparable with its future in other lines.   As was shown in the
chapter on topography and climate, the shores of Alaska are
washed by an ocean current that sweeps across the Pacific from
the coasts of Japan, and, in consequence, southern Alaska and
much of the coast district has a climate comparable with that
which makes, for instance, the British Isles remarkable for their
fertility.

Sitka is no farther north than Edinburg, and the northernmost point of Sweden is nearer the North Pole than the northernmost point of Alaska.   The great warm current that tempers

the climate of the Alaskan coast makes it, it is claimed by many, a country in which agriculture may be followed as successfully as in many of the older countries of the world, where the climate is not essentially different.

Simply Lacks Tilling.

It is claimed by many that all that is lacking near the coast is for the soil to be tilled, and that it can be made to produce practically the same products that grow in Norway, Sweden and Great Britain. That the extreme northern plains, where the mercury often falls to 80 or 90 degrees below zero, and where, even in midsummer, the ground only thaws out two or three inches, can be transformed into an agricultural region, there are few to believe. But most people who have visited the country believe there are fertile regions enough to support millions of people.

Baranof, in the early days of the Russian occupation of the country cleared fifteen kitchen gardens. He ripened barley and potatoes and common vegetables, What is more, this has been done every year since. If Alaska is a glacier-abounding and snow-clad country, it is nevertheless true that fine grasses spring up naturally on any clearing. Wild timothy and coarser grasses commonly grow from three to four feet high, and clover thrives about as luxuriantly as it does in more southern latitudes.

In the neighborhood of Vancouver the natives cultivate potatoes and a sort of tobacco. Each family has its little plantation sheltered away in some nook. Here they plant their tubers and sow their grain. Even in the barren regions of the north, Dawson City, Circle City and Klondike, it is a common practice of the miners to grow turnips on the house tops. There the sun, even in the depth of summer, only thaws out the ground two or three inches, but by putting a generous covering of soil on the house tops, so that it gets the heat from the dwelling beneath, little trouble is experienced in maturing vegetables. Apparently what is lost in intensity of heat is made up by the length of the period which the sun shines.

Garden Vegetables Raised.

Since the United States occupation of the country it has been a common practice of residents in the more settled parts to raise radishes, lettuce, onions, cauliflowers, cabbages, peas, turnips, beets, parsnips and celery. Single potatoes have been produced weighing as much as a pound and five ounces. Hay is commonly cured throughout the entire southeastern portion of Alaska, and this has been done since 1805. It is said that by adopting Norwegian methods larger and better crops could be cured.

By way of comparison it may be stated that wheat is cultivated in Norway as far north as the 64th degree ; rye as far north as the 69th degree ; barley and oats as far north as the 70th degree. Apples, plums and cherries come to maturity there up to the 64th and 65th degrees, while raspberries, strawberries, currants and gooseberries thrive well at the North Cape, which is 71 degrees 10 minutes. It is an often forgotten fact that throughout Southern Alaska, at least, there are two or three weeks of really hot weather, when the mercury rises as high as 92 degrees.

Dr. John G. Brady, a Presbyterian missionary at Sitka, expresses the belief that the country has an agricultural future. Says he :

"The Kake Indians furnished the Russians with potatoes. Some of the natives at Wrangel are clearing off garden patches this year. Much can be done in this direction, for Alaska will furnish vegetables for a teeming population. There are several thousand acres in the neighborhood of this place upon which the finest vegetables may be raised with certainty. The soil for the most part is a vegetable mould mixed with sand.

" Mr. Smiegh, of this place, has had a garden for the last seven years. He says he has grown cabbages weighing twenty-seven pounds. He has tried peas, carrots, leeks, parsnips, turnips, lettuce, radishes, onions, potatoes, parsley, celery, horse radish and rhubarb. He has also tried cucumbers and beans, but they did not do well. Cauliflowers and celery surpassed any he raised in other places.

" The wild black currants abound in the woods. The tame currants do well. Gooseberries do well and have a delicate flavor. The cabbages grow wild and are six or eight inches in diameter. Mr. Burns, who has had a garden for the last three years, agrees with Mr. Smiegh. The strawberry grows wild near Mount Edgecombe."

Missions in the Wilderness.

Dr. Sheldon Jackson, Commissioner of Education, who had spent many years traveling the Alaskan Territory, was asked, after the Klondike fever broke out and the grave difficulty of supplying the mining colony with suitable food became a vital problem, of his views of the agricultural possibilities of the country. It was Dr. Jackson, by the way, who, in company with Mrs. McFarland, took the initial steps in establishing Presbyterian missions in the wilderness.

His residence in Alaska was protracted and his work as a missionary took him to so many parts of the country that he had ample means to observe climatic conditions and the most desirable places for agricultural enterprise. He thoroughly agreed with those who had the interest of the miners at heart that it was a matter of prime importance to take immediate steps for supplementing the mining activities with agricultural enterprises that

22

would limit the possibilities of suffering and disease. Said he :

" The warmest friends of Alaska do not claim that it is rich in agricultural resources, or that it will agriculturally bear comparison with the rich valleys of the Mississippi River ; but they do claim, that while there are large areas of mountains and unproductive land agriculturally, yet there are valleys and plains where with suitable care many of the earlier vegetables, fruits and grains can be raised.

Gardening is Common.

"On Kadiak, on adjacent islands and on the shores of Cook's Inlet, where there are small Russian Creole settlements, they have for three-quarters of a century supplied themselves with vegetables and potatoes raised in their own gardens. During recent years the government and mission teachers in Southeast Alaska have in some instances had good vegetable gardens.

In Northern Alaska, less than 100 miles south of the Arctic Circle, the teachers of the Swedish Evangelical mission at Unalaska in 1891 cleared four acres of ground, on which they raised seventy bushels of potatoes. As that region has a frozen subsoil covered with a heavy coating of moss, the removal of the moss and the cultivation of the ground will cause the soil to thaw out at a greater depth than it would otherwise. So that years of cultivation will cause the ground to yield much more plentifully than when first cultivated."

Dr. Jackson gave some interesting illustrations of experiments that have been tried in various parts of the country, all going to prove that, difficult and unsatisfactory as agricultural experiments for a time might be, they would ultimately prove a success and be a great blessing. Continuing he said :

"In 1887, on the site of Lake Labugo, on the headwaters of the Yukon, over 2000 miles from Behring Sea, a missionary, passing along, saw ten heads of volunteer wheat, nearly ripe, on the twenty-second of August, in a place where some miners had camped the year before and dropped the seed.

"Not only in the mild belt of Southern Alaska, but also in the Arctic and subarctic belt of Northern Alaska, various wild berries grow and ripen in profusion (cranberries, currants, raspberries, huckleberries, blackberries, strawberries), and there is no question that if the government places Alaska on an equal footing with the other States and Territories in the establishment of one or more experimental stations, it will be demonstrated that sufficient vegetables can be raised for the consumption of its people. And if there is found a section so far north that the profitable raising of vegetables and grains becomes impossible, that region can be utilized by the introduction of herds of domestic reindeer."

Would Introduce Reindeer.

Dr. Jackson is an ardent advocate of the introduction of reindeer into Alaska, as a means of solving the transit difficulties. Up to the present time, practically the only means of transportation on leaving the coast, is either to go up the rivers during the brief summer months, or to take the overland trails during the remaining nine months of the year, using dogs as pack animals, and as steeds for sledges. On the matter of introducing reindeer into the country, Dr. Jackson said:

"Taking Norway and Sweden, where complete statistics are to be had, as a basis of calculation, and applying the same average to Alaska, it is found the country is capable of sustaining 9,200,000 head of reindeer, which will support a population of 287,500 living like the Laps of Lapland.

"The stocking of Alaska with tame reindeer means the opening up of the vast and almost inaccessible central region of Northern and Central Alaska to white settlers and civilization, and the opening up of a vast commercial industry.

" Lapland, with 400,000 reindeer, supplies the grocery stores of northern Europe with smoked reindeer hams, smoked tongues, dried and tanned hides, and 23,000 carcasses per annum to the butcher shops.   On the same basis, Alaska, with its capacity for 9,200,000 head of reindeer, can supply the markets of North America with 500,000 carcasses of venison annually, together with tons of delicious hams and tongues and finest leather. Surely the creation of an industry worth from $83,000,000 to $100,000,000 where none now exists is worthy the attention of the American people."

Testimony of Mr. Ogilvie.

The testimony of William Ogilvie, who made an official report to the Dominion Government of the characteristics of the country, its resources and its possibilities, is of importance, and extracts are here given from that portion of the report bearing upon feasibility of agricultural enterprises.   Mr. Ogilvie is not an enthusiast, and his statements may be taken as an impartial account of the country by one who, trained in methods of observation, combines good judgment with the expedients of enforced policy.   As to the Yukon River and its valley Mr. Ogilvie says :

" The agricultural capabilities of the country along the river are not great, nor is the land that can be seen from the river of good quality.   When we consider further the unsuitable climatic conditions that prevail in that region, it may be said that as an agricultural district this portion of the country will never be of any value.

" Many meteorological records show over 8 degrees of frost on August 1st, over 10 on the third, and four times during the month the minimum temperature was below freezing.

"Along the east side of Lake Bennett, opposite the Chilkoot or western arm, there are some flats of dry, gravelly soil, which would make a few farms of limited extent.   On the west side, around the mouth of the Wheaton River, there is an extensive flat of sand and gravel, covered with small pine and spruce of stunted growth.

Coarse Grass for Cattle.

"Along the western shore of Tagish Lake there is a large extent of low, swampy flats, a part of which might be used for the production of such roots and cereals as the climate would permit.   Along the west side of Marsh Lake there is also much flat surface of the same general character, on which I saw some coarse grass which would serve as food for cattle.   Along the east side the surface appeared higher and terraced, and is probably less suited to the requirements of the agriculturist.

"Along the head of the river for some miles below Marsh Lake, there are flats on both sides, which would, as far as surface conformation goes, serve as farms.   The soil is of much better quality than any heretofore seen, as is proved by the larger and thicker growth of timber and underbrush which it supports. The soil bears less the character of detritus, and more that of alluvium, than that seen above.

" Some miles down the lake an extensive valley joins that of the lake an the west side.   This valley contains a small stream. Around this place there is some land that might be useful, as the grass and vegetation is much better than any seen so far.

On the lower end of the lake, on the west side, there is also a considerable plain which might be utilized ; the soil in parts of it is good.   I saw one part where the timber had been burned some time ago ; here both the soil and vegetation were good, and two or three of the plants seen are common in this part of Ontario, but they had not the vigorous appearance which the same plants have here."

Mr. Ogilvie had not a little to say on the forestation of the country and its possibilities in the line of lumber. Speaking of the timber lands in the district considered in the passage just quoted, he says:

" Northward from the end of the lake there is a deep, wide valley, which Dr. Dawson has named ' Ogilvie Valley.' In this the mixed timber, poplar and spruce, is of a size which betokens a fair soil; the herbage, too, is more than usually rich for this region. This valley is extensive, and, if ever required as an aid in the sustenance of our people, will figure largely in the district's agricultural assets.

" Below the lake the valley of the river is not as a rule wide, and the banks are often steep and high. There are, however, many flats of moderate extent along the river and at its confluence with other streams. The soil of many of these is fair.

"About forty miles above the mouth of the Pelly River there is an extensive flat on both sides of the Lewis. The soil here is poor and sandy, with small open timber. At Pelly River there is a flat of considerable extent on which the ruins of Fort Selkirk stand. It is covered with a small growth of poplar and a few spruce. The soil is a gravelly loam of about eight inches in depth, the subsoil being gravel, evidently detritus. This flat extends up the river for some miles, but is all covered thickly with timber, except a small piece around the site of the fort."

An Experimental Station.

There is every likelihood to believe that in the near future the United States government will have an agricultural experimental station in the valley of the Yukon. The desirability of such an experimental farm growing out of the necessities and the hardships of the mining populace was suggested by P. B. Weare, of the North American Transportation and Trading Company. A meeting was held in Chicago early in August, 1897, at which the development of the agricultural resources of Alaska was exhaustively discussed.

Secretary Wilson was present and pledged himself to work for the immediate establishment of such an experimental governmental farm in the Yukon valley. He expressed it as his belief that there would be little trouble in getting Congress to appropriate at least $15,000 for this purpose. So far as he knew, there was no reason why a trial in the line of developing agricultural industries in Alaska should not be made early in the spring of 1898.

Mr. Weare's plan contemplates the sending of a body of experienced farmers from the older and better settled States, and putting into their hands every possible means for testing what can be done in raising grains, fruits and vegetables. Secretary Wilson was entirely in accord with Mr. Weare, and the belief was expressed that within a few years there will be thousands of acres under cultivation at no great distance from the gold fields in the Yukon valley.

Views Thought Utopian.

Many to whom this plan of establishing a government farm was broached thought the views of Mr. Weare and Secretary Wilson a little too Utopian. They thought it might be possible to make a great success of farming in Southern Alaska, say, in the neighborhood of Sitka, but considered that the climate was too rigorous and the summer season too short for farming to be a success along the Yukon and Klondike rivers. It was generally conceded, however, that it would be a long step towards the solution of the food problem if agriculture could be developed to a large extent in the southern portion of the territory, so that the matter of transporting provisions to the camps would not be so costly.

After the decision to establish the experimental farm had

been made, Secretary Wilson expressed himself as follows:

" I am greatly interested in the development of Alaska.   With the aid of three experienced men, who are now in the Yukon country, the Department of Agriculture is making extensive investigations, with a view of learning the value of the agricultural resources of the principal valleys, and it is certain an experimental farm will be established within a year near the junction of the Yukon and Tanano rivers, or in some other favorable location."

Projects of Individuals.

The same all-important work which the United States government will take upon its hands and push will probably receive great assistance from private enterprises.   Scarcely had the Klondike fever broken out, and reports as to the difficulty of getting good wholesome food at the mining camps had been brought south, when Swan Frederickson, a hardy Norseman, who had served for years with the Hudson Bay Company, came forward with a proposition for a company to be called The Alaska Settlement Company, whose aim it should be to encourage imigration and foster agriculture in the country immediately south of the Yukon.

Frederickson said that he had lived too long in Alaska not to know what he was about, and that he was satisfied that with ample capital and judicious methods of procedure the population of the territory could be greatly increased and thousands of acres, that now are of no use whatever, could be reclaimed and made to subserve the comfort and happiness of the people.   He said it only wanted pluck, enterprise and perseverance to make Alaska from the southern limit virtually to the Yukon River one of the happiest agricultural regions in America.

With a capital of $100,000 Frederickson is positive he can start some thrifty settlements of Norsemen farmers, and the com-

pany will make plenty of money by a monopoly of town site
and commercial privileges. He insists that a good business can
be done in raising beef, mutton, hardy vegetables and horse fodder for the thousands of miners who are pouring into Alaska.
The number to be fed will increase rapidly from now on, and
Frederickson waxes enthusiastic in discussing the possibilities of
his scheme ; but there is no capital yet in sight for starting the
work.

Farming Not Enticing.

Farming in Alaska does not sound like a particularly enticing
proposition, but there are other enthusiasts besides Frederickson
who are pushing the idea. They not only maintain that grains
and grasses can be raised in some parts of the Territory, but
even talk about vegetables and fruits. What's more, they quote
Joaquin Miller's letters in support of their scheme.

Ranch booming in southern California in its palmiest days
never had more earnest advocates than these men who are trying to develop the agricultural and horticultural possibilities of
Alaska. They have no land to sell there, but they want to go
into the farming business under the shade of Mount St. Elias or
some other favored spot, and would like some capital to make a
start on, with big profits later on for all interested parties.

As a berry-growing region Alaska has greater promise than
would be supposed for a country part of which lies beyond the
Arctic Circle. At present it is reported there is but one fruit
tree growing in that climate, it being a wild crabapple, which is
not palatable. Whether or not the hardier forms of apples
growing in the Northern States will thrive and the fruit come to
maturity on the prairies along the Yukon is a question. But a
great variety of berries do grow, and many of them grow wild.

Strawberries, cranberries, gooseberries, raspberries and huckleberries not infrequently attain great size. A berry unknown in

southern regions, the roseberry, which grows on a species of rosebush, abounds in the Alaskan valleys. These berries are said to be delicious. They grow in large quantities in Russia, where the natives make preserves that they prize most highly. For some time large invoices of cranberries grown in Alaska have been received and sold in the markets of San Francisco.

It is reasonable to suppose that when small fruits grow wild in such abundance they can be easily cultivated and produce a profitable crop. Indeed, it is believed that more money can be made in raising berries there than in mining gold—at all events there is less risk of loss. Turnips, radishes, potatoes, and cabbages can be raised in the climate, it is believed.

Industries Largely Transformed.

An enterprise was proposed early in August, 1897, with the purpose of making the raising of dogs a distinct and separate enterprise or industry in Alaska, somewhat on the line of Dr. Jackson's proposition to introduce reindeer as a means of solving the transportation problem. The enterprise grew out of the scarcity of sledge dogs on the overland routes.

A kennel owner offered to furnish a stock of draft dogs and take in payment part cash and the rest in the stock of the company which he proposed to organize. There was no intention of introducing any of the breeds of dogs commonly found in the Southern States. These it was said would be wholly worthless for the purpose for which animals are needed in Alaska. On the contrary, dogs used in Siberia and other countries too cold for horses, would be imported and bred in such numbers as to supply the demand and make the enterprise a success from a financial standpoint.

Short as is the history of Alaska, it will be seen that its industries and its commercial enterprises have been practically transformed since the first days of Russian occupation, and it will also be seen that there is every prospect that the transformation will be still greater during the two or three years, following the discovery of gold in the Yukon Valley. Of the first commercial enterprises carried on in the country practically only one survives to-day in a hopeful and remunerative way. Seal fishing, as has been shown, has had its day of rise and decadence. The time was when hundreds of thousands of valuable skins in periods of glutted market were thrown into the sea for the mere purpose of keeping up the prices. To-day while sealing is still carried on, it is carried on in a way so limited as to contrast strangely with the former days of intense activity in this industry.

The Seal Fisheries.

The same is true of hunting and trapping on the mainland. The yearly output is now in no wise comparable with that of the palmy days of the Russian Fur Company and the American Fur Company. The falling away in sealing is due to the wholesale slaughter of the animals, for whose preservation the Government was obliged to take the strictest measures. It is altogether probable that with a wise policy in limiting the number of seals killed for their furs, sealing may in future years be as profitable as ever. It is not deemed probable that hunting and trapping wild animals on the mainland for their furs will ever be what it once was.

The fishing industry on the coast and along the rivers is bound, it is said, to continue, not merely holding its own, but developing into ever increasing enterprises. There is much to be hoped for in the timber districts, for despite the adverse reports that have been made on the forestation of the country, it must be remembered that there are whole regions where the white man has scarcely set foot. What these unknown regions may contain is now a mere matter of conjecture. The history of lumbering in the United States shows that this industry is a mere growth dependent upon exploration and subsequent enterprise. It is not unlikely that lumbering in the wilds of Alaska will develop into something which even the most sanguine to-day little suspect.

Mining and Agriculture.

In view of the excitement incident to the discovery of gold in the Yukon Valley and the impetus it has given, not merely to the work of prospectors and miners, but to that of scientific investigators, the probability is that the leading industry of Alaska for many years to come will be that of mining. And directly connected with and dependent upon it, there is likelihood also of a marked development of agricultural pursuits.

Until early in 1897, when travelers returned from Alaska and were asked what the chief occupations of the people were, they would say, of course, fishing and hunting. But the mere fortunate discovery of golden treasure in the ground will likely give a new trend to the entire development of the country. To shoot and trap and fish was naturally both the amusement and the employment of the Indians and Esquimeaux and such white men as ventured into the country on trips of exploration.

But with the white man as a hunter for gold instead of for animals, it was a different matter. He came, he saw, he dug, and in digging he found riches. The glittering gold greeted his eyes and the fever of gold fell upon the whole country. It is the common belief that this malady, engendered by good fortune, will shape the destinies of Alaska, and transform it from an unknown wilderness of plains and valleys and mountain peaks and glaciers into one of the most remarkable and important mining and agricultural regions of the world.

CHAPTER XII.

Resources and Wealth.

Record as a Fur Country—State of Development Twenty Years Ago—How the Golden Treasures were Discovered and Developed—Report of Geological Survey Expert Spurr—Professor Elliott's Review—Alaska Richer than Klondike—West of the Coast Range—Mint Director Preston's Views—United States Leads the World in Gold Production—From the Alaska Mining Record—Value of Yukon Gold—Cook's Inlet Diggings —Some Scattered Streaks—Experts in the Field—John W. Mackey Quoted—Other Mineral Resources—Canadian Report.

IN 1867 most people who freed their minds had only hard things to say of " Russian America," which the policy of William H. Seward had just incorporated in the territorial area of the United States. Seven millions, even in those days of " war prices," seemed a large sum to throw away, and all but a few long-headed men regarded as clearly thrown away money used to acquire that reputed ice-locked land of bergs and glaciers. They were certain no good thing could come out of it, and their expectations of returns on the nation's investment were circumscribed by estimates of the interest on the purchase price which the fur industry would probably pay. That there was or ever would be anything in the " great country" except fur, was not a canon of the popular faith. And faith was the largest ingredient in the logic with which Seward supported his project—faith in the still hidden treasures of that vast terra incognita, which, it seems, has waited thirty years for justification.

Repellant to the immigrant as Alaska has seemed for most of three decades, it would appear likely that the region is about to be shown as one of the rich areas of the nation. The gold craze on the Canadian Klondike has not only served to stimulate the news of other gold discoveries in the adjacent United States

349

territory, but has brought to light before the public the existence of other wealth producing resources within the old Russian colony which have hitherto been known or guessed at only by a few, and which promise well for development.

What are the resources of Alaska?

First, of course, in present importance are the mineral deposits and here gold is at the head of the list. There is silver, too, as usual, associated with the more precious metal. Besides these there are copper, iron, lead, plumbago, marble, coal, sulphur, bismuth, kaolin fireclay, gypsum and petroleum.

Allied to these minerals are many gems, among them the famous Alaskan diamonds, garnets, amethysts, zeolites, agates and cornelians. Fossil ivory is frequently found, and it has been claimed by scientific men that the ivory finds in the frost beds of Siberia might probably be duplicated in Alaska as the result of systematic prospecting for these treasures of extinct pachyderms.

Vegetable and Animal.

The resources of Alaska in the vegetable kingdom cover a long list of valuable woods, the cedars especially being unsurpassed. Small fruits are plentiful in the southeastern or Sitkan portion, and experiments within a few years give hopes that agriculture and stock-raising are not impossible industries, but they lack the confirmation of extensive experience.

In the animal kingdom furs from amphibious and land animals are the principal sources of wealth. The whale fisheries have hitherto been profitable industries, but the extermination of the " right " whale by the hunters and the market for oil and " bone " have latterly reduced the value of this industry as a resource. The salmon, which abound in Alaskan waters, have developed two great industries in canning and salting, and the cod fisheries on the great banks along the Aleutian chain are important. There are many other food fishes, also, ample for local consumption, but of a commercial value not yet ascertained. It has been said there is "more fish than water in Alaska;" but this may be taken as hyperbole. As to the food animals, a project is under way to introduce reindeer into the country for the value there may be in their hides and meat, but the scheme is still in the experimental stage.

In any estimate of either the resources or the native and natural wealth of Alaska, it should be borne in mind that no systematic development has yet taken place along any lines except the fur and salmon industries. Except in the Sitkan region, the exploitation of the gold area has been more accidental than designed, and comparatively no attention has been paid to the other minerals. There are no statistics from which to compile comparative tables, and all statements must perforce partake of the nature of generalities. The Russians had no use for Alaska except for its furs, and for ten years after the territory had passed from the dominion of the double-headed eagle to that of the one-headed bird of Uncle Sam the new owners had no definite idea that they had bought anything more valuable than fur seals and sea otters.

Twenty Years Ago.

In 1877 Henry W. Elliott wrote as follows of the new Territory:
"At present, however, beyond the fur trade, there is nothing doing whatever in Alaska—no settlers, no mines, no mills. If we ever utilize the spruce and fir timber on the Sitka coast we must encourage and foster the effort in the line of ship-building, for this timber is too gummy and resinous for the ordinary use of house-building and furniture-making. If gold or silver is discovered in Alaska it must be of unusual richness, or it will never support any considerable body of men up there, so far away

from the sources of necessary supply. The reputed Alaska gold mines are not in Alaska at all."

Mr. Elliott was a noted and shrewd observer, and he had had ten years acquaintance with Alaska, but Birch Creek and Forty-Mile were then unheard of, and even the auriferous riches of Douglas Island were not dreamed of.

Gold on Douglas Island.

Gold-bearing rock was discovered on Douglas Island in 1880, and the next year the famous Treadwell mine was located there in the largest solid body of ore on the Coast. The deposit is a mountain of gold-bearing quartz, worked from the surface like an ordinary stone quarry. The ore only runs from three dollars to seven dollars to the ton, but as it costs one dollar and a quarter or less a ton to mill it, the property is considered one of the most profitable mines in the world. The largest stamp mill in the world, running 240 stamps, handles the output.

Following the location of the Treadwell mines other gold areas were discovered, and it soon became well known on the coast that there was yellow dust in many portions of the "Pan-Handle," and also in the Yukon Basin, though the rigor of the climate and the remoteness of the diggings from bases of supplies long kept the country from being developed in response to the impulse of the discoveries.

Then came the placers around Circle City and Fort Cudahy, and hard after them the marvelous strikes in the Klondike just across the border, and the golden future of Alaska was an established fact.

The report on the Yukon gold region by Josiah Edward Spurr, the geological survey expert, who headed a party that made a thorough investigation in Alaska last summer, gives new facts about the interior. It says as to the Forty-Mile gold district that in the latter part of 1887 gold was struck in Franklin Gulch, and ever since it has been a constant payer. The discovery of Davis Creek and a stampede from Franklin Gulch followed in the spring of 1888. In 1891 gold mining in the interior, as well as on the coast, at Silver Bow Basin and Treadwell, received a great impetus. The event of 1892 was the discovery of Miller Creek. In the spring of 1893 many new claims were staked, and it is estimated that eighty men took out $100,000. Since then Miller Creek has been the heaviest producer of the Forty-Mile district, and until recently of the whole Yukon. Its entire length lies in British possessions. The output for 1893, as given by the Mint Director, for the Alaskan creeks, all but Miller Creek being in American possessions, was $198,000, with a mining population of 196.

The total amount produced by the Yukon placers in 1894 was double that of the previous year. In 1895 the output had doubled again.

Forty-Mile district in the summer of 1896 is described in the report as looking as if it had seen its best days, and unless several new creeks are discovered it will lose its old position.

Large Profits Reported.

The Birch Creek district was last summer in a flourishing condition. Most of the gulches were then running, miners were working on double shifts, night and day, and many large profits were reported. On Mastadon Creek, the best producer, over thirty miners were at work, many expecting to winter in the gulch. As to hydraulicking, the report says: "Some miners have planned to work this and other good ground supposed to exist under the deep covering of moss and gravel in the wide valley of the Mammoth and Crooked creeks by hydraulicking, the water to be obtained by tapping Miller and Mastadon creeks

23

near the head.   It will be several years before the scheme can be operated, because both of the present gulches are paying well and will continue do so for at least five years."

Expert Spurr's report on the Klondike district is as follows :

" With the announcement of gold here in the winter of 1896–97 there was a genuine stampede to the new region.   Forty-Mile was almost deserted.   But 350 men spent the winter on the Klondike, in the gulches and at the new town of Dawson. The more important parts of the district are on the Bonanza and Hunker creeks.   According to the latest information 400 claims have been located up to January 1, 1897 ; about half as many on Hunker Creek.   There is plenty of room for many more prospectors and miners, for the gulches and creeks which have shown good prospects spread over an area of 700 square miles. The estimated Alaskan gold production for 1896 is $1,400,000.

Professor Elliott Again.

It is interesting at this point to see how Professor Elliott's views have changed between 1877 and 1897.   Here is what he said last summer of Alaska :

" My experience in the Klondike region leads me to believe that while there is considerable gold in the crevices and along the rivers, washed down for ages from the mountains by attrition and the glacial displacements, the ' pockets ' in which large quantities are to be found, including nuggets and much pure gold, are comparatively few.   One man may find a ' pocket,' and get thousands from it, while hundreds of others may toil near by for a few dollars' worth of metal a day.   I understand there are now about 7,000 people in the Klondike region seeking for gold, while hundreds of others are flocking there as fast as possible.   Mark my word, you will hear of a lot of disgusted men returning to the States next spring, having failed to ' strike it rich,' as they had hoped.   I would advise no man who is established in business here, who is married, or who has any responsibility resting upon him, to go to the new gold fields.

"Alaska is a healthful country, there being no malaria or mountain fever.   A curious fact is that any one afflicted with neuralgia or rheumatism is completely cured of it in that climate.   The clear, dry atmosphere and the rapid changes of the body's tissues doubtless account for this.   One's appetite is tremendous in that climate.   A man will eat four times as much food as he does here and not feel uncomfortable.

"There is plenty of fuel, poplar, beech and fir trees lining the numerous streams.   Of course, the culling and hauling of timber make it very expensive.   Houses are nothing but log huts, two or three feet of which are below ground, with earth banked about the sides and even over the roofs.   Eight or ten miners will lie down to sleep on the rude bunks within these cabins, wrapped in their heavy blankets."

Alaska Richer than Klondike.

A scientific expert of the Coast Survey, who knows what he is talking about from experience, believes Alaska is richer than the Klondike.   He sums up his reasons thus :

"A study of the map convinces me that the greater part of the gold fields of the extreme Northwest will finally be found within the limits of our territory.   I went through Alaska as a member of the boundary commission, and am very familiar with the valley of the Yukon and the surrounding country.   The greatest activity in placer mining is now in the British possessions, about forty miles east of the 141st meridian, which is our boundary.   But if you look at the map and see where gold has been found, you will observe that all the lodes seem to lead into Alaska.

"There is a certain regularity about gold findings.   South of

the Klondike in British Columbia is the Cariboo region, which
was the scene of a former gold excitement.   Crews on vessels
deserted, and there was the same sort of a rush, on a smaller
scale, that we have seen in the Klondike.   Then directly east of
the ' Pan-Handle ' of our Alaska territory is the celebrated Cassiar country.   Here are the headwaters of the Pelly River, and
the confluence of the Lewis and the Pelly makes the Yukon.   The
richness of the Cassiar country has long been known, and it belongs to the same general trend, geologically speaking, as the
Klondike.   This trend is parallel to the west coast of the continent.   Wherever the tributaries of these rivers have been prospected gold has generally been found.   Forty-Mile Creek, Sixty-
Mile Creek and Birch Creek are instances in point.   The
headwaters of all these streams are in a group of mountains, the
area of which is probably a thousand square miles.   It is mostly
unexplored, but largely within the territory of the United States,
and it is probably rich in gold.   Of the country farther north we
know little as yet, although it is well watered, and belongs to the
same mountain range.   It is entirely likely that placer mining
can be carried on through this country for a distance of 500
miles.

West of Coast Range.

" Besides this trend of gold country parallel to the west coast,
it will be observed that there is another remarkable region west
of the coast range, which converges into the same Alaskan
territory.   Beginning at Juneau there is a great deal of quartz
mining and near that town the largest stamp mill in the world has
been built.   The ore is a low grade, yielding only about $2.50 a
ton, but it can at that figure be very profitably worked.   At
Yakutat Bay, right under Mount St. Elias, there is considerable
placer mining, and at Cook's Inlet, farther north, still more.
Compared with the region in Alaska, which now seems likely

to be rich in gold, the California territory was very small.

" I am much impressed with the opportunities for profit in other things in Alaska besides this gold. The fisheries of the coast are most remarkable, and when fully developed may yield larger returns than the mines. Then the coal, now that a population is going into the country which will want to use it, is a very important thing. Some system of easy transportation across country, from one river to another, might be profitably established. The inhabitants of the Yukon Valley will always have to draw their food supplies from the outside. That is one of the most desolate regions on the face of the earth. Game is very scarce. The Indian population is slight, which proves how difficult it must be to get food."

Rich Finds in Alaska.

F. G. H. Bowker, one of the returned Yukoners, who brings back nearly $40,000 in gold dust, the result of six months' work, is authority for the statement that on the American side of the international boundary placer fields have been found which even put those of the Klondike into the shade.

When his party was descending the Yukon on the return from Dawson City the steamship was intercepted by a man who desired to send letters and papers back to civilization. This man was one of a party who had gone down the river from Dawson in the hope of locating rich beds of which Indians in the vicinity had been telling. The members of the party were well known to the Yukoners and full credence is given to the story.

Bowker and his associates were told that just across the Alaska boundary, on the American side, the party had found placer fields fabulously rich in gold. They had staked out claims and begun to work them.

" Every one of us has taken out thousands of dollars in dust and nuggets already," said Bowker's informant, "and there seems no limit to the gold in sight. It is more abundant than on the Klondike and easier to work, the gold being very near the surface of the ground. We are all rich already, but we are going to stay through next winter."

Further information was conveyed that there were only white men in the new district, and they had the field practically to themselves. They advised Bowker and his companions to forsake Klondike claims on their return from the States and take claims in the new diggings.

The point at which the fortunate treasure-hunters are working is northwest of Dawson and but a few miles west of the boundary. Their claims are in a valley of one of the numerous creeks emptying into the Yukon.

Mint Director's Report.

Director of the Mint Preston, in a report on the gold of Alaska and the adjoining Klondike territory, which may fairly be considered at the same time as the Alaskan auriferous area, since the lodes and placiers of one are for practical purposes precisely similar to those of the other, says:

"That gold exists in large quantities in the newly discovered Klondike district is sufficiently proven by the large amount recently brought out by the steamship companies and miners returning to the States who went into the district within the last eight months. So far $1,500,000 in gold from the Klondike District has been deposited at the mints and assay offices of the United States, and from information now at hand there are substantial reasons for believing from $3,000,000 to $4,000,000 additional will be brought out by the steamers and returning miners sailing from St. Michael's the last of September or early October next. One of the steamship companies states that it expects to bring out about $2,000,000 on its steamer sailing from St. Michael's on September 30th, and has asked the government to have a revenue cutter act as a convoy through the Behring Sea. In view of the facts above stated, I am justified in estimating that the Klondike District will augment the world's gold supply in 1897 nearly $6,000,000.

Richness of the Klondike.

" The gold product of the Dominion of Canada for 1896, as estimated by Dr. G. M. Dawson, Director of the Geological Survey of that country, was $2,810,000. Of this sum the Yukon placers, within British territory, were credited with a production of $355,000. The total product of that country for 1897 has been estimated at $10,000,000, an increase over 1896 of $7,200,000. From this the richness of the newly-discovered gold fields of the Klondike is evident.

" Of all the gold-producing countries, of course, the Klondike is at present one of most absorbing interest. It strikes the imagination to-day as California did the minds of the forty-niners. It will add in 1897 possibly $6,000,000 to the gold treasure of the world.

" Now as to the influence of such addition to the world's gold, the influence it will exert depends mainly on how many years the Klondike District shall continue a producer and how large its annual increment to the world's existing stock of gold shall be. There is every reason to believe that Alaska and the adjacent British territory are possibly as rich in gold as was California and Australia when first discovered. I have estimated that the Klondike district will in 1897 produce $6,000,000 worth of gold. It will add to the product from year to year probably for a minimum of one or two decades."

Mr. Preston calls attention to the fact that the United States leads the world in gold production. He estimates the gold production of the world for 1896 to have been $205,000,000, of which the United States contributed over $53,000,000. For 1897 it is believed the world's gold product will reach at least $240,000,000, an increase of $35,000,000 over 1896. He says : "As an indication of the increase in the world's gold product for 1897 the following table, showing the product of the United States, Australia, South Africa, Russia, Mexico, British India and Canada for 1896, and the probable output of these countries for 1897, is given :

	1896.	1897.
United States	$53,000,000	$60,000,000
Australia	46,250,000	52,550,000
South Africa	44,000,000	56,000,000
Russia	22,000,000	25,000,000
Mexico	7,000,000	9,000,000
British India	5,800,000	7,000,000
Canada	2,800,000	10,000,000
Total	$180,850,000	$219,550,000

"That the world's great product will continue to increase for a number of years to come," says Mr. Preston, "is self-evident, as new mines will be opened up in all parts of the world, and with the improved appliances and methods for extracting the gold contained in the ores it is believed that by the close of the present century the world's gold product will exceed $300,000,000.

From the Mining Record.

The Alaska Mining Record, in a summary of the business of 1896, gives some interesting figures, as follows, about the gold output :

"The output of the mines of Alaska is difficult of estimation. The vastness of the mining territory, the extremely migratory characteristic of the population and the entire absence of reports and statistics from a great part of the smaller camps render it difficult to arrive at a statement approximating correctness except by careful study and watchful attention to every detail. The following estimate is the result of just such work, and is believed to be as nearly correct as is possible, and still represent fully, yet conservatively, the production of gold in Alaska during 1896:

Total output of quartz mines	$2,355,000
Lituya Bay placer mines	15,000
Cook Inlet placer mines	175,000
Birch Creek district, Yukon mines	1,300,000
Other Yukon districts	800,000
From several small creeks in various parts of the territory, worked by arrastas	25,000
Total output	$4,670,000

" This is an increase over 1895 of $1,670,000. At the same time the number of new discoveries which promise well has been great. These will be more or less productive during the next year, and a corresponding increase is assured. Two new mills of ten stamps each have been erected during the past year, and sixty-five stamps have been added to mills already operating, bringing the number of stamps now dropping in Alaska to 549, of which all but ninety-four are in continuous operation, these latter being closed down by climatic severities during the winter season. As development is carried forward, however, steps are taken to overcome this, and it is but a question of a short time when all our mines will run regardless of climate or season. It is quite likely that during the coming summer no less than 250 stamps will be added to the present number."

Value of Yukon Gold.

Assistant Weigher W. A. Underhill, of the Selby Smelting Company, of San Francisco, says the gold from the Yukon is not as valuable as that produced in California. He states his point in these words:

"It is a fact that the Yukon gold is not as valuable as that produced in this State. The nuggets from the Yukon are worth $17 and $18 per ounce, and the finer gold dust is worth from $16 to $17 per ounce. The California gold value is about $1 an ounce more. Its nuggets run from $18 to $19, and gold dust never less than $17 per ounce."

There would seem to be no doubt that gold exists in paying quantities in many other portions of Alaska than in the quartzite veins of Douglas Island or the placers around Circle City. "Color," in fact, is a characteristic of the whole Yukon basin and of a great number of valleys and gulches in other parts of the Territory.

At Cook's Inlet.

George Hall, a Cook's Inlet miner, has this to say about that region :

"I want to deny the stories told by 'tenderfeet' sheep herders and grape pickers, who say that there is no gold in Cook's Inlet. I'll wager that from $400,000 to $500,000 will be taken out of the Sunrise City district this summer. On Cañon Creek, Mills Creek, Gulch Creek and Bear Creek the various mines are working from five to twenty men, each at $4 a day, and they are taking out at least $20 a day to the man. Of course, this is not doing as well as the Klondike, but it is a mighty sight better than nothing.

"The Pelly Mining Company took out $45,000 last year, and is working ten men this year, who are averaging $20 a day to the man. Wages on the Pelly mine are $4 a day and board.

"An old practical miner who went to Link Creek, which had been prospected time and again by 'tenderfeet' and pronounced valueless, took out $10,000 last fall, and is now working twenty men. There are three or four other claims on Link Creek paying equally as well. Claims on Gulch Creek, which was discovered by a man named Shuffler, were averaging $20 a day to the man on July 4th.

"We have a prosperous community at Sunrise—about 200 population, two general merchandise stores, two saloons and a hotel. It is no country for men who expect to pick up gold by the handful, but is good for practical, hard-working miners."

Told by a Kadiaker.

Dr. C. F. Dickenson, a resident of Kadiak, recently wrote :

"In my opinion there are just as good placer diggings to be found at Cook's Inlet as in the Klondike region. There is not a foot of ground in all that country that does not contain gold in more or less appreciable quantities. There is room there for thousands of men, and there is certainly no better place in the world for a poor man."

George F. Becker, in an unpublished report made to the geological survey of his investigation in 1895 of the coastal gold districts, says that most of the islands of the Alexander Archipelago contain gold deposits, yet unworked, that would probably repay very handsomely well-directed efforts of placer mining. These deposits are in the neighborhood of Sitka, and generally on Baronoff and Admiralty Islands, and the beaches of the adjacent mainland. Another fairly promising region is in a group of deposits on the Kenai Peninsula, on the southeast shore of Cook's Inlet, and at Yakutal Bay and the beaches of Kadiak Island.

Gold and silver have been discovered in the extreme northern portion of the Territory, but no systematic prospect has ever been conducted, and the value of the deposits cannot be estimated.

In the region of Lake Clark, a newly discovered body of water in the Southwestern mainland, the census agent reported "pay" gold in the creek beds, but said the dust was as fine as flour, and would require special apparatus for working.

Professor G. F. Wright, of the Chair of Geology at Oberlin College, wrote of the general prospect to the *New York Journal,* as follows :

" As to the ultimate yield of the mines or the prospect of finding more, we have nothing but conjecture to go upon. The geologist who have visited the region were not the ones who discovered the gold. What the prospectors have found points to more. The unexplored region is immense. The mountains to the south are young, having been elevated very much since the climax of the glacial period. With these discoveries and the success in introducing reindeer, Alaska bids fair to support a population eventually of several million. The United States must hold on to her treaty rights with Great Britain for the protection of our interests there."

Experts in the Field.

Samuel C. Dunham, expert of the Federal Bureau of Labor, left for Alaska early in August, under Government direction, to investigate the gold belt and report this coming winter. His inquiry will cover the extent of the deposits, opportunities for business, for investment of capital, labor, wages, cost of living, climate, best means of reaching the gold fields and kindred subjects.

The Government at Washington will send a mining expert into the Klondike country next spring to make an estimate as to the probable amount of gold in this region.

In reaching this conclusion the Treasury Department is following the precedent established in the case of the gold discoveries in the Rand, South Africa. When those discoveries were reported the Rothschilds sent Hamilton Smith, of New York, to estimate the value of the fields, and he reported $3,000,000,000 as his estimate.

Mining experts doubted the correctness of Mr. Smith's conclusions on account of the smallness of the space occupied by the mines, and the German Government sent Bergath Schmeiser, a noted mining engineer, to make a report. The government of the United States followed Germany's example by sending George F. Becker.

John W. Mackey Quoted.

John W. Mackey, the last of the Bonanza Kings—now president of the Commercial Cable Company and of the Postal Telegraph System, and one of the world's great capitalists—knows more, probably, about the vicissitudes of gold hunting and placer mining than any man in America. He spoke of the reports of the marvelous richness of the Alaskan and Klondike gold fields, as follows :

"I have no reason to doubt them. I have had great confidence in the mining possibilities in British Columbia and Alaska —have always believed that those frozen, almost inaccessible regions contain heavy deposits of precious metals. Some enormous ' finds ' of gold have undoubtedly been made there, and yet we know little or nothing of the possibilities of the country. Think of Williams' Creek, for instance, in the Caribou region in British Columbia. As long ago as 1860 something like fifty millions of gold were taken out. It was placer mining there, just the same as the Klondike.

" The gold is right on the surface. It is a mountainous country, overrun with lava at some remote age, and centuries ago, probably, the great forces of nature were at work and melted the gold in a natural crucible.

" The particles of gold are now washed out by the waters, and are generally found along the course of mountain streams. You will always find the best placer gold near the banks of streams and barren water courses. Scientific mining preserves a much larger portion of gold dust than formerly, and I presume it destroys a great deal of the individuality in a working miner. Thus far the Klondike region has seen only old-fashioned, primitive mining, the men groveling in the dirt with their hands and washing out the gold dust in a simple pan, picking nuggets with their fingers.

Modern Mining Methods.

" In time modern mining methods will be carried up to the Yukon country. The recent discoveries prove that it is immensely rich. All parts of the country will be opened. Capital will always go where there is a chance for legitimate investment, and transportation facilities will increase as rapidly as the travelers.

" Whether interest in the Alaskan mines will increase depends on future reports. I see in it something like the excitement of the early '50s over the gold discoveries of the Pacific coast region. The reports of rich individual finds are likely to continue, and the arrival of every ship loaded with fortunate gold hunters will stimulate the imagination, hopes and desires of the would-be gold hunters. We hear nothing of the failures, you know. One man who is lucky is more talked about than a thousand who fail.

" My experience is, I think, that about one man in ten used to get on in the mining days in California. I do not mean that one man in ten became a millionaire. I mean made a living and a little more. The thriftless and careless ones go to the wall, while the hard workers, who have a definite purpose in view and who cling tenaciously to it, succeed in mining as in other occupations.

" But, as I said, in placer mining there is a good deal of luck in locating the claim. One man will take out a great deal and another man nothing. As to the limits of British Columbia mining I cannot say, but I think there are immense gold deposits yet to be found."

Henry Ellsworth Haydon, former Secretary of Alaska, speaks of the gold production as follows :

" From many places in the Pacific coast States miners have been drifting Alaskaward for years, locating pay quartz and placer claims in southeastern Alaska and along the Yukon River and its tributaries, and feeling assured all the time, from every indication, that the wind-blown snow plumes on the mountain tops waved above crowns of gold.

A Happy Surprise.

" Long prior to 1887 Juneau and a comrade went prospecting in Alaska. They were hunting quartz. Paddling along the coast in a canoe, they saw far up a mountain side, which skirted a lonely bay, the glimmer of white outcroppings from the dull gray of the surrounding rocks. They beached their canoe, and after a hard climb reached the spot. The rock was worthless, but the summit was not far off, and desiring to see what was on the other side, they pushed onward until they stood where they could look down into a ravine, through which a mountain stream rushed tumultuously toward the sea. They noted that the bed of the stream was strewn with big white boulders, and curiosity and hope led them to descend to it and investigate. Joe told me he was breathless when he got there, and they both sat down on the banks and wondered if it were true.

" Before them, where the crystal water babbled, they saw white rocks veined with gold and inlaid with nuggets, many as large as a thumb nail. They stayed there while their provisions lasted, a few days only, gathering together $14,000 in virgin gold.

" In the rear of Juneau, on the mainland, is Silver Bow Basin, where some rich placer mines are being worked. Placer mining is carried on in at least eight districts, viz. : Silver Bow Basin,

near Juneau ; Sum Dum and Shuck, some distance south ;
Latuya Bay, on the coast north of Cross Sound ; Yakutat,
Kenai Peninsula ; the Fish River district, on Norton Sound,
at Cook's Inlet, and the Yukon district, including the rivers flowing
into the Yukon.

Placers in Yukon Basin.

"In the absence of statistics it is difficult to obtain reliable
information, but work in these placers continues, which is evidence
of success.   For ten years at least men have worked placers in
the Yukon district.   Leaving Juneau early in the spring, they have
gone out over the Chilkat Pass and down the little chain of lakes
on the other side, making long portages, it is true, and enduring
some hardships, to the Yukon River.   They have returned to
Juneau in the fall, year after year, bringing with them from $2000
to $3500 each in gold dust, the product of the summer's work.
But they are improvident, these men who win gold from the beds
of rivers, and when the spring comes they are stranded financially, many of them without a grub-stake, but they 'win out'
some way and go back again to return—unless they have crossed
the divide forever—and repeat the same old story of excess and
extravagance.

"They never grow money wise, these grizzled veterans of the
rocker, the gold pan, the pick and the shovel, but after all they
are of God's people, and I like them.

"Quartz lodes are worked in ten or more districts, some of
which are large and contain many district claims.   The ten districts referred to are as follows : Sheep Creek region, which
yields ore containing silver, gold, and other metals ; Salmon
Creek, near Juneau, silver and gold ; Silver Bow Basin, mainly
gold ; Douglas Island, mainly gold ; Fuhter Bay, on Admiralty
Island, mainly gold ; the Silver Bay Mining District, near Sitka,
gold and silver ; Besner Bay, in Lynn Canal, mainly gold ; Fish

River Mining District, on Norton Sound; Unga District and Lemon Creek."

The undeveloped and almost unthought of mineral resources of Alaska, other than gold, deserve a passing glance. Another year or two will, perhaps, give some statistics of deposit and production which are lacking now.

Copper promises to be a valuable and important resource of the territory. It is found pure or "virgin" in many places and has given its name to the little known Copper River. A valuable deposit of bronze copper has been worked for years.

An expedition has been organized to go out from Tacoma and Port Townsend to explore a rich copper field, in which there is believed to be also much gold, which is known to exist along the Copper River. For many years past gold, copper and furs have been brought out of that region by Copper River Indians, and exchanged with traders for firearms and food. The Copper Indians are a ferocious tribe, and during the last few years have become well equipped with guns and ammunition. Knowing the value of their rich stakes, and that the ingress of white men would mean their retirement, the Indians have steadfastly refused to permit a single white man to explore their country. Every man making the attempt has been told to keep out, and when he persisted has been killed.

After the Copper.

The Copper River tribe numbers nearly 1000, and as they have been well able to carry out their threats, no attempt to molest them has been made in recent years. Now, however, it is intended to teach these natives that white men must eventually be allowed to prospect and take out the mineral riches of their domain.

One hundred men, heavily armed, will compose the expedition.

24

They will be led into the Copper River section by Judge Joseph Kuhn, who has been collecting data regarding Copper River for years, and was the originator of the project. The Indians will not be molested unless they attack the exploring party. Traditions of the last sixty years have ascribed great mineral wealth to the Copper River country. At Sitka, it is said, that in 1831 a Russian trader invaded that section with eight men. They were killed when within two days' march to the seacoast.

Coal of fair quality exists in good quantity in several parts of Alaska. At Coal Harbor an ample supply of a rather poor quality of lignite has been worked in a spasmodic way for some time. A semi-bituminous lignite is mined along the northern coast by whalers for use on the spot. It makes steam quickly, but the quantities of ash and cinder are something of an objection. A glossy, semi-bituminous lignite, which steams well and is mined without much labor, is found near Kilisnoo, and good coal exists on Silkinak Island. A new coal mine has just been opened six miles from Fort Cudahy, and will be promptly developed to supply fuel to the river steamers. Coal is also mined in the Pelly River country.

Lead and Other Riches.

Lead is found on Whale Bay and Kadiak Island, and there are indications of paying deposits in the interior. A mine on the Fish River has been opened for working by a San Francisco company.

Graphite abounds about Port Clarence. Marble exists in inexhaustible quantities.

Petroleum has been found in what are believed to be paying quantities on a lake near Kadmai Bay. Samples sent down for analysis were of marvelous richness, and a company has been formed to handle the product for the Alaskan mining camps.

A San Francisco expert, just returned from Alaska, sums up the resources as follows:

"There are other discoveries awaiting the pioneers of Alaska than that of gold. Iron and coal abound in these rugged mountains, and the necessity of development will be immediatly apparent. The source of a new commerce will be established. An impetus will be given to the manufacturing interests of the Pacific Coast, and the community wealth will receive a more substantial benefit than could possibly accrue from individual accumulation of riches."

Canadian "Blue Book."

The Canadian Government has issued a "Blue Book" on the Klondike, extracts from which deserve a place here. It says:

"It is beyond doubt that a considerable number of pans of the dirt on different claims have turned out over $200 worth of gold, while those which run from ten dollars to fifty dollars have been very numerous. In the line of these finds further south is the Cassiar gold fields, in British Columbia, so the presumption is that we have in our territory along the easterly watershed of the Yukon a gold-bearing belt of indefinite width and upward of 300 miles long, exclusive of the British Columbia part of it."

"Gold is not the only mineral wealth of the Yukon, it appears. Mr. Ogilvie states that copper has been found on the Ton-dac Creek, above Fort Reliance, and several small veins have been found in the vicinity. With better facilities it may become, he says, a valuable feature of the country. A small seam of asbestos was also found a short distance from Fort Cudahy, and as there is quite an area of serpentine in that neighborhood, asbestos of commercial value may yet be found.

"Still another valuable feature is the coal fields which the district possesses. On Coal Creek, about seven miles up, overlying a coarse sandstone and under drift clay and gravel, a seam of twelve feet six inches has been discovered. It is certain that coal extends along the valley of the Yukon from Coal Creek, ten or twelve mile down, and from Coal Creek up to Twelve-Mile Creek, which flows into the Yukon about thirty miles above Fort Cudahy. Coal is also found in the upper part of Klondike and on other creeks."

Gold-bearing quartz, the report states, has been found in Cone Hill, which stands midway in the valley of the Forty-Mile River, a couple of miles above the junction with the Yukon. The quantity in sight surpasses that of the famous Treadwell mine on the coast, and the quality is better. Were it on the coast the Treadwell would be diminutive beside it.

Not far from Cone Hill a ledge had been found last spring on the Chindindu River (known in the district as the Twelve-Mile Creek), by an American expert prospecting for the North American Transportation and Trading Company, which the expert said he had never read of or seen anything like in the world. He had spent years of his life in the best mining districts of the United States, and he assured Mr. Ogilvie that this section of country promised better than any he ever saw before, and he was going to spend the rest of his life there.

By Governor McIntosh.

Governor H. C. McIntosh, of the Northwest Territory, which includes the Canadian Yukon, says the Klondike diggings will reach $10,000,000 in the season of 1897. In a recent interview about the new camp, Governor McIntosh said:

" We are only on the threshold of the greatest discovery ever made. Gold has been piling up in all these innumerable streams for hundreds of years. Much of the territory the foot of man has never trod. It would hardly be possible for one to exaggerate the richness, not only of the Klondike, but of other districts in the Canadian Yukon. At the same time, the folly of thousands rushing in there without proper means of subsistence and utter ignorance of geographical conditions of the country should be kept ever in mind.

" There are fully 9000 miles of these golden waterways in the region of the Yukon. Rivers, creeks and streams of every size and description are all rich in gold. I derived this knowledge from many old Hudson Bay explorers, who assured me that they considered the gold next to inexhaustible.

" In 1894 I made a report to Sir John Thompson, then premier of Canada, who died the same year, at Windsor Castle, strongly urging that a body of Canadian police be established on the river to maintain order. This was done in 1895, and the British outpost of Fort Cudahy was founded.

Prospect in Other Streams.

" I have known gold to exist there since 1889, consequent upon a report made to me by W. Ogilvie, the government explorer. Many streams that will no doubt prove to be as rich as the Klondike have not been explored or prospected. Among these I might mention Dominion Creek, Hootalinqua River, Stewart River, Liard River and a score of other streams comparatively unknown.

" It is my judgment and opinion, that the 1897 yield of the Canadian Yukon will exceed $10,000,000 in gold. Of course, as in the case of the Cariboo and Cassiar districts years ago, it will be impossible accurately to estimate the full amount taken out.

" There is now far in excess of $1,000,000 remaining already mined on the Klondike. It is in valises, tin cans and lying loose in saloons, but just as sacredly guarded there and apparently as

safe as if it were in a vault.   Already this spring we have official
knowledge of over $2,000,000 in gold having been taken from
the Klondike camps.   It was shipped out on the steamships
Excelsior and Portland.

"Incidentally I may say we have data of an official nature
which lead us to believe that the gold output of the Rossland
and Kootenai districts for 1897 will be in excess of $7,000,000.
I should have said, and I have no hesitancy in asserting, that
within the course of five years the gold yield of the three districts named will exceed that of either Colorado, California or
South Africa."

A more complete statement of the seal and salmon industries
will be found in another chapter.

Adds to our Knowledge.

In these days when every scrap of information regarding
Alaska and the gold discoveries is eagerly sought, and the greed
of gold is leading many to almost certain destruction, it is well
to consider what is a redeeming feature of the gold craze.   The
finds in the upper Yukon country can at best benefit only a
limited number of people in a direct manner, while the educa-
tional value of the gold discoveries to all civilized nations really is
unlimited.   Only a few weeks ago Alaska in general and the
Klondike region in particular were comparatively unknown.   The
maps contained only indefinite outlines of the more important
streams and mountain ranges, and as to places of human abode,
with the exceptions of a few in Southern Alaska, none was recorded.   Look at the change now.   Chilkoot Pass, Dyea, Lake
Linderman, Bonanza Creek, Circle City, Fort Cudahy, St. Michael's, Dutch Harbor, etc., are on everybody's lips, and many who
could not locate St. Louis accurately on the maps talk of the
Klondike River as familiarly as of the Mississippi.

CHAPTER XIII.

Gold Mining in Alaska.

Antiquity of Placer Mining—How Nature has Filled the Gravel with Gold
—Selecting a Locality—Building a House—Out Prospecting—Thawing
the Ground—How to Distinguish Gold from other Minerals—Pyrites,
Mica, Black Sand—Mechanical Assay—Locating the Claim—Local
Customs—Commissioner Herrman's Digest—Getting Out the Gold—
Mining in Winter—Work Along the Yukon—Sluicing for Gold—Dry
Placer Miners—Dredging for Gold—Old Miner's Advice—Gold-bearing
Quartz—How Gold Came to Klondike—Banks and Banking.

NO history has recorded, nor has tradition handed down, whether the first gold which excited man's admiration and afterward his cupidity was a nugget of the virgin metal or only glittering, yellow dust. Probably it was the former and quite likely the lump was a large one. But since that primitive time the thirst for gold then created has grown more insatiable till famishing mankind in the search for the precious metal has literally changed the face of nature over a good portion of the known world.

Probably the first man to make a " strike " valued the nugget mainly because it was large and bright, but smaller bits of the same brilliant substance came ere long to have a recognized value proportioned to their size, and when at length some unusually long-headed antedeluvian hit upon the fact that a pound of gold dust could be made into one lump just as large and just as brilliant as a nugget of the same weight the day of " dust " had dawned. And the day of dust was the day when men began to " wash " the golden sands of the ancient river beds and lay up for themselves treasures on earth.

Placer mining, in which the gold found " free " in the gravel beds is washed clean of earthy dross, is essentially " poor

man's " mining.   It needs few tools and little capital, and there
is no hindering patent on the process.   It has been follo.ved
from the earliest times and in much the same manner in all
parts of the world.   Nations which had nothing else in common
were alike in their methods and tools for placer mining.   The
pans and panning described by Mungo Park were practically
identical with those of the " days of '49," and the prospector of
'97 in the Klondike needs no other types of tools than are in
use by the rude native miners of every gold bearing region on
earth.

In the shallow diggings or placers nature has for ages been
performing the work for which the quartz miner must invent all
manner of machinery and employ a vast amount of capital and
skilled labor—the disintegration of the gold-bearing rock and
the concentration of the metal.   Consequently, the unskilled
laborer, whose capital is his own strength and a few of the sim-
plest tools, is able to extract, on a remunerative scale, immense
quantities of gold which, under its original condition, spread
through quartz and other hard rocks, would have needed vast
amounts of capital and much machinery for its elimination, and
in many instances would not have repaid the outlay.   It is easy
to see why placers are " poor men's " mines.

Exhausting the Surfacings.

The exhaustion of the shallow placers of the older gold fields
is fast approaching, that class of mining being abandoned in
those regions in America almost entirely to the patient Chinese.
Yet it should not be forgotten these shallow washings have often
led the miner to the very door of vast storehouses of wealth in the
veins in the hills and mountains.   In California, in New South
Wales and in Victoria deep leads were nearly all discovered by
prospecting the surfacing.   From this the Alaskan miner will

understand that however rich his placer claim may be, it is, more than that, the likely guide post to a still vaster treasure, and he will be able to understand why " Lucky " Baldwin intends to turn his great experience and ample resources to the locating of the " mother lode."

But the majority of the men now in or going into the Alas-

HYDRAULIC MINING.

kan diggings or the Klondike have neither taste, time nor means to hunt for the " mother lode." They have taken it for granted that nature has extracted the yellow metal from the rocks for them, and they want the benefit of her bounty in a hurry, and all they can get of it.

The first thing for the prospector to do is to pick out a likely

locality to prospect. Judgment and technical knowledge and
experience all count for something in making this choice, but
they are not infallible. The novice may have better luck than
the old-timer, and it is worthy of note in this connection that old
miners are firm believers in "luck." The experiences of the last
two years in the Yukon Basin would seem to go far to confirm
their faith.

A man just back from Dawson City with $100,000 in dust to
his credit told this story :

" Men who had scarcely one dollar six months ago are now
bonanza kings carrying $50,000 in gold dust and owning claims
that they would not sell for that amount. It is simply chance
or luck and nothing else. Dozens of worthy fellows have
worked hard and not " struck " anything yet, while others have
literally stumbled into their good fortunes. Last November a
man went out on the creek with others to stake a claim. He
was so drunk that he scarcely knew—much less cared—where
he was or what he was doing, but he staked. Now, he can command his hundreds of thousands."

Building a House.

Having selected a locality the next thing is to build a house,
or hut, for the daily life of a prospector or miner on the Yukon
is rough and hard, and a warm home is absolutely essential to
the health and cheerful spirit without which he cannot hope to
succeed. If there are four men in the party, the building need
not take more than a day. Architecture is all " out of the
same log " in that region, and any house will do for a model.
Four log walls well chinked with the abundant moss, a dirt roof
and a chimney are the main essentials.

Then, out for " color."

Prospecting in this land of long winters is generally conducted in the season when everything is locked in frost. During the short summer the streams are full of rushing water, and prospecting except along the banks is difficult and often impossible. The absence of water might be deemed a drawback in winter prospecting, but the novice will quickly learn that it takes but a little water to wash out a sample pan, and that amount can easily be obtained by melting snow or ice. Moreover, to an expert placer miner, water is not a necessity. He pans dry. The Alaskan "dust" is very coarse averaging nearly a wheat grain in size. This makes easy panning.

Mrs. Frederick Schwatka gives a none too alluring picture of this stage of the Yukon miner's experience in these words:

"There isn't very much said about the kind of ground that the gold hunters have to prospect over in the river regions. It is frightfully hard to travel. In the winter it is all ice and in the summer it is buried deep with drift wood and débris from the spring floods till it is almost impassable. All the rivers are flooded every spring and fall and the waters carry off huge pieces of frozen banks."

But the Alaska argonaut knew all this before he started, so he is not disheartened.

Thawing the Ground.

In hunting for gold prospectors dig a hole down to bed rock, which is generally found at a depth of from fifteen to eighteen feet. In the Yukon Basin they have to melt the ground, a few inches at a time, as they dig. The first twelve feet or so of earth is non-auriferous. Under it lies a stratum of coarse gravel three feet or more in thickness, which is rich in the precious metal, most of it being in the shape of small nuggets or grains. It is called "dust," but it is much coarser than the dust found in other parts of the world. Some of it is so large that a big percentage can be picked out by hand as the gravel is brought up out of the hole, but the general practice is to sluice or pan wash it.

The feeble suns of the short summer do not thaw out the frozen ground to its full depth in the Yukon Basin, and it has to be softened by building huge fires, which are kept going night and day until the earth is in such shape that the miners can force their way through it with picks. This done, a number of holes are dug on each claim, but even then when the gold gravel is taken out it is in frozen chunks resembling small masses of concrete. By making these holes in the summer the miners are enabled to work underground a portion of the winter and thus prepare for an early wash-up when the spring thaw comes in June. To take advantage of this the gravel which has been dug out during the winter has to be again softened by fire before it can be put through the sluices or pans and the gold separated.

The gravel is packed in a kind of clay, which makes a conglomerate like concrete, through which, when frozen, the strongest man cannot force a pick. When this gravel is thawed it is broken up with picks and thrown in a big heap with shovels. It varies in depth from fourteen to twenty feet, and it is richest in gold close to the bed rock. This is because gold is heavier than gravel and settles toward the bottom of any bar or bank in which it has accumulated. It is almost unnecessary to say that in sinking the holes or shafts every foot of the ground must be prospected for " pay dirt." This part of the prospecting consists simply in washing out pans of the gravel or sand ; if gold is found the claim should be " located " or staked out at once.

How to Tell Minerals.

It is necessary to remind the novice that all is not gold that glitters. Since the days when the earliest Virginian explorers sailed back to England with a ship-load of yellow sand under the delusion that they had a cargo of gold, " tenderfeet " have been easily misled, when seeking gold, by iron and copper pyrites and by mica. How to distinguish these natural counterfeits is worth knowing.

Iron pyrites, or bisulphide of iron, is a brass-yellow mineral occurring in small cubical crystals. It is easily discriminated. When strongly heated it is attracted by the magnet, while gold never becomes magnetic. Gold is malleable and iron pyrites brittle. Gold may be cut in flakes, pyrites not. Heated in nitric acid pyrites dissolves with effervescence and abundant red fumes, gold is unaffected. The specific gravity of gold is about four times that of iron pyrites. Mercury absorbs gold dust, but not iron pyrites.

Copper pyrites, or yellow copper ore, the principal source of copper, is a deep brass-yellow colored mineral with a strong metallic lustre. Its primitive crystalline form is the regular tetrahedron. It crumbles freely under the hammer, and yields to the knife; but instead of giving a solid chip as gold would, produces only dust. Heated on charcoal before the blowpipe it loses its yellow color and fuses into a dull black globule. Mixed with carbonate of soda and a little borax and subjected to the blowpipe it will yield a button of metallic copper.

Mica is a yellow, glistening mineral of foliated structure, and semi-metallic luster. It is much lighter than gold and becomes flakey when heated to redness and loses its lustre on cooling, whereas gold would remain unchanged.

Black Sand.

In assaying the gold sand of rivers, streams, and beaches of the Pacific coast, some difficulty is occasionally met with from the specular and titanic iron known technically as black sand. Platinum and iridium are often found in the same sands. Following are convenient methods of testing these sands :

For Atwood's test, take 100 to 1000 grains and attack with aqua regia in a flask ; cool for thirty minutes, dilute with water and filter.   If gold is present it will be in solution in the filtrate. Evaporate the filtrates to dryness, add a little hydrochloric acid and redissolve the dry salt in warm water ; add to the solution so formed, protosulphate of iron, which will throw down the gold as a fine, dark precipitate.   Dry and burn over the lamp.   Mix residuum with three times its weight of lead, fuse, scarify and cupel.

Mechanical Assay.

The mechanical test or assay of auriferous sands is of the utmost practical value, and may be thus described as scientifically performed, it being understood this is only a working test, and does not give all the gold as shown by a careful fire assay : Put 2000 grammes in a pan or, better, in a batea, and wash carefully until the gold begins to appear.   Use clean water, and when the pan and the small residue are clean, pour off most of the water and drop in a globule of pure mercury and a piece of cyanide of potassium.   As the cyanide begins to dissolve, impart a rotary motion to the dish—best done by holding the arms stiff and moving the body.   As the mercury rolls over and ploughs through the sand, under the influence of the cyanide, it will collect all the particles of free gold.   When all has been collected, transfer the mercury carefully to a small porcelain cup or test tube, and boil with strong, pure nitric acid.   When the mercury is all dissolved, the acid is poured off, more nitric acid is applied cold and rejected, and the gold is then washed with distilled water and dried.   The second washing with nitric acid is to remove any nitrate of mercury.

The resulting gold is not pure, but has the composition of the natural alloy.   To purify it, melt it with silver, hammer it out thin, boil twice with nitric acid, dry and heat it to redness.   To calculate the assay, take each of the original 2000 grammes to mean a pound and decimals of a gramme to mean decimals of a pound. Multiply the value of gold by the fraction of a gramme produced, and the result will be the value of the gold in a ton.

In this same connection it may be noted that it is important, in estimating the value of purchased gold dust to examine carefully to see if there is any counterfeit or " bogus " dust present. If all from the same locality the dust will have a uniform color. A fair sample of the whole lot of dust under inspection should be placed in an evaporating dish and nitric acid poured upon it If any reaction takes place there is foreign matter present.

Locating the Claims.

If the prospects indicate a claim that will pay for working, the miner's first step is to locate the claim.

The manner of locating placer mining claims differs from that of locating claims upon veins or lodes. In locating a vein or lode claim, the United States statutes provide that no claim shall extend more than 300 feet on each side of the middle of the vein at the surface, and that no claim shall be limited by mining regulations to less than 25 feet on each side of the middle vein at the surface. In locating claims called " placers," however, the law provides that no location of such claim upon surveyed lands shall include more than 20 acres for each individual claimant. The supreme court, however, has held that one individual can hold as many locations as he can purchase and rely upon his possessory title ; that a separate patent for each location is unnecessary.

A patent for any land claimed and located may be obtained in the following manner: "Any person, association or corporation authorized to locate a claim, having claimed and located a piece of land, and who has or have complied with the terms of the law, may file in the proper land office an application for a patent under oath, showing such compliance, together with a plat and field notes of the claim or claims in common made by or under the direction of the United States surveyor general, showing accurately the boundaries of the claim or claims, which

GUARDING HIS CLAIM.

shall be distinctly marked by monuments on the ground, and shall post a copy of such plat, together with a notice of such application for a patent, in a conspicuous place on the land embraced in such plat, previous to the application for a patent on such plat; and shall file an affidavit of at least two persons that such notice has been duly posted, and shall file a copy of the notice in such land office ; and shall thereupon be entitled to a patent to the land in the manner following : The registrar of said land office upon the filing of such application, plat, field notes, notices and application, shall publish a notice that such application has been made for a period of sixty days, in a newspaper to be by him designated, as published nearest to such claim ; and he shall post such notice in his office for the same period. The claimant at the time of filing such application, or at any time thereafter, within sixty days of publication, shall file with the registrar a certificate of the United States surveyor general that $500 worth of labor has been expended or improvements made upon the claim by himself or grantors ; that the plat is correct, with such further description by reference to natural objects or permanent monuments as shall identify the claim and furnish an accurate description to be incorporated in the patent. At the expiration of the sixty days of publication, the claimant shall file his affidavit, showing that the plat and notice have been posted in a conspicuous place on the claim during such period of publication.''

If no adverse claim shall have been filed with the registrar of the land office at the expiration of said sixty days, the claimant is entitled to a patent upon the payment to the proper officer of $5 per acre in the case of a lode claim, and $2.50 per acre for a placer.

As to Local Customs.

The location of a placer claim and keeping possession thereof until a patent shall be issued are also subject to local customs, about which the wise miner will thoroughly inform himself. In Alaska the holder of a claim is required to do at least $100 worth of work on his claim every year for five years to get an absolute title to it. He has the privilege of doing the entire $500

25

worth of work at once if he chooses to do so, and on proof of it
may get his patent.   The man who locates a claim is allowed a
full year before he puts up his location notice for working the first
assessment, during which time his right is absolute and is also
negotiable.   A purchaser fulfilling the obligation entered into
by the discoverer enjoys the same rights.

In Alaska and in the Klondike the first miners in a district
hold a meeting and fix the size of the claims, and also agree as
to how much work shall constitute an assessment.   The miners
also elect a register.

The size of a claim, as fixed by agreement among the miners
of any particular locality, is a section of the creek of a certain
length—sometimes 200 feet—and it extends from rim to rim in
width.   The reason of this variableness in the size of claims on
the different creeks is that on some a greater length is required
to make them worth a man's while to work them.   The paying
deposits may be scattered so a man could make wages only by
working here and there over a large territory.   Of course, the
conditions surrounding the first discovery made on a creek are
the basis for fixing the size of a claim on that stream.   The discoverer of a new field is allowed two claims, while others are
permitted to take but one at a time.   However, when a locator
has worked out his assessment of a few days' work he is at
liberty to take another.

Commissioner Herrman's Digest.

Commissioner Herrman, of the United States Land Office at
Washington, briefly digested the law bearing on placer claims as
follows :

" When you patent a claim it is necessary for you to be a citizen of the United States or to have declared your intention of
becoming one.

" This law, however, is of little consequence when placer digging is concerned. Under our laws anybody is privileged to
dig out gold wherever it is found. When it comes to taking out
a patent for the land the miner will have exhausted the superficial supply of gold and moved on.

" There is practically no need of taking out patents for placer
mining. The miner comes along, sees a likely piece of ground,
digs up a few panfuls, extracts the gold, if there is any, stays
there till he has obtained as much as he can from that piece of
ground with his primitive implements, and then moves on to
another likely piece.

" Pretty soon along comes the quartz miner with his machinery and takes out a claim for a piece of ground which the placer
miner may have worked superficially."

As to locations on the Klondike, see the chapter an " Mining
Laws."

Getting Out the Gold.

Now comes the hardest part of the miners' work—getting out
the golden treasure.

In summer in Alaska about the only tools required in the
placers are a pick, shovel and gold pan, about the size of a small
dish pan and made of copper or white enameled iron, preferably
the latter because the relief enables the miner to see the gold
more distinctly especially when it is in fine specks. The miner
squats beside the water, dips water into the pan, oscillates it with
a motion that can only be acquired by experience, and gradually
sloughs out the water, dirt, gravel, etc., retaining the gold in the
pan. Gold being the heaviest substance it is, of course, the
easiest to retain in the pan. If it be in the shape of nuggets,
the miner picks them out of the pan with his fingers ; if the
gold be in small particles, fine gold or " flour " gold, he dries
the pan in the sun and carefully brushes the deposit into a

piece of buckskin or other material used for carrying the precious metal.   Some miners prefer the cradle to the pan for getting gold.

It is nearly always desirable, but not always possible, to have a sluice.   This sometimes is very primitive.   It may be only a gully bottomed with cobblestones, or plank troughing, with riffles or cleats at intervals across the bottom.   In either case, the gold-bearing dirt or gravel is thrown in while water is running through the sluice.   The current is supposed to carry away the worthless rocks and dirt, allowing the gold to sink to the bottom.   If the gold is in finely divided particles, the sluice is made tight and quicksilver is placed above the riffles, which envelops and holds the gold dust.   No two mines are exactly alike, and the manner of working them has to be varied to suit the circumstances.

Mining in Winter.

In placers in winter in Alaska and in the Klondike, practically all the year round, it is necessary to melt the frozen auriferous gravel by means of huge fires in order to make it possible to work it with a pick.   Formerly miners used to thaw out the whole area of their claims down to bed rock.   Now they sink a shaft to the bottom of the gravel, and tunnel along underneath in the gold-bearing layer.   As the tunnel is all the way through the solid frozen earth, no shoring is required, and the only expense for timber is for fuel.

The way in which the tunneling is done is interesting, as it has to be carried on in cold weather, when everything is frozen. The miners build fires over the area which they wish to work, and keep them lighted for the space of about twenty-four hours. Then, at the expiration of this period, the gravel will be melted and softened to a depth of perhaps six inches.   This is then taken off and other fires built, until the gold-bearing layer is reached. When the shaft is down so far fires are built at the bottom, against the sides of the layer, and tunnels made in this manner. Dry wood is piled against the face of the drift, and then other pieces are set slantwise over the heap of fuel. As the fire burns, the gravel falls down from above and gradually covers the slanting shield of wood. The fire smoulders away and becomes charcoal burning. It is when it reaches this confined stage during the night that its heat is most effective against the face of the drift. Next day the miner finds the face of his drift thawed out for a distance of from ten to eighteen inches, according to conditions. He shovels out dirt, and if only part is pay dirt he puts only that on his dump. Thus, at the rate of a few inches a day, drifting out of precious gravel goes on, and the dump is slowly added to until spring, when the torrents come down, and the washing and sluicing and cradling begin.

Work on the Yukon.

The mines of the Yukon are of a class by themselves, and it is necessary to follow new methods for getting the gold. To begin with, the ground is frozen. From the roots of the moss, which often is more than a foot thick, to the greatest depth that ever has been reached, the ground is as hard as a bone. The gold is found in a certain drift of gravel, which lies at varying depths, often as far down as twenty feet. Only that portion of the gravel just above hard pan—by which is usually meant clay—carries gold in any quantity, and in favored localities this particular gravel is extraordinarily rich. In fact, there is more free gold found within the same space, taking the whole district through, than ever was found anywhere in placers. Toward the heads of the creeks, and likewise toward the original source of the mineral, the gravel is found nearer the surface than at places further down the streams. It is also coarser gold, but, on the other hand, it covers a narrower strip of the valley. Going down the creeks, the deposit is spread out over a much wider area, and is deeper in the ground. The gold is in smaller particles, but the quantity may be as great as anywhere. As in nearly all placer mines, the low places of what has formerly been the bed of the creek are the richest, the deposits decreasing in quantity toward the outer edges.

Another Description.

Land Surveyor Ogilvie gives the following description of a method of placer mining in vogue across the border :

"The process of placer mining is about as follows : After clearing all the coarse gravel and stones off a patch of ground, the miner lifts a little of the firmer gravel or sand in his pan, which is a broad, shallow dish, made of strong sheet-iron ; he then puts in water enough to fill the pan and gives it a few rapid whirls and shakes ; this tends to bring the gold to the bottom on account of its great specific gravity. The dish is then shaken and held in such a way that the gravel and sand are gradually washed out, care being taken to avoid letting out the finer and heavier parts that have settled to the bottom. Finally all that is left in the pan is whatever gold may have been in the dish, and some black sand which almost invariably accompanies it. This black sand is nothing but pulverized magnetic iron ore.

"Should the gold thus found be fine, the contents of the pan are thrown into a barrel containing water and a pound or two of mercury. As soon as the gold comes in contact with the mercury it combines and forms an amalgam. This process is continued until enough amalgam has been formed to pay for ' roasting' or 'firing.' It is then squeezed through a buckskin bag, all the mercury that comes through the bag being put back into the barrel to serve again, and what remains in the bag is placed in a retort, if the miner has one, or, if not, on a shovel, and heated until nearly all the mercury is vaporized. The gold then remains in a lump with some mercury still held in combination with it. This is called the 'pan' or 'hand' method, and is never, on account of its slowness and laboriousness, continued for any length of time when it is possible to procure a 'rocker,' or to make and work sluices.

Sluicing for Gold.

" Sluicing is always employed when possible. It requires a good supply of water, with sufficient head or fall. The process is as follows : Planks are procured and formed into a box of suitable width and depth. Slats are fixed across the bottom of the box at suitable intervals, or shallow holes bored in the bottom in such order that no particle could run along the bottom in a straight line and escape running over a hole. Several of these boxes are then set up with a considerable slope, and are fitted into one another at the ends, like a stovepipe. A stream of water is now directed into the upper end of the highest box. The gravel having been collected, as in the case of the rocker, it is shoveled into the upper box, and is washed downward by the strong current of water. The gold is detained by its weight, and is held by the slats or in the holes mentioned ; if it is fine, mercury is placed behind the slats or in these holes to catch it.

" In this way about three times as much dirt can be washed as by the rocker, and consequently three times as much gold can be secured in a given time.

" A great many of the miners spend their time in the summer in prospecting, and in the winter resort to what is called 'burning.' They make fires on the surface, thus thawing the ground until the bedrock is reached. The pay dirt is brought to the surface and heaped in a pile until spring, when water can be obtained. The sluice boxes are then set up and the dirt is

washed out, thus enabling the miner to work advantageously
and profitably the year round."

Captain J. F. Higgins, of the steamer Excelsior, one of the
Alaska boats, wrote to a friend in San Diego the following story
of good luck in the Yukon placers :

" There is about fifteen feet of dirt above bedrock, the pay
streak averaging from four to six feet, which is tunneled out
while the ground is frozen.   Of course, the ground taken out is
thawed by building fires, and when the thaw comes and water
rushes in they set their sluices and wash the dirt.   Two of our
fellows thought a small bird in the hand worth a large one in
the bush and sold their claims for $45,000, getting $4500 down,
the remainder to be paid in monthly installments of $10,000
each.   The purchasers had no more than $5000 paid.   They
were twenty days thawing and getting out dirt.   Then there was
no water to sluice with, but one fellow made a rocker, and in
ten days took out the $10,000 for the first installment.   So, tun-
neling and rocking, they took out $40,000 before there was water
to sluice with."

Dry Placer Miners.

Machines known as " dry placer miners " are in use in various
southern diggings and may be expected to make their appearance in Alaska and the Klondike soon, where it is believed they
would be peculiarly well adapted to the conditions imposed on
mining by the climate.   A feature of some of these dry washers
is that, unlike sluicing or hydraulicking, they will effect a separation of the gold from the black sand.

The principle in these dry washers is that of the air blast removing or blowing the fine sand or dust from the finely pulverized material which is fed upon a panning table of perforated
metal covered with cloth and crossed by copper riffles.   The
sand and earthy dust are blown away, the gangue rolls down

the incline over the riffles, and is discharged as tailings, and the gold settles on the cloth behind the riffles and is removed in the daily "clean up." A small size of dry washer is made for prospectors.

A combination sled and gold "rocker" is being largely sold. It is about six feet long, eighteen inches wide and the runners stand up about ten inches. The "bed," when taken off, constitutes a "rocker" of a form approved by miners. It is claimed that 300 pounds of provisions, besides a miner's outfit of tools can be carried on it.

Dredging for Gold.

One of the new schemes for getting the gold out of the Yukon is to dredge the river bed. A company has been formed to carry out the work, and intends beginning work in the great river in the spring. The promoter argues that the gold deposits of the rivers and creeks are the results of the washing down by high waters and the carrying down of ice floes. Upon this assumption the argument is made that in the deeper channel the gold has sunk lower, and, as the dredgers will work down to bed rock, the belief is that the result of pumping from the bottom will be proportionately richer.

An experiment is being conducted in Frazier River in the use of centrifugal pumps on barges to pump up the earth along the bottom of the river and wash out the gold that has been deposited there for ages. The nozzles of these pumps, which are screened to prevent big bowlders from being taken in, are forced to the bottom of the river, and as the sand and water reach the top of the barge they are carefully screened, so that all the gold is secured. If the experiment proves a success it will revolutionize placer mining.

A report on the Birch Creek district, issued during the summer of 1897, says :

"Some miners have planned to work this and other good ground supposed to exist under the deep covering of moss and gravel in the wide valley of the Mammoth and Crooked Creeks, by hydraulicking, the water to be obtained by tapping Miller and Mastadon Creeks near the head."

A machine has recently been invented, intended to use Alaska petroleum if it can be had in sufficient quantities, and if not, oil brought from the States or from Ontario, by means of which it is expected to thaw the frozen gravel and drift in the placer beds, and vastly cheapen and expedite the process of gathering the gold. The machine is so light that one man can easily handle and move it from place to place.

The fuel oil is contained in a tank which is mounted on wheels, and is provided with a blower to force air into the tank and oil out. A lead of pipe runs under a piece of sheet iron, usually three feet long by twenty inches wide, which has beveled sides. Beneath the cover is a coil of perforated pipe through which the oil makes its escape and is burned. It is so arranged there is always a downward draft, and the force of the flame is continually against the ground.

Old Miner's Advice.

Here is some good advice by an old miner to "tenderfeet," who are apt to stampede easily and be led to run after false gods :

"If you have once got a claim that is paying a fairly satisfactory amount of gold stick to it. You are just about as apt to strike a rich pocket there as anywhere else, and it is much better to be taking out even a comparatively small sum regularly than to spend your time roving from one place to another, and getting next to nothing anywhere. You have got to have perseverance, and be willing to plod in this pursuit, as well as in any other, if you want to succeed in it."

It is advice worth pondering and heeding.

HYDRAULIC MINING—WASHING OUT THE GOLD.

395

Placers, wherever found, are indications of gold-bearing veins in the neighborhood. Alaska is believed to be no exception to the apparent rule. That rich quartz will be found in the highlands of the Territory there seems to be no good reason to doubt, and the day when the subterranean mining industry will be the principal resource of the " Seward Purchase " may not be far distant. As usual, the first craze was over the placers, but the extraordinary richness of the surfacings attracted the attention of men of capital, and their agents are already in the field prospecting for gold-bearing quartz. The sequence of development in new gold fields is always the same—first, the men with pans to gather the riches on the surface ; next, miners with " long Toms " ; third, hydraulicking, and then, quartz mining underground. Alaska may break the record for getting into the fourth stage.

How Gold Came to Klondike.

Professor Frederick Wright, writing of " How Gold Came to the Klondike," says :

" Little is known about the geology of the Yukon River, where the Klondike mines have been found. Being placer mines, the gold may have been transported many miles. The means of transportation are both glaciers and rivers. The Klondike region is on the north side of the St. Elias Alps. Alaska was never completely covered with glacial ice. The glaciers flowed both north and south from these summits. Dawson and Professor Russell both report well-defined terminal moraines across the upper Yukon Valley. The source of the Klondike gold, therefore, is from the south.

Placer mines originate in the disintegration of gold-bearing quartz veins or mass like that at Juneau. Under subaerial agencies these become dissolved. Then the glaciers transport the material as far as they go, when the floods of water carry it on still further. Gold, being heavier than the other materials associated with it, lodges in the crevasses or in the rough places at the bottom of the streams. So to speak, nature has stamped and panned the gravel first and prepared the way for man to finish the work. The amount of gold found in the placer mines is evidence not so much perhaps of a very rich vein as of the disintegration of a very large vein.

" The ' mother lode' has been looked for in vain in California, and perhaps will be so in Alaska. But it exists somewhere up the streams on which the placer mines are found. The discovery of gold in glacial deposits far away from its native place is familiar to American geologists.

".It is evident, however, that in Alaska the transportation of gold has not gone so far."

General Duffield, Superintendent of the Coast and Geodetic Survey, also inclines to the glacier view. He says :

" The gold has been ground out of the quartz by the pressure of the glaciers, which lie and move along the courses of the streams, exerting a tremendous pressure. This force is present to a more appreciable extent in Alaska than elsewhere, and I believe that as a consequence more placer gold will be found in that region than in any other part of the world."

Dr. Everett's Views.

Dr. Willis E. Everett, of Tacoma, says :

" Alaska was once under glaciers, and the gold now found undoubtedly comes from glacial action, primarily, which has been going on for many centuries. The miners are finding, however, that what they usually consider bed rock is only a false bed rock, and that underneath there is still another bed rock, with larger lumps of gold than are found on the first. I believe that the country in the interior, back of Klondike, will furnish enormous

quantities of gold, and that the rich strikes already made are but
a small beginning.   The district will prove to be about 300 miles
square."

This theory of Dr. Everett would seem to be borne out by
the experience of a young Chicago "tenderfoot" who, being un-
learned in miner's traditions, not only dug down to hard-pan,
but went straight on through the clay and found a fabulously
rich deposit of "dust" and nuggets.   Had he been an old miner
he would have stopped at hard-pan and the treasure would not
have been uncovered.

Professor Emmon's Theory.

Professor S. F. Emmons, of the Geological Survey, says:
"The real mass of golden wealth in Alaska remains as yet untouched.   It lies in the virgin rocks, from which the particles
found in the river gravels now being washed by the Klondike
miners have been torn by the erosion of streams.   These particles, being heavy, have been deposited by the streams, which
carried the lighter matter onward to the ocean, thus forming, by
gradual accumulation, a sort of auriferous concentrate.   Many
of the bits, especially in certain localities, are big enough to be
called nuggets.   In spots the gravels are so rich that, as we
have all heard, many ounces of the yellow metal are obtained
from the washing of a single panful.   That is what is making
the people so wild—the prospect of picking money out of the
dirt by the handful literally."

Gold-bearing quartz is plentiful in the southeastern portion of
Alaska, around the great Alaska-Treadwell and Alaska-Mexican
mines and their smaller likenesses.   Such quartz has been found
in Cone Hill, midway in the valley of the Forty-Mile, and vague
reports of quartz finds worth working have come in from other
sections which the winter's prospecting is. expected to verify.

And in the spring, too, " Lucky " Baldwin starts out to find the
" mother lode." There is no doubt that lode mining will be
carried on in the Alaskan mountains when the country is settled.

Banks and Banking.

After the miner on the Yukon has dug and panned out his
gold, although the country is full of naturally honest men and
of others as honest as a wholesome fear of Judge Lynch can
make them, his next thought will be where he can stow it away
and keep it safe till he gets ready to carry it back to civilization.
Heretofore he has deposited it, if he banked it all, with Captain
Healy in his safe at Circle City. Next year he will have banking facilities of approved pattern at his very door.

The North American Transportation and Trading Company
has decided to carry out the plan of establishing five, and possibly six, banks on the Yukon, at Dawson City, Fort Cudahy
Circle City, Fort Get There and St. Michael's. W. H. Hubbard,
of Chicago, went into the basin via the Chilkoot Pass in August
to complete the arrangements for opening the institutions. Before leaving for Alaska, he said :

"The banks will be primarily banks of exchange. We shall
accept gold dust and sell exchange on Chicago, New York and
San Francisco for it. In Chicago we shall accept currency and
issue letters of credit to those going into the mines.

"As I understand it, gold dust is the only ' currency' in the
interior of Alaska. It passes current for $17 an ounce, its
market value being a trifle more than that amount. Gold dust
is used even in petty transactions, as there is not enough silver
for change. A miner going into a saloon for a drink takes out
his bag of dust, lays it on the bar, and the saloon-keeper weighs
the fifty cents or one dollar and hands back the change. All
supplies are paid for in like manner.

"Loans by the banks will be a later consideration. No doubt traders will flock in and all kinds of business established. The merchants there as elsewhere probably will need accommodations, and where their standing warrants it we shall let them have money. The banking business is in embryo. My work will be to establish it at the five posts which the North American Company has founded."

The Canadian Government has under consideration a project for the establishment of a "treasure house" at Dawson City in which will be stored the miners' gold and for which they will receive drafts on United States or Canadian banks for the full market value of their "dust."

If the gold is stored in a central place, under this proposed plan, the officials of the law will find the task of preserving order greatly simplified, for the miners will not be under the necessity of carrying arms, nor will the rougher sort likely spend as much gold in riotous living. It will, of course, be necessary for the government to take great precaution to insure the safety of the gold, but the presence of fifty or a hundred mounted police and three or four Maxim guns will be a great deterrent to the envious and greedy.

Wells, Fargo & Co. will likely establish an office in Dawson City in the spring.

Effects of Discovery.

Touching the effect of the discovery of gold in Alaska, Director of the Mint Preston, said :

"It is too early to determine. We cannot expect to see any material effect in the London market, where gold is quoted every day, until a year or two have passed.

"I should judge from all accounts that the discoveries of the Klondike region would add a tremendous amount of gold to the world's stock. The tendency of this will be, of course, to increase the value of silver, but I doubt if it will very greatly raise its market value. At any rate, we must wait from one to two years to determine that.

" It is unfair to assume that the increase in the value of silver resulting from the discovery of gold in Alaska will be anything like that which resulted in the early '50s from the discoveries in California and in Australia. At that time the supply of silver in the United States was almost nil, and there was very little silver coinage. At the present time, however, there is so much silver that the world, as the market has indicated this week, does not know what to do with it. There cannot be expected, therefore, a very high jump in the price of silver under any discovery of gold."

26

CHAPTER XIV.

Resume of Mining Laws.

Law and Order—Fees for Mining—Rights of Miners—Quartz Mining—
Surveys and Reservations—Voice of the Press—Penalties Imposed—Call
for United States Troops—Size of Claims—Canadian Laws.

IN gold mining the law may be the survival of the fittest, but
it is not the rule of the strongest. Every phase of the
work is hedged around by legal enactments, and the miners
are obliged to observe as much red tape, away out in the wilderness, thousands of miles from civilization, as a citizen would in
New York or Chicago.

On the American side of the boundary line all mining operations are subject only to the United States mining laws and the
general laws of the State of Oregon, as they existed in 1884,
when the law providing a civil government for Alaska was
passed.

That law provided "that the general laws of the State of
Oregon now in force are hereby declared to be the law in said
district, so far as the same be applicable and not in conflict with
the provisions of this act or the laws of the United States."

Thus the laws of Oregon in force May 17, 1884, are the laws
of Alaska. As a matter of fact, however, little attention to
niceties of detail is ever paid. In a large sense, the law of
the miners is an unwritten code, but that code is kept within the
legal statutes.

On the Canadian side of the boundary—that is, in Klondike—
the mining laws of British Columbia are in force. For the convenience of readers who may contemplate trying their fortunes
in the great Northwest a digest of the mining laws of both countries is herewith given.

402

The Placer Mining Law of the United States, from the Revised Statutes, provides as follows :

The term " placer claim " as defined by the Supreme Court of the United States, is : " Ground within defined boundaries which contains mineral in its earth, sand or gravel ; ground that includes valuable deposits not in place, that is, not fixed in rock, but which are in a loose state, and may in most cases be collected by washing or amalgamation without milling."

The manner of locating placer mining claims differ from that of locating claims upon veins or lodes. In locating a vein or lode claim, the United States Statutes provide that no claim shall extend more than 300 feet on each side of the middle of the vein at the surface, and that no claim shall be limited by mining regulations to less than 25 feet on each side of the middle of the vein at the surface. In locating claims called " placers," however, the law provides that no location of such claim upon surveyed lands shall include more than 20 acres for each individual claimant. The Supreme Court, however, has held that one individual can hold as many locations as he can purchase and rely upon his possessory title ; that a separate patent for each location is unnecessary.

Proof of Citizenship.

Locaters, however, have to show proof of citizenship or intention to become citizens. This may be done in the case of an individual by his own affidavit ; in the case of an association incorporated by a number of individuals by the affidavit of their authorized agent, made on his own knowledge or upon information and belief ; and in the case of a company organized under the laws of any State or Territory, by the filing of a certified copy of the charter or certificate of incorporation.

A patent for any land claimed and located may be obtained in the following manner : " Any person, association or corporation authorized to locate a claim, having claimed and located a piece of land, and who has or have complied with the terms of the law, may file in the proper land office an application for a patent, under oath, showing such compliance, together with a plat and field notes of the claim or claims in common made by or under the direction of the United States Surveyor General, showing accurately the boundaries of the claim or claims, which shall be distinctly marked by monuments on the ground, and shall post a copy of such plat, together with a notice of such application for a patent, in a conspicuous place on the land embraced in such plat, previous to the application for a patent on such plat; and shall file an affidavit of at least two persons that such notice has been duly posted, and shall file a copy of the notice in such land office ; and shall thereupon be entitled to a patent to the land in the manner following :

Publishing of Notices.

"The registrar of said land office upon the filing of such application, plat, field notes, notices and affidavits, shall publish a notice that such application has been made, for a period of sixty days, in a newspaper to be by him designated, as published nearest to such claim ; and he shall post such notice in his office for the same period. The claimant at the time of filing such application or at any time thereafter, within sixty days of publication, shall file with the registrar a certificate of the United States Surveyor General that $500 worth of labor has been expended or improvements made upon the claim by himself or grantors ; that the plat is correct, with such further description by reference to natural objects or permanent monuments as shall identify the claim and furnish an accurate description to be incorporated in the patent. At the expiration of the sixty days of publication, the claimant shall file his affidavit showing that the plat. and notice have been posted in a conspicuous place on the claim during such period of publication."

If no adverse claim shall have been filed with the registrar of the land office at the expiration of said sixty days, the claimant is entitled to a patent upon the payment to the proper officer of $5 per acre in the case of a lode claim, and $2.50 per acre for a placer.

The location of a placer claim and keeping possession thereof until a patent shall be issued are subject to local laws and customs.

It will be seen from the following that the Mining Laws of British Columbia differ somewhat in detail from those of the United States, but are designed to cover essentially the same points and subserve the same purpose. The Canadian Statutes make these provisions :

Placer Mining—Registration and Fees.

At the close of the second sitting of the Canadian Cabinet it was announced that the Government had decided to impose a royalty on all placer diggings on the Yukon in addition to $15 registration fee and $100 annual assessment. The royalty will be 10 per cent. each on claims with an output of $500 or less monthly, and 20 per cent. on every claim yielding above that amount monthly. Besides this royalty it has been decided in regard to all future claims staked out on other streams or rivers, that every alternate claim should be the property of the Government, and should be reserved for public purposes and sold or worked by the Government for the benefit of the revenue of the Dominion.

For " bar diggings "—A strip of land 100 feet wide at highwater mark, and thence extending into the river at its lowest water level.

For " dry diggings "—100 feet square.

For "creek and river claims"—500 feet along the direction of the stream, extending in width from base to base of the hill or bench on either side. The width of such claims, however, is limited to 600 feet when the benches are a greater distance apart than that. In such a case claims are laid out in areas of 10 acres, with boundaries running north and south, east and west.

For "bench claims"—100 feet square.

Size of claims to discoverers or parties of discoverers—To one discoverer, 300 feet in length; to a party of two, 600 feet in length; to a party of three, 800 feet in length; to a party of four, 1000 feet in length; to a party of more than four, ordinary sized claims only.

New strata of auriferous gravel in a locality where claims are abandoned, or dry diggings discovered in the vicinity of bar diggings, or vice versa, shall be deemed new mines.

Rights and Duties of Miners.

Entries of grants for placer mining must be renewed and entry fee paid every year.

No miner shall receive more than one claim in the same locality, but may hold any number of claims by purchase, and any number of miners may unite to work their claims in common, provided an agreement be duly registered and a registration fee of $5 be duly paid therefor.

Claims may be mortgaged or disposed of, provided such disposal be registered and a registration fee of $2 be paid therefor.

Although miners shall have exclusive right of entry upon their claims for the " miner-like " working of them, holders of adjacent claims shall be granted such right of entry thereon as may seem reasonable to the superintendent of mines.

Each miner shall be entitled to so much of the water not previously appropriated flowing through or past his claim as the superintendent of mines shall deem necessary to work it, and shall be entitled to drain his own claim free of charge.

Claims remaining unworked on working days for seventy-two hours are deemed abandoned, unless sickness or other reasonable cause is shown, or unless the grantee is absent on leave.

For the convenience of miners on back claims, on benches or slopes, permission may be granted by the superintendent of mines to tunnel through claims fronting on water courses.

In case of the death of a miner, the provisions of abandonment do not apply during his last illness or after his decease.

Acquisition of Mining Locations.

Marking of locations—Wooden posts, four inches square, driven eighteen inches into the ground and projecting eighteen inches above it, must mark the four corners of a location. In rocky ground stone mounds three feet in diameter may be piled about the post. In timbered land well-blazed lines must join the posts. In rolling or uneven localities flattened posts must be placed at intervals along the lines to mark them, so that subsequent explorers shall have no trouble in tracing such lines.

When locations are bounded by lines running north and south, east and west, the stake at the northeast corner shall be marked by a cutting instrument or by colored chalk, " M. L. No. 1 " (mining location, stake number 1). Likewise the southeasterly stake shall be marked " M. L. No. 2," the southwesterly " M. L. No. 3 " and the northwesterly " M. L. No. 4." Where the boundary lines do not run north and south, east and west, the northerly stake shall be marked 1, the easterly 2, the southerly 3 and the westerly 4. On each post shall be marked also the claimant's initials and the distance to the next post.

Application and affidavit of discoverer—Within sixty days after marking his location the claimant shall file in the office of the dominion land office for the district a formal declaration, sworn to before the land agent, describing as nearly as may be the locality and dimensions of the location. With such declaration he must pay the agent an entry fee of $5.

Receipt issued to discoverer—Upon such payment the agent shall grant a receipt authorizing the claimant, or his legal representative, to enter into possession, subject to renewal every year for five years, provided that in these five years $100 shall be expended on the claim in actual mining operations. A detailed statement of such expenditure must also be filed with the agent of Dominion lands, in the form of an affidavit corroborated by two reliable and disinterested witnesses.

Annual renewal of location certificate—Upon payment of the $5 fee therefor a receipt shall be issued entitling the claimant to hold the location for another year.

Rules for Partnerships.

Working in partnership—Any party of four or less neighboring miners, within three months after entering, may, upon being authorized by the agent, make upon any one of such locations, during the first and second years, but not subsequently, the expenditure otherwise required on each of the locations. An agreement, however, accompanied by a fee of $5, must be filed with the agent. Provided, however, that the expenditure made upon any one location shall not be applicable in any manner or for any purpose to any other location.

Purchase of location—At any time before the expiration of five years from date of entry a claimant may purchase a location upon filing with the agent proof that he has expended $500 in actual mining operations on the claim and complied with all other prescribed regulations. The price of a mining location shall be $5 per acre, cash.

On making an application to purchase, the claimant must deposit with the agent $50, to be deemed as payment to the government for the survey of his location. On receipt of plans

IN THE HANDS OF A VIGILANCE COMMITTEE.

and field notes, and approval by the surveyor general, a patent shall issue to the claimant.

Reversion of title—Failure of a claimant to prove within each year the expenditure prescribed, or failure to pay the agent the full cash price, shall cause the claimant's right to lapse and the location to revert to the crown, along with the improvements upon it.

Rival claimants—When two or more persons claim the same location, the right to acquire it shall be in him who can prove he was the first to discover the mineral deposit involved, and to take possession in the prescribed manner. Priority of discovery alone, however, shall not give the right to acquire. A subsequent discoverer, who has complied with other prescribed conditions, shall take precedence over a prior discoverer who has failed so to comply.

When a claimant has in bad faith used the prior discovery of another and has fraudulently affirmed that he made independent discovery and demarcation, he shall, apart from other legal consequences, have no claim, forfeit his deposit and be absolutely debarred from obtaining another location.

Rival applicants—Where there are two or more applicants for a mining location, neither of whom is the original discoverer, the Minister of the Interior may invite competitive tenders or put it up for public auction, as he sees fit.

Transfer of Mining Rights.

Assignment of right to purchase—An assignment of the right to purchase a location shall be indorsed on the back of the receipt or certificate of assignment, and execution thereof witnessed by two disinterested witnesses. Upon the deposit of such receipt in the office of the land agent, accompanied by a registration fee of $2, the agent shall give the assignee a certificate entitling him to all the rights of the original discoverer. By complying with the prescribed regulations such assignee becomes entitled to purchase the location.

Regulations in respect to placer mining, so far as they relate to entries, entry fees, assignments, marking of locations, agents' receipts, etc., except where otherwise provided, apply also to quartz mining.

Nature and size of claims—A location shall not exceed the following dimensions : Length, 1500 feet ; breadth, 600 feet. The surface boundaries shall be from straight parallel lines, and its boundaries beneath the surface the planes of these lines.

Limit to number of locations—Not more than one mining location shall be granted to any one individual claimant upon the same lode or vein.

Mill sites—Land used for milling purposes may be applied for and patented, either in connection with or separate from a mining location, and may be held in addition to a mining location, provided such additional land shall in no case exceed five acres.

General Provisions.

Decision of disputes—The Superintendent of Mines shall have power to hear and determine all disputes in regard to mining property arising within his district, subject to appeal by either of the parties to the commissioner of dominion lands.

Leave of absence—Each holder of a mining location shall be entitled to be absent and suspend work on his diggings during the "close" season, which "close" season shall be declared by the agent in each district, under instructions from the minister of the interior.

The agent may grant a leave of absence pending the decision of any dispute before him.

Any miner is entitled to a year's leave of absence upon proving expenditure of not less than $200 without any reasonable return of gold.

The time occupied by a locator in going to and returning from the office of the agent or of the superintendent of mines shall not count against him.

Additional locations—The minister of the interior may grant to a person actually developing a location an adjoining location equal in size, provided it be shown to the minister's satisfaction that the vein being worked will probably extend beyond the boundaries of the original location.

Forfeiture—In event of the breach of the regulations, a right or grant shall be absolutely forfeited, and the offending party shall be incapable of subsequently acquiring similar rights except by special permission of the minister of the interior.

Trouble Over Mining Laws.

It was natural to expect that in a mining region so remote from districts in which there was an established order of affairs, in two countries between which there was a boundary line dispute of long standing—and in governments, or nominal governments, laws in unsettled regions are bound to be more or less dead letters—where mining was done under different systems of regulations and requirements, there should be more or less jealousy, friction and trouble.

Those who predicted a clash—and there were many such on the first news of the discovery of gold in the Klondike wilds reaching southern cities—were not disappointed.

Differences did arise almost immediately. These were due partly to a misunderstanding or an ignoring of the existing mining laws and partly to the greed of Great Britain in seeking to make a rich thing of the find by imposing exactions on the miners who crossed the real or alleged boundary line and staked off claims on the territory claimed by Canada.

The Canadian government lost no time in taking official action and there was a prospect of international hostilities.

On July 30, 1897, the Dominion Cabinet reached an important decision as to the imposition of a tax in the Yukon district on all

American miners.   This perhaps is best told in a telegraphic report from Ottawa, which was sent out at the time.   Says this report :

" Under the regulations recently issued the fee for registering a claim was fixed at $15, while an annual assessment of $100 was to be paid by the holder.   Now, in addition to this, a royalty of 10 per cent. will be levied upon the output of all claims yielding $500 and under to each claim, and 20 per cent. upon each claim yielding over that amount.

"Among those posted the opinion is freely expressed that it will be impossible to so supervise the output of these thousands of individual claims as to collect royalty upon the exact yield. Another obstacle is the fact that the mines all lie within a comparatively short distance of the boundaries.   There is nothing to prevent the miner from carrying the bulk of his gold dust, on the quiet, down the river to the boundary line, and once in American territory he is out of the jurisdiction of the Canadian tax collector.

Reservation of Grounds.

" In addition to the royalty every alternate claim in all placer grounds is to be reserved as the property of the government. These government reserves are to be sold or worked by the government for the benefit of the revenue of the Dominion. This is considered a startling departure from all the traditions of placer mining the world over.

" Two customs officers will be dispatched to a point near Lake Tagish, where all goods sent in by the Taiya route (Chilkoot Pass) can be intercepted.   At this point also a strong mounted police post will be erected, and the strength of the Yukon police will be augumented by an additional detachment of eighty men.   Small police posts will be established about fifty miles apart up to Fort Selkirk.   These will serve as stations for the dog trains carrying mails, and also for the relief of such travelers as may make the journey overland during the winter.

" There will be established a regular monthly mail service between Taiya and Fort Selkirk. The government has also determined to test the feasibility of connecting Dawson City with Taiya by means of a telegraph line. Should it be found impracticable to construct an ordinary overhead system a species of land cable may be employed to convey the wire laid on the surface.

" In the meantime the survey for a route overland from Taiya will be pushed, and upon the surveyors' report will depend the carrying out of the proposal of constructing a wagon road through the country at least to the head of uninterrupted navigation on the Yukon River. Diplomatic communication will be entered into with the United States authorities for the purpose of establishing a *modus vivendi* so as to give the Canadian Government the right of way through the country."

The miners summarily condemned the action of the Dominion Cabinet and rose up almost to a man against the payment of the tax. They denounced the step as rank robbery and declared that the Dominion officers would have a high time in collecting the monies levied.

Much indignation was aroused not less in the press than among the public, as the following newspaper comments show :

Press Is Indignant.

Bulletin : Canada cannot very well hold on to all the gold in the Klondike, but the Dominion Government will put a royalty on claims and gather in as large a share as possible. Let the Dominion statesmen go on if they think there is no such thing as manifest destiny.

Evening Report : The news about the imposition of a mining tax by the Canadian Government suggests that a war vessel be sent to Dawson City without loss of time.

Chronicle : The Dominion Government has thrown fairness and caution to the winds and gone to the unexpected length of imposing a royalty on all placer diggings on the Yukon, besides a $15 registration fee and $100 annual assessment. The royalty named is 10 per cent. on claims with an output of $500 or less monthly, and 20 per cent. on every claim yielding above that amount. Additionally, the government will reserve every alternate claim in any new gold district that may be found, and will impose a heavy tariff upon all goods coming in from the American side.

With the latter proviso we do not, of course, find fault, but the proceeding as a whole shows an intent to keep American miners out of the field in which they were pioneers and where they have uncovered the richest finds.

The Canadian government, however, apparently meant business, and it proceeded to cloister the tax it had imposed with a certain amount of terror in the way of penalties. According to the amended regulations issued, any miner who defrauds the government will be made liable to the confiscation of his claim and the withdrawal of his right to have any holding in the future. The penalty for the trespassing clause reads as follows :

Penalties are Imposed.

" Entry shall only be granted for alternate claims, the other alternate claims being reserved for the crown, to be disposed of at public auction or in such manner as may be decided by the Minister of the Interior. The penalty for trespassing upon a claim reserved for the crown shall be the immediate cancellation by the gold commissioner of any entry the trespasser may have obtained for a mining claim, and the refusal of the acceptance of

any application which the trespasser may at any time make for a
claim.   In addition to such penalty the mounted police, upon

LYNCH LAW IN KLONDIKE.

requisition from the gold commissioner, shall take necessary steps
to eject the trespasser."

A scheme was likewise devised by the Canadians to prevent or
limit the flow of gold to this country.   This move also met the
bitterest opposition, from the fact that a large percentage of the
miners in the Klondike district were Americans who went there,

braving perils and hardships, on a mere chance of making fortunes, and who resented being taxed for the privilege in the first
place, and, in the second place, having restrictions placed upon
them as to the disposition of their finds.

The scheme was devised by Captain Strickland. Following
is a report of his plan :

" Captain Strickland said the plan which he has already suggested, and which the Dominion government was inclined to favor,
provided they had a large enough police force to be assured of carrying it out, was to pass a law prohibiting the export of gold except
by Dominion officials. The gold dust brought in by the miners
of all nationalities would be carefully weighed by officials of the
Canadian government. A fixed value would be placed on the
metal, according to assayers' estimates, and this value would be
paid in money of only local value."

Klondike a Free Country.

In official circles in the United States the manifestos of Canada
were deemed " amusing literature." Said one of the leading
officials of the State Department at the time :

" The gold fields are free to all. Of course it is possible for
Great Britain to pass an alien law which would keep citizens of
the United States out of the new gold fields, but the result would
be that it would keep their own people out as well, for, while it
is true that the fields already explored seem to be on Canadian
territory, they cannot be reached at all except by passing through
the American territory of Alaska. It is well nigh impossible to
make the journey overland from British Columbia to Forty-Mile
Creek or any of the headwaters of the Yukon. It is necessary
to go through Alaska to get to the gold fields, and the gold
which is taken from there must go through Alaska to get to
civilization.

27

"The Canadians have been talking of establishing custom houses to levy some kind of a toll on the importation of supplies. There has been no talk of any prohibition of mining by American citizens, for if that were done all we would have to do would be to prevent the transit of Canadian miners across our territory, and thereupon the gold fields would have to be abandoned.

"Up to the present time no mortal man can say exactly where the boundary line between the American and the British posessions runs. The meridian fixed by the treaty has not been determined astronomically. The preliminary surveys show that the new gold fields are on Canadian soil, but the margin is so slight that neither government would care to assert authority where there is nothing to be gained by it. The miners themselves have established a local government, as is the case in all mining fields, but when the proper time comes the British Government, which is the best equipped in the world for looking after far-away dependencies, will take care of its own. American miners can go there without fear of interference on the part of Canada, but the information in our possession goes to show that many of those who do go will never return, for a famine in the Yukon country during the long winter season seems to be almost inevitable."

John Sherman Talks.

In the matter of an alien law, Secretary of State, John Sherman made the following statement:

"We have an alien law of our own. We have never enforced it against gold miners. Canadian citizens have been free to come into the United States and mine for gold under the same terms that our own citizens did. There has never been any friction over the matter.

"Where a man has taken up a land claim for the purpose of residence and cultivation we have always insisted that he be a citizen.   The same has been done under the Canadian Government.

"Where a man has simply prospected for gold with the intention of digging into the ground a little ways and taking what he could find from land against which there was already no claim, he has never been interferred with on our side of the boundary.   I do not think that the Canadian Government will change that course of procedure.   If they do it may lead to fully as much embarrassment to them as to our miners.

Through Clinched Teeth.

Canadians, however, continued to talk through clinched teeth, and, on an intimation being made that the United States would look out for the interests of its citizens, spoke with satisfaction of the policy of backing up the Dominion's claims with guns.

"It is hardly necessary," says the Toronto *World*, "to reply to the threats of Americans in the matter.   The government of Canada has already made its reply, and that reply is based on action, not on words.   A large force of mounted police and two Maxim guns are now on the way to the Klondike country, and if the miners whom the United States journals are inciting to revolt only make the attempt, they will perhaps meet with a reception warmer than they anticipated.

"Surely it is time that the people of this country, and especially the party in power, began to consider the relations of Canada with the United States from an entirely new standpoint. Hitherto the Liberal party has regarded this people as a friendly neighbor, from whom Canadians might expect fair treatment, at the least, while our habit has been to yield to them over much, and rather to supplicate such treatment from them than demand it as of right."

The United States government meant to stand by its word and

protect its people, though.   There was a call for troops, and on July 26, 1897, the following telegram was sent:

" Washington, D. C., July 26, 1897.—Shafter, Commanding Department of California : Can you spare a full company of infantry for the establishment of a post at Circle City this season for the protection of American interests ?   Men may be selected for duty from various commands.   Answer immediately.

" ALGER, Secretary."

General Shafter answered in the affirmative, and as a result of orders Captain Patrick Henry Ray, Eighth United States Infantry, stationed at Cheyenne, Wyo., was instructed to take a detachment of troops to the Yukon district.   The troops sailed from Seattle—six officers and fifty-six men—on August 5th, by way of St. Michael's for Circle City, and the thousands who were on their way or who intended to go to the gold fields had the assurance that they and their interests would be protected.

Limited Size of Claims.

Early in August of 1897, too, the Canadian government took a new tack in the matter of mining regulations by restricting the size of claims that would be allowed.   Instead of allowing 500 feet, as the regular law provided, the Dominion decided that it would fix the limit at 100 feet.   This decision was made on August 9th, to go into effect immediately.   This was designed to revolutionize the old plan of operations, which is thus described by Thomas Cook, an old miner who spent years in the region :

" In Canada the placer mines are, as a matter of course, close to the water and every man when he makes his prospect is allowed to stake off about what he considers 500 feet on each side of the place up and down the river.   That gives him the width of his claim 1000 feet, and this width extends from the river back to the foot of the mountain, whether it is a cañon or a plain.

" Then he puts up his stake and the government surveyor comes along and sets off the 500 feet each way exactly. Every man must pay a license of $15 a year and he must put in three months' work on the claim during the year. If the work is not done, there are plenty of men ready to report him and take the claim.

"Americans like the Canadian laws better than the laws of the United States, because they know their claims are better protected, and there is no claim-jumping so long as a man abides by the laws. The government follows up the miners by building roads. I don't want to say anything against our own laws, for I am an American, but it is a fact that we get better protection and the government takes more interest in helping the miners along in Canada."

The new mining enactment passed by the Dominion expressly forbids the " grub-staking " of prospectors or prospecting by proxy. In the future if any man wants a lawful share of the riches of the Klondike region he must work with pick, shovel and gold pan.

Slap at the United States.

The law, it was said, is clearly a slap at the United States. It is intended to restrict the immigration of American miners. By the provisions of the act it is unlawful for any person or corporation to prepay transportation " or in any way assist or encourage the importation or immigration of any foreigner or alien into Canada."

All such contracts are declared void and unlawful, and the penalty attached is $1000 for each and every offense, and all parties to the contract are individually liable.

The " exemptions " from the act include nearly all classes of labor except mining and prospecting. Informers are to receive 50 per cent. of the penalties collected.

CHAPTER XV.

Gold Crazes of Other Days.

Mining Excitements in Other Countries—Australia and South Africa lay the Old World under Tribute—Outbreaks of the Fever in America—Early Case in North Carolina—Stampede of '49—" Pike's Peak or Bust "—Recollections of the Argonauts—The Rocky Belle Camp Craze—Rush to Stevens' Claim—Excitement About Tombstone—Placers in Baja, California—Harqua Hala Diggings—Randsburg and Its Boom—Comparisons with Klondike—What the Early Stampedes Cost in Cash and Life.

FROM the far-away days of the Scriptural land of Havilah, the world has been subject to going crazy over discoveries of gold. A large part of history is a record of events for which gold has been more or less directly responsible. Most of the wars of invasion have been waged to gain gold, or its equivalent in transmutable form. Gold lured the Spaniards to the Antilles and the Englishman to Virginia. Lust for gold cost the Aztecs an empire and enslaved the Incas. Gold hunters gave Australia and New Zealand and South Africa to civilization. Gold has never had but one rival as a civilizer—religion—and, to produce a stampede, not even plague or famine ever equalled it.

Though Australia and South Africa had some gold excitements which laid the Old World well under tribute for the bravest and sturdiest, as well as the greediest of its population, America, and especially the United States, has had more gold fevers and had them harder than any other region on the globe. There was as much of a craze as the new country could stand, probably, when gold was discovered in the Carolinas, when the nation was a youngster, and there were some other relatively minor outbreaks of the auriferous malady in other sections early in the century; but it was not until the war with Mexico had given both the opportunity and the hardy men to take advantage of it, by stimu

422

lating the spirit of Western exploration, that America began in real earnest to show of what it was capable when the gold fever "struck in."

California, Pike's Peak, Washoe, Salmon River, Frazer River, Montana, Black Hills, Leadville, Tombstone, Kootenai, Cariboo, Randsburg, Alaska—every one a stampede.   Gold has made no other history like it.   Monte Cristo was a poor fellow in comparison with the heroes of those stampedes ; Ophir and Golconda were poor "streaks" beside the treasure houses in the mountains of those days ; and Mungo Park and Rider Haggard prosy tellers of true stories, beside the masters of golden fiction, that America produced or imported during the latter half of the nineteenth century.

When the gold fields of California were discovered in the " days of '49," the eastern half of the continent began to depopulate itself at a rate which brought a new State into the Union in three years.   The news of Major Sutter's wonderful strike in the Sacramento sands crossed the ocean and European adventurers joined in the rush to the Pacific slope.

Perils of '49.

Yet it was no child's pastime, that journey to the golden valleys of the Sierras nearly fifty years ago.   Two thousand miles of wilderness, partly a desert of perils, partly stern mountain chains, bleak and impassable, had to be traversed and almost every foot of the way was beset by blood-thirsty Indians or marauding white renegades.   Or else the argonaut risked the hazards of the sea and either crossed the Isthmus of Panama and dared its deadly fever, which too often undermined his health for all time, or spent six months or a year in the monotonous voyage " around the Horn."   Anyway he went, it cost time and money unstinted to reach the land of gold.   And when they got there they were out of the world. Everybody else was across the mountains or the sea, mails were few, expensive and uncertain, and it sometimes cost the total proceeds of a day's hard work in the placer and took a year's time to get a letter to the old home "in the States" and an answer from the dear ones back again.

"This Alaska is a regular parlor game to what we had to undergo in '49 and the early '50's," was how President Addison Ballard, of the Forty-niners Association in Chicago put it. "Cold! why we had to cross mountain tops that were covered with ice and snow as cold as any ever produced in Alaska. We had not only that to contend with, but also the blazing heat of the tropics, the thousand and one dangers and trials of the plains, the sufferings and privations of the most barren and sterile and forbidding deserts ever crossed by man. Savage beasts and still more savage men besetting every mile of our way and that way was a trail across trackless plains through a country undeveloped, unopened and unknown.

Only Locomotive a Mule.

"All of this had to be contended against at a time when the resources of civilization were comparatively primitive. We had no railroads then, our only train was the prairie schooner, our only locomotive a mule team or a span of oxen. We had no tinned meats, condensed milks or preserved fruits in those days ; we had to do with the roughest food, sometimes furnished by our rifles, and oftentimes that in scanty quantities. Then there was the sickening, saddening oppressive sensation of being cut off from the rest of the world and the possibility of never being again brought in touch with home and friends and civilization."

George W. Custer, Auditor of the Board of Education, Chicago, another '49er, who went overland in 1850, remembered the hardships well enough to shudder as he talked of them.   He said :

"It was the fourth day of April, 1850, that my father made up his mind to go to the California gold fields, and started with his family across the country to where we were told men could dig up nuggets with their heels right out of the soft surface mold all over the peninsula of California.   I shall never forget our experiences on that trip.   Hundreds of people started out without sufficient money or provisions, and as a result they perished of hunger and thirst on the great American desert of the Salt Lake district, through which their path lay.

Fourth of July in the Desert.

"Our family formed a portion of the caravan known as the Patterson Rangers.   It was composed of twelve wagons, forty-seven men and a boy (myself).   We ate dinner on the Fourth of July, 1850, right in the heart of the desert, and on that evening we practically ran out of provisions.   It was the poorest Fourth of July dinner I ever remember to have eaten.   I remember it well.   We each had a small piece of smoked meat and a biscuit.   My father, who had smuggled a small jar of sweet jelly with him, smeared a little of it over my dry biscuit in honor of the occasion.

"Our trail was littered with the remains of other caravans of pioneers who had preceded us across the deadly waste.   The skeletons of men and animals dotted both sides of the trail, and wagon wheels, old arms, rusty swords, broken rifles and other relics of the victims of that terrible summer were lying around in profusion.   The value of the material that lay there decaying on the desert would, I believe, if fairly computed, run up into the hundreds of thousands of dollars."

And these were not even fair samples of the experiences of

hardship and peril of the California argonauts. Yet the craze lasted and men by the thousand kept rushing West by land and sea to the placers of the Pacific slope.

Then the Australian gold fever came on in 1851 and 1852, and right on top of that the Colorado discoveries—" Pike's Peak or Bust "—and it seemed for a time as if all the civilized world that was not already at the mines was pushing and crowding to get there. Stories of disappointments and disasters to those who had " gone in " did not deter those who were going ; it was according to the ethics of gold hunting that bad luck was individual and good luck only was " catching." And so they rushed in, and where one " struck it rich " nine " went broke." The world had seen nothing like it since the Crusades.

The Rocky Belle Craze.

Arizona supplied some good samples of the gold fever in the seventies. Probably the wildest and craziest stampede ever known in the Southwest was that to the Rocky Belle Camp in Northern Arizona, in the region of the Moqui Indian reservation, in December, 1874. The region is 8000 feet above the sea level and lies among snow-clad mountains. It was an unusually cold winter when the news went abroad that Hank Binford and his companion had struck a whole mountain of gold rock that assayed over $900 to the ton.

A week more and over 2000 miners from every part of Arizona and Southern California were moving day and night, scarcely stopping for food and sleep, toward the Rocky Belle Camp. Hundreds of men traveled 700 and 800 miles on foot and with mules and donkeys to the new diggings, and nearly all traveled across desert and mountain for a distance of 250 to 300 miles. As the multitude journeyed on, the report of the riches of Hank Binford's find grew until it seemed as if wagon loads of rich gold ore awaited the travelers. Merchants and professional men in Maricopa and Tucson, and that part of Southern Arizona became imbued with the spirit of the miners, and, turning their business over to others, joined in the movement on Rocky Belle.

The hardships that the fortune seekers suffered in the mountains will never be fully known. A large number of men coming out of the warm, balmy air of the semi-tropic valleys lost their lives among the snowbanks and ice in the mountains, and many a man was made an invalid for life because of exposure to the biting cold during the stampede. A severe blizzard raged in the mountains for several days while the miners were slowly trudging through them. In one party of over 100 men from New Mexico, four men were frozen to death one morning, and it is thought that fully twenty more died in the same way in the mountains at that time. To this day there are in California and Arizona gray-haired miners who lack a finger, a toe, or an ear, lost in the terrible cold of that stampede.

When the last of the Rocky Belle diggings were reached it was soon seen that there was no ore in the district worth the digging except in the claims held by Hank Binford and his friends, and that the reports of their find had been exaggerated beyond all reason. Binford's own mine petered out a year or two later, and he got only a few thousand dollars from it.

Stevens Starts a Stampede.

Along in the summer of 1878 a miner named Stevens wrote to a friend in Phoenix that he had found a claim that beat anything in mining outside of the Comstock lode in Nevada, and that with a common iron mortar and pestle he had pounded out from $70 to $100 worth of gold dust a day. The claim was located 120 miles northeast from Kingman, near the since famous Harqua Hala mining region, and there was a chance, so Stevens wrote, for other men to strike it rich up there.

Of course, such news could not be kept quiet. It traveled with miraculous speed through every camp in the Salt River valley and over to Prescott. In less than two weeks all that part of Arizona was deeply stirred by the reports, which no one seemed to have time to investigate, of the richness of the mines that Stevens had found. A thousand or more miners caught the fever so badly that they started on foot across the country for Stevens' camp without delay. It was a hot, dry summer and the journey entailed several weeks of severe physical labor, torturing thirsts and the endurance of a temperature that usually stood over 110 degrees in the shade. A dozen men died from fever and in wild delirium under that awful sky, and as many more miners never recovered from disorders caused by the privations of that stampede across the desert of Arizona.

Having arrived at the Stevens' camp the excited men realized that there were claims worth working by about 100 men. Several hundred claims were staked out in less than a day after the excited miners got to the scene, but in a fortnight the camp population fell from 1200 to less than 400. In a month more about 100 persons were left to do all the mining. The camp was abandoned entirely ten years ago.

Mad Rush to Tombstone.

With the possible exception of the rush to the Leadville mining district in Colorado, there has been none anywhere in forty years attended with excitement that followed the news of the finding of great deposits of gold and silver in Tombstone in 1879. Miners from every part of the Pacific coast caught the fever for gold, and as week after week samples of the Tombstone rock were more widely circulated, and rumors went forth concerning the fortune this or that man or company was getting out of the hills and mountains about the new camp, thousands started for Tombstone.

Hundreds of young men and youths in the older States were wild with zeal to hasten to the new Eldorado and started across the continent with little or no preparation. In less than four months after Gird and the Hawkinses began getting several thousand dollars a day from their mines, there were over 6000 persons in the camp, and several months later Tombstone had a population of over 10,000 men and 200 women. There never was another camp in the Southwest like that at Tombstone in 1879 and 1880. Indeed, there have been very few similar communities in the world.

Wealth and Death Indiscriminately.

For over seven months the daily output of precious metal averaged about $50,000. Over a dozen men went there penniless and came away worth over $500,000 in less than a year, and six or seven men struck it rich and sold out for over $1,000,000 each. Fully half the population walked hundreds of miles to get there. No railroad ran through Southern Arizona in those days, and the awful Colorado and Mojave desserts had to be crossed in wagons or on foot by the multitudes of fortune seekers from California. Desert sandstorms were encountered and for days travelers to Tombstone endured a temperature of over 130 degrees in the shade. Many a man died on the hot, sandy plains. Miners on their way to the new camp from the East and South toiled across the Arizona alkali plains through immense cactus areas, and risked their lives in the then hostile land of the Apache Indians. But hardship, pain, suffering and risk of life were all secondary to an early arrival in Tombstone and the location of a mining claim.

When Tombstone was reached there were new privations and more physical distress, for the greater number, especially for those who had hastened from offices, stores, shops, clerkships and the pastor's study.   Over one-third of the men in camp had very little money, or none at all, and knew no way of earning it except by the hardest kind of manual labor, to which they were unused.   It cost $1 a night to sleep in a dirty, rough pine bunk. Water sold at 20 cents a gallon, a small dish of beans at 50 cents, tallow candles at 2 bits (25 cents), common overalls at $5 each, smoked hams at $12 each, and cowhide boots were disposed of as fast as they could be hauled to camp across the desert from Los Angeles and Yuma for $35 a pair.   It was a ground-hog case with these commodities for the first ten months of Tombstone—take them at the price asked or go without.

Placer Mines in Lower California.

In the last ten years there have been four or five stampedes to mining camps in the Southwest.   In the middle of the winter of 1890 California, as far north as San Francisco and Arizona, as far east as Prescott and Phoenix, were stirred up as they had not been for several years by the news that rich placer mines had been found by Mexicans in Lower California, seventy miles south of San Diego.   That was one of the most spontaneous stampedes known in that region.

Samples of the pay dirt were sent to San Diego to be analyzed one Sunday afternoon.   The assayer found it would run over $400 to the ton.   Somehow the secret got out and was telegraphed up the Pacific coast.   The telegraph operators in San Francisco spent the next two days and nights in sending and receiving messages about the new diggings.   Before Thursday morning 6000 to 7000 men and youths were on their way by cars, wagons, horses, coasting vessels and foot, to San Diego and Lower California.   The hardware stores in Los Angeles and San Diego, and in every village for 100 miles around, sold every pickax,

shovel, tin dripping pan, wash dish and milk and bread pan they
had on hand to persons who equipped themselves for placer
mining and started in a day for the mines.

The boom had a short life and almost died a-borning. For a
week little was talked of in the California cities but the discovery
of gold in Baja, California, and the prospect of another edition
of the days of '49. Then, when the first victims of the fever
who had been down to the mines returned to San Diego, declar-
ing the stories of wealth there to be lies, and the excitement
only a manufactured imitation of the genuine article, the old
miners who had not time to get out shook their heads at the
other fellows and said, " I told you so."

Harqua Hala Diggings.

Thousands of people will never forget the rush for the Harqua
Hala diggings in the spring of 1892. The mines were found
in the Northwestern part of Arizona, close to the Colorado
River and the boundary lines between Arizona, California and
Nevada. For several months in the winter of 1891–92 there
came almost every week news of the big prospects that a half
dozen miners, who had been moving from one camp to another
in the territories, and in Mexico, for nearly a generation, had at
last come across at Harqua Hala. Along in March and April
quantities of gold dust and nuggets from the mines came into
the hands of bankers in San Bernardino and Los Angeles.

Newspapers published reports as to the prospects at Harqua
Hala, and in a week or two there was another general rush for
the diggings. The railroads did a land office business for several
weeks in carrying men as far as the Colorado River. From there
the travelers to Harqua Hala packed themselves on little river
steamboats at exorbitant rates of travel. Hundreds of miners
who had hardly a dollar tramped over the mountains 150 and

200 miles to the mines.   And then they all tramped back again, wiser and poorer.

And then there was Randsburg—that little cluster of claims and grog shops that sprang into existence in the heart of a California desert on the strength of bags of specimens flashed by a few highly-imaginative prospectors.   There is no denying there is gold and a good deal of it in the vicinity of Randsburg—but it is a good plan to stop the denying right there.

In a general way Randsburg was a forerunner of the Klondike affair.   As soon as the newspapers gave up their columns and pages to stories and illustrations, everyone who could make or scrape together the necessary sum to reach the mines got a prospector's outfit and marched for Randsburg.   Some stayed there and some came back to civilization to tell of what they didn't earn.   Those who stayed, as a rule, went to work for the syndicates that practically control the claims.   If anyone is making money out of these diggings, it is the syndicate in charge.   So far as the lone prospector is concerned, he is a dead one.   He may pan out enough to keep body and soul together and lend strength to his thirst for conquest, but there he stops.

Randsburg and Klondike Contrasted.

Something else there is about Randsburg that may have a bearing on the Alaskan fever.   It is regarding the personality of the army of prospectors.   Frequently the characteristics of a few daring individual spirits lend a color to an entire community.   The news of the Randsburg Eldorado had hardly been taken from the ticker when the gambling element, which had been browsing about the State in an aimless sort of fashion, determined to introduce the illusive, yet seductive, pea, monte, the wheel and any number of other devices for the purpose of separating the curious from their good money.   In addition to all this, there

was a flourishing dance hall, roof garden, and all-around vaude-
ville show, so dear to the early novels of Bret Harte.   The few
cents the sydicate didn't get away from the pick-and-shovel brigade floated into the pockets of the " sure-thing " men before pay
day entered on its second childhood.

Randsburg and Klondike tales and events have much in common.   The stories of the rivers, hills and valleys of gold have
already been told and set the blood of the imaginative tenderfoot
boiling.   Horses, lots and even personal effects have been disposed of on all sides at a great sacrifice for the purpose of obtaining the wherewithal to reach the Eldorado of the pole.   Some
have already started on their perilous journey ; others are about
to hurl themselves into the Klondike maelstrom, and yet a third
class are still looking about them in search of an opportunity to
join hands with their brethren and one or two of the sisters who
are braving the tortures of a polar winter in the mad hunt for
fortune.

Some Tough Characters.

A number of the dispatches and all the statements issued by
the Alaska Commercial Company and other transportation concerns have been to the effect that the people going north are all
good, square, honest and upright miners.   They have said nothing about the sure-thing men, the army of thugs, ex-prize
fighters and general disturbers who are turning heaven and earth
to reach the Klondike.   Yet such is the actual state of affairs.
There is a brigade of muscular young men who have drifted to
the coast since the revival of pugilism.   These fellows would
sooner go to jail than work at any honest trade or occupation.
It is almost an impossibility to get them to train for a go in their
dearly beloved prize ring.   Yet there is nothing they won't do,
outside of work, to get hold of a piece of money.   The sandbag
is their favorite method.

28

These gentlemen are all going to Alaska. On the surface they are sincere in their claims that they are going to work. And they will work if they can find the gold lying on the bank of a babbling brook. The actual state of affairs in a nutshell is this: These fellows will take any chance under the sun to get money. They will stop at nothing. A man's life is no more to them than a snowflake to a storm. If things come their way they will, within certain limits, conduct themselves in accordance with the law, but as soon as they see they are "up against it" they will cast reserve and all scruples to the winds and begin tearing things wide open.

Honesty Versus Starving Idleness.

Just think of the number of people who have already gone and those who are determined to go to the Klondike who have nothing more than the mere price of getting there! Now you can take it for granted that these men are, as a rule, good, honest fellows, willing to do a fair day's work for a fair day's pay and take a chance of striking it rich on the side. But it is not every man who can remain good, honest and square under certain conditions. These men have gone and are going to Alaska under a delusion. They imagine they are going to get $15 a day whether school keeps or not. Naturally $15 a day looks like a great deal of money to men who have been making $2 and $3 a day. And so it is. But you do not actually get the $15, or anything like a tenth part of it, for an average day's work under the most favorable conditions in the Klondike. Of course you may be handed the $15, or its equivalent, on the completion of a day's labor, but how about the expense of living? If you get $15 a day for your work you may rest assured that $14 of it will go for board and lodging, and as a rule you board and lodge with the man or syndicate for whom you work.

Then there is another important factor to be taken into consideration which has been given the general overlook in the newspapers. A man does not and, in fact, cannot, even under the most favorable conditions, work the entire year round. There are months when you are compelled to remain indoors, rolled in skins if you are fortunate enough to have them, with nothing to eat but a bit of dried bacon, providing you are sufficiently wealthy to be able to afford this luxury. So you see a man who is not his own boss runs an excellent chance of working a season and winding up the year by being over head and heels in debt to his employer.

The gold stories from Alaska are by no means new. Some years ago there was a general exodus to the Yukon. The small army who went northward at that time have not yet returned laden down with yellow metal. A few fortunate ones have come back with a fair return for their labor and a library of romance that puts the professional writer of fiction to the blush. But what has become of the 2000 or 3000 who went up at the same time and practically have not been heard of since? How about those private graveyards in the ice fields and the unfortunates who will never return to tell the tale of hardship and suffering that accompanies an Alaskan winter?

In a way, the fever of '49 has a bearing on the fever of '97. The pioneer days of California form a basis of comparisons and enable those who will to draw conclusions.

Has Faith in Prospectors.

Hear the Argonaut Auditor Custer again:

" These Alaska prospectors are doing better than the '49ers did. I notice that those who have gone to the front are telling the truth and not sending back exaggerated reports, or painting the roseate pictures that the first of the California pioneers made

in the first flush of the western gold find.   It was the false reports made by some of the early California gold hunters that led
so many people unprepared into the western wilds, and filled the
great plains with the bones of unfortunate immigrants.   The
people are now being warned of the hardships and privations
which await them in their quest for fortune, and of the means
with which they must be provided to overcome them.

 " Our party made no money in California, and came back in a
year.   Two thousand others did the same. Of course, thousands
made their pile, though tens of thousands were disappointed.
But that came from expecting too much.   I don't think that will
be the case with the Alaska gold campaign.   The boys who
have gone out first are apparently moderate in their statements,
and I believe it will prove a great place for hardy and adventurous men to seek fortune and find it.   The California gold fever
did much to open up and build up this country, and I believe the
Alaska gold fields will also be a great benefit to this country and
its people."

" Go to Alaska, Young Man."

President Addison Ballard felt like Mr. Custer, only more so.
 " This Alaska gold discovery is great," he said.   " I don't believe there is any great exaggeration in the stories told.   I am
not surprised at all at them, for I have always held that along
that vein of territory clear to the North Pole the earth is full of the
precious yellow metal, and not only of gold, but of silver, copper
and other metals of value.   Why, if I was a young man to-day,
I would be off to Alaska just as quick as I could get my kit together.   I wouldn't stay around this town one minute longer
than it would take me to get my tools and other necessaries in
shape for transportation."

Mr. Ballard's hair is white as befits a man who went " overland " forty-eight years ago, but his eye sparkled with the argo-

naut spirit, and he looked like a second Jason setting out for the fleece as he spoke.

"I'll tell you that the man who loafs around here in Chicago out of work, flat broke or toiling for starvation wages these days is a pesky fool," he continued. "Of course, I would not advise men in very poor circumstances and with large families to take care of, to rush off there unprovided and expect to pick the gold up in handfuls right off the face of the earth. We didn't pick it up in nuggets out of the dust at our feet in California. You don't get gold anywhere without you work for it, and the gold hunters of Alaska, as well as those of California, will have to dig for it if they are to get it. The men who go up there in those regions after wealth and fortune could not do better than to bear in mind the little ditty so often sung by the California gold miner:

> "They told us of the heaps of dust,
> And the lumps so mighty big;
> But they never said a single word
> How hard it was to dig.

Easy to Get There.

"Now, what is the case with this Alaska business? Why, they have the railroad trains to carry them right to the very foot hills where the precious metal lies concealed. They have a country thoroughly explored, the geography of it thoroughly understood and comparatively quick means of communication. I tell you the pioneer of Alaska will be a featherbed pioneer compared to the old forty-niner, when the history of both comes to be told. And yet, if it was all to be done over again, not all the dangers and discomforts of the 'overland route,' the horrors of the sea voyage and the 'weathering of the Horn,' the fever of the Panama, the hunger and thirst of the desert would deter me from starting once again.

"No; I wouldn't be deterred by any little hardships such as they are talking about in connection with this Alaska business, and while it can never confer the lasting benefits upon the country that the pioneers of California did, for it was the pioneer of the diggings who opened up the far West and brought State after State into the Union till it reached from ocean to ocean, the Alaska gold find will, in my opinion, be a good thing for the whole country and enrich great numbers of our citizens.

> " Ho for California,
>   That's the land for me ;
> Away to Sacramento,
>   With my washbowl on my knee."

Fruit Belt Versus Arctic.

Yet it may be well to remember that in the days of the rush to the gold fields of California, it was almost impossible to get the worst of a venture to that part of the Pacific coast. Starvation was almost out of the question, save in the northern and mountainous districts, and a comfortable bed could always be found on the hillside of the land of eternal summer. There were no huge ice and snow fields practically destitute of bird and beast. On the contrary, there were streams full of fish, anxious to be caught, and forests inhabited by flocks of birds that have since acquired reputations for high prices in city eating houses. Again, the argonauts of California and Nevada were almost exclusively hard headed, painstaking and sober minded men, who were willing to brave hardships and privations providing they ultimately obtained independence for their pains.

There are a great many people woefully ignorant of the true condition of affairs in the Alaskan country. Even among the enthusiasts will be found few, if any, who are conversant with the subject in general, let alone in detail. The greater number of men who have already started for the Yukon, and the vast army who are ready to march forward at a moment's notice, know nothing about the actual condition of affairs. For them this book is published.

The Black Hills.

The rush to the Black Hills of Dakota differed from some others in that the primary placers gave place quickly to lode mining, and the perils from climate and human enemies were minimized from the start. Gold was discovered in 1874 and the great stampede to the diggings began to culminate in 1875. The auriferous land was on an Indian reservation, and United States soldiers protected the white trespassers and throttled the remonstrant redskins until the United States government made a forced purchase of the territory, and the miners thenceforward had things their own way.

The ores of the Black Hills are refractory and it required much capital to develop the mines. Mills began to spring up in 1876, and to-day the Homestake Company controls 580 stamps in this rich district. The total stamps running number 685.

CHAPTER XVI.

Side=Lights.

THOUGH there is a dark side to the Klondike craze, silhouetted in blasted hopes, physical misery, wrecked fortunes and even death, there is a humorous side as well, rather grim at times and often having the comedy, trenching perilously close on tragedy, but still pregnant with a realizing sense of the grotesque, and apt to jar a smile out of the most disagreeable situations. A siege of the gold fever offers unlimited opportunity for the display of idiosyncracies, and what passes for humor in new societies is most often only the discovery of unexpected traits in the hap-hazard assemblage. The experiences of a mining craze are prolific of the absurd and the ridiculous,—the craze itself has a humorous phase in that it is a craze, and the gay recklessness with which men chase golden phantoms is only the absurd antithesis to the faith in human gullibility with which schemers bait hooks for gumptionless suckers and play and land their foolish prey.

The Klondike craze, both in and out of the diggings, has run the gaunt of the jester's part. Sometimes in its brief duration it has been a question who were the crazier, those who rushed to the placers or those who stayed behind to laugh at the reckless argonauts. Some of the queer features of the '97 fever are

440

worth recording for the digest of human nature there is in them
—" What fools these mortals be ! "

A Kalamazoo man announced his intention of establishing a
balloon route to the Klondike. When the air was full of hor-
rowing stories of the awful perils of the passes and the " sure
death " which lurked in the maelstrom-like rapids and the bleak
and ice-locked marches of the river trail, he came to the rescue
with a rose-hued story of and air-ship he was building, which
would sail over anything, carry a ton of supplies and make the
trip to the gold fields and back in a fortnight. People wrote to
him from all over the nation to secure passage, offering ridiculously large sums for even a " berth in the steerage." One
Illinois man (perhaps forgetting for a moment he lived in the
sucker State), sent a draft for $500 for a round trip ticket. To
the credit of the air-ship navigator, be it said, he returned the
draft to the sender.

The balloonist announced at the outset that he could take
only two men besides himself and that the party intended to stay
in the Klondike only long enough to locate two or three million-
dollar claims and then scud home to the celery town to spend
the winter. Like Orpheus C. Kerr's famous machine-gun which
would have killed a thousand men a minute if the crank would
have turned, there was only one defect in the Kalamazoo air-
ship—it would not sail, and the great trans-continental air-line
was never opened.

Bicycles for Argonauts.

Some New Yorkers figured out a scheme for taking their
party into the Klondike on bicycles. Every detail of the
machines was thoughtfully considered and worked out. So
successful was it considered the " bike " route was sure to be,
that a syndicate was formed to manufacture the special wheels
for the market, and the promoters declared the day of Indian

packers, burros, dogs and reindeers was waned almost to sunset. The wheel was designed especially for use via the Chilkoot Pass, though it was likely to prove as useful by any other land route. The prospectus said:

" Every miner who goes to the gold fields must take with him about 1000 pounds of supplies, and the only way to transport them is for him to carry them on his back. The most that a man can carry for any distance is 200 pounds. The method now in vogue is to carry one load about five miles, hide it so that it will not be destroyed by animals, and then go back for another load. In this tedious way the goods are finally transported to their destination.

Style of the Wheel.

" The Klondike bicycle is specially designed to carry freight, and is in reality a four-wheeled vehicle and a bicycle combined. It is built very strongly and weighs about fifty pounds. The tires are of solid rubber one and a half inches in diameter. The frame is the ordinary diamond, of steel tubing, built, however, more for strength than appearance, and wound with rawhide, shrunk on, to enable the miners to handle it with comfort in low temperatures. From each side of the top bar two arms of steel project, each arm carrying a smaller wheel, about fourteen inches in diameter, which, when not in use, can be folded up inside the diamond frame.

" Devices for packing large quantities of material are attached to the handle bars and rear forks, and the machine, it is estimated, will carry 500 pounds.

" The plan is to load it with half the miner's equipment, drag it on four wheels ten miles or so. Then the rider will fold up the side wheels, ride it back as a bicycle, and bring on the rest of the load."

At last accounts no one had gone to Dawson City by bicycle.

The syndicate had overlooked the one thing besides a good wheel necessary to successful country riding—good roads. General Coxey had never been to Alaska.

A sledge and boat company exploited a sectional steel vessel, which was to serve the double purpose of water craft and land conveyance. Oars and sails would propel it in the water, while on land the argonauts would pull it along easily after a couple of plates at the sides were let down so as to form a flat surface under the keel. It was to be fitted with air chambers and burglar-proof compartments for storing the precious gold dust. This transportation scheme, needless to say, fell flat.

Stock in Dogs.

Hearing there was a scarcity of dogs in Alaska, a kennel owner tried to organize a stock company to furnish a supply of canine draft animals. The fact that such dogs as could be furnished from the States would be valueless in Alaska, for sledge drawing did not worry the brainy fancier at all, if, indeed, he ever thought of it. But others thought of it, and the company was never formed.

The North American Transportation and Trading Company offered miners a way of getting into the Klondike, which beat the balloon and "bike" and other easy modes of transportation, though there was an arduous side to it which kept many from taking advantage. The company needed wood in readiness for its Yukon steamers, as soon as ice goes out in the spring and navigation opens ; and it proposed to pay each passenger whom it transported as far as Hamilton's Landing, four dollars a cord for chopping wood during the eight winter months, the scene of activity to be between the Landing and Fort Yukon. It was estimated a good chopper could get up three cords of spruce or hemlock a day in the Alaska climate, which would enable the prospector to reach the Klondike with a comfortable "stake" in his pocket and his muscles seasoned for the hard labor of hunting for "pay dirt."

Grub-stakers proved one of the most ample crops of the craze. They sprung up everywhere, and all they wanted was an "angel." A grub-staker is a man who wants somebody to stake him with grub, and "grub" is Klondike for beans, bacon and tea. An "angel" is one who advances, loans, or in any manner puts money in the hands of the grub-staker. The grub-stakers were all willing to go to the Klondike and endure hardships and face death and locate a million-and-a-half dollar gold mine, if somebody would advance the money for the grub and the transportation. Then the "angel," when the mine was located, would reap the reward of his childlike trust and implicit faith, for, by mining law, the "angel" receives one-half of all the grub-staked one discovers.

Grub-stakers haunted railroad and steamship offices in the great centres and in the ports of the coast, and offered every man with money who could not go himself, a chance to go by proxy, and, astonishing as it may seem, many an "angel" let go of his savings to send to the diggings a man without credentials or residence, and whose very name was often suggestive of the probability that neither man nor money would ever be heard of again.

Schemes of Prospectors.

An Eastern argonaut, who was awaiting "steamer day" in Seattle, wrote home of his experience with grub-stakers in these words :

"Broken down prospectors, who have been unable to make a strike in the West, offer their services in trying to find gold for other people in Alaska. Few of them pretend to know anything about the Yukon country, but they are all sanguine of being able to go direct to the right spot and unearth a valuable placer deposit. The only requisite is clothes, food and money, especially the latter. Thus equipped these prospectors will go to the Klondike and send back at once half the gold they find. Odd tales are told about some of these fellows. If reports be true, some of the grub-stake money finds its way at once into the till of the nearest saloon, and the only prospecting done is that entailed in a hunt for new innocents.

"Men who have just come back from the gold fields, as they assert, offer bargains in the way of partnerships in claims. They proudly exhibit bottles of gold dust in proof of the rich strikes they have made, and then name prices which would be ridiculously cheap for bona fide properties of the kind described. It is pretty difficult to trace an Alaska claim at this distance from its location, and there is no satisfactory way of establishing its existence, dimensions, or worth. When the mining fever is on a man, however, he overlooks such minor things as these, and jumps in haste to close what he calls a good bargain. He doesn't stop to consider the risk he is running, and goes away to make room for another customer, who will buy the same claim right over again."

Clairvoyants on Deck.

Clairvoyants put in their bid to be recognized as factors in the Klondike development. Something in the nature of a grub-stake company was formed by a number of spiritualists in Chicago and an advance agent or prospector sent out to locate the rich claims which a well-known "medium" professed to be able to discern clairvoyantly across the vast intervening distance. Some of these claims were said by the "spirit guides" to be fabulously rich and all of them well worth the finding. Maps were drawn and explicit directions given and a new field for "prospecting" duly opened.

Anything with the name " Klondike " on it, especially if it was mining stock, was a pretty sure seller after August 1st. All that was necessary was that the price should be cheap and terms easy. Plenty of shrewd men took early advantage of this and some printing presses were kept working overtime getting out the prospectuses and certificates for these "mining companies." How many were "bitten" by these sharpers and how many hundreds of thousands of good money they absorbed will never be known, but it is certain that a very small percentage of those who invested in Alaska companies will ever see even the " first annual report " of the concern's announcing that they must be revivified by a ten per cent. assessment or shut up shop.

Magic in "Klondike."

The magic word " Klondike " seemed to be ample indorsement in the estimation of the general public for any kind of an Alaskan proposition, no matter how wild or ridiculous its scope. Railways running for hundreds of miles over wastes of ice and snow were minutely laid out on paper and their earning capacity soberly computed by men accredited with the possession of business ability. Electric light plants were advocated for Dawson City and similar mining towns. Development of the coal beds as fuel for great central depots for piping heat to the gulches to thaw the frozen gravel was seriously talked about. Had some gold lunatic proposed the sawing of the Alaskan ice into railroad ties or telegraph poles for use where timber was scarce, it would have caused no more than a ripple of surprise, to judge from the bare-brained schemes which really enlisted financial backing. Everything was possible in Alaska, according to the promoters.

One of the oddest things brought to light was an attempt to organize a barber's syndicate to invade the upper Yukon country. One winter's experience in the Arctic region satisfies nearly every man that it is safer and more comfortable to keep his face free from hair. Moisture from the breath freezes mustache and beard into cumbersome and dangerous chunks of ice in that cold climate, and in trying to remove them pieces of frozen flesh are liable to be torn off. Safety lies in clean-shaved faces. Many men cannot shave themselves and many of those who can, have no razors fit to use. The result is a demand for barbers. Knowledge of this led one sanguine young shaver to broach the idea of taking a party of brother workmen to the Klondike and there was considerable enthusiasm over the scheme.

An amateur detective set seriously about organizing a stock company to send himself and a corps of trained sleuths to the Klondike, where he believed there is a rich gold mine in arresting many criminals for whose capture large rewards are offered. He was morally certain Willie Tascott, and a lot of other badly wanted men were there masquerading as miners under the Arctic Circle. He regarded the scooping in of these men, and the prize money appertaining to them, as a vastly easier and more lucrative way of making a fortune than burning down to bed rock through eighteen feet of frozen gravel. But the police laughed at him.

Samples of Argonauts.

How little many would-be argonauts knew of the Klondike, or anything connected with it, was illustrated in a New York railroad ticket office. A well-dressed man pushed his way through the crowd, and throwing a big roll of bills on the counter, cried out:

"Give me a first-class, and a lower berth."

"Where to?"

"Klondike."

He was indignant when the ticket seller tried to explain that sleepers were not run regularly over Chilkoot Pass.

A man bought an "outfit" at a Seattle store, and found his bill was forty dollars over his funds.

"Never mind; I'll pay you at Dawson," he said to the cashier, and seemed dumbfounded when he learned the clerk was not going to the Klondike.

All sorts of men wanted to do all sorts of things in the diggings, beside dig for gold.

One man wanted to practice law at Dawson, or any other place on the Yukon, and wanted the agent's advise as to the size of library he had best take along. His feelings were hurt when he was told a hot milk route would probably pay better.

Another advertised for parties to form a company to send a stock of "ladies' and gents'" ready-made garments to the Alaskan gold fields. The venture may be a success if the supplies are limited to those for men, as "ladies" who are among the best people of the Alaskan wilds, show a preference for white bearskins and walrus oil overknit wear.

Women at the Camps.

Another season may change this, however, for there is a chance that women with white skins of their own will be much in evidence in the camps in 1898. Several promoters have already arranged to establish matrimonial agencies in the Klondike. One of them says:

"Thousands of poor but thoroughly respectable girls even in this State are looking for honest employment, and would go to Alaska to get it if they were assured they would be properly cared for. In the towns and villages of New England the number of women is so far in excess of the men and employment so hard to get that thousands would be willing to go to Alaska under proper conditions. I propose to secure places in advance for companies of, say, 100 girls, and have their employers advance money for their transportation from the States and recompense me for my trouble besides. No girls will be accepted except such as can bring the highest recommendations as to character and respectability. Arriving at the gold district each one will be assigned to her place, but all will be located within a short distance of each other, so that they may have association and be able to counsel each other.. Under their influence the camp would take on a homelike appearance, and the miners would not feel that sense of isolation which sends so many to their graves. They would be served with well-cooked food, and the general health of the camp would be vastly improved."

Charlotte Smith, the Eastern sociologist, wants to transplant 4000 or more working women from sweatshops and factories to Klondike camps. Hers is not a money-making scheme—she is laboring solely in what she thinks the best interests of humanity. Transportation from a life of drudgery, with a bare pittance in the way of wages, to homes in Alaska would, in Miss Smith's opinion, be a blessing which thousands of women would be glad to embrace.

Bogus Employment Bureaus.

Employment bureaus to engage miners to work in the Klondike made their appearance with the first signs of the craze. Several of them flourished in the coast cities for some time, and the proprietors accumulated quite a fund from gullible and impecunious victims of the fever before the police swooped down and arrested the sharpers.

Some of the miners coming back with a " pile " had as hard a time to keep their gold from the sharpers as the tenderfeet had to keep their greenbacks.

Shortly after the arrival of the last ship from Alaska at San

29

Francisco a number of the Yukoners had a reunion at a private hotel on California street hill. There was everything on hand to make the function pleasant, and the evening passed rapidly. Then there was an adjournment to a music hall on the edge of the "tenderloin," and there was more of the wine, women and song business. The Yukoners found that whisky at ten cents a glass was a more potent liquor than they had met even at Forty-Mile.

There was no limit to the orders, for the men were in for a good time. Some of them, with considerable foresight, placed their sacks in the safe of the saloon. When they did this they had more confidence in the integrity of the strong box than in their own capacity for liquids, but their confidence was misplaced, according to reports. One of the party, who was at one time a leader of the Yukon pioneers, deposited a sack containing $400 in the safe. When he called for it he found that some one else had broken into the safe and had taken one-half of the stuff that was in the receptacle. One man lost, according to his statement, $214, and his companion about $100. The party broke up about the time the cars began to run in the morning, and when the sacks were demanded there was a scene.

Accused of Robbery.

This was nothing, however, to what occurred the night following the orgie. Those who lost their money met in the refreshment room of the hotel in which they were staying, and each accused the other of being accessory to the robbery. Had it not been for the intervention of several policemen, called by the proprietor, there might have been several owners of rich claims lying on the slabs of the morgue the next morning.

Gamblers reaped a harvest in the coast cities as long as miners were returning with their dust. Gaming was the only pastime

at the diggings and it was easy for the card sharps to find and
fleece their victims among the home-coming argonauts. Play
had been relatively as high as fair on the Yukon and before the
pioneer discovered he was made a victim, he had generally been
well "plucked." The supply of these easily duped miners ran
out after a time, however, and then the professional gamblers
started for the fountain head at Dawson City. It speaks well
for the caliber of the '97ers that while many of the blackleg
fraternity undoubtedly got through the outposts, many more
were turned back on their journey to the mines with some short,
stern advice not to make another attempt to get in.

Type of a Miner's Paper.

One of the oddities of the craze was a little three column-
folio sheet purporting to be published at Dawson City, and which
gained much notoriety during its brief day of novelty. The
Klondike *Morning Times* may be taken with as many grains of
salt as the reader may see fit, but, as an antitype of frontier mining journalism, it is worthy of the days of Bret Harte.

The editor seems to have started the paper, because he
needed money. This may be inferred from the subscription
price, which is announced without any attempt at extenuation as
$7.50 a single copy or $350,000 a year, payment to be made in
nickels, nuggets or stamps. Some concessions are made for
club orders, the editor offering 1,000,000 copies for $30,000.
The subscriber is advised to read the paper quickly, or he'll not
believe all there's in it.

The sensation of the day was a disturbance in the Dirty Dog
saloon the night before. The editor at once grasped the news
value of the story, recognizing its " human interest " at a glance.
He played it under a " scare " head consisting of the expressive
monosyllable " Biff," followed by three-line pyramids and " cap "

lines in which the various features of the story were strikingly indexed.

The story in vernacular is as follows :

" There was a hot time in the old town last night, as the frequenters of the Dirty Dog saloon will testify.

" In the course of a quiet little poker game there was a clash between Bonanza Bill, formerly of Circle City, and a half-breed Indian known in the diggings as Chilkoot Charley.

" The stakes were large. Over $2,000,000 in nuggets glittered on the table when all played dropped out excepting Bill and Charley.

" Charley finally weakened and called his antagonist.

" Bonanza Bill proudly displayed a pair of fours.

" ' No good,' said Charley, as he began to rake in the shining pot, ' I've got sevens.'

" ' Stop!' roared Bonanza, and with a quick movement he seized the cards from Chilkoot Charley's hand.

" Charley had a pair of deuces only.

" Piqued at the idea of being played for a good thing by a half-breed Indian, Bonanza Bill lost his temper and, seizing a cast iron cuspidor, he brought it down upon Charley's head with great emphasis. Skull and spit-box were both wrecked by the force of the collision.

" The Indian was buried in a snowbank at the foot of Easy Street at 2.30 A. M."

The prospects of the Dawson City and Elsewhere Railroad are flatteringly exploited, and the enterprise and liberality of the editor are revealed in a voting contest for the most popular faro dealer, the winner to get a free trip to Juneau.

CHAPTER XVII.

Camp Life and Morals.

Mining Towns in the Alaskan Wilderness Similar to Other Rude Communities, with such Peculiarities as are Born of Climatic and Topographical Features—All Have Their Social Amenities—The Bible and Shakespeare Appeal to the Literary Tastes of the Fortune Seekers—Watching of Property Early a Necessity—Sharpers Lose no Time in Getting in Their Work—Gamblers also Flock Toward the Yukon to Intercept the Returning Miners and Fleece Them—Whiskey Trade Flourishes in the Wilds.

THE mining camps of the Yukon Valley resemble the mining camps of all other gold diggings the world over, with such minor differences as are born of the characteristics of the country. Their life is a rude life, a life of hardship, a life of temporary expedients, and yet a life that has a bright side for every dark side it presents. The Yukon valley is well worthy of a Bret Harte to recount its pretty romances, its heroism, its humble joys, its pathos and the strong traits of character it develops or brings to notice.

Situated as the camps are, thousands of miles from civilization, it would be strange did their life not present oddities and striking features of exceptional interest to new comers. There is the absence of conveniences usually to be found in such place; the same tendency to recklessness and improvidence; the same summary execution of unwritten law; and in fact everything that tends to make a mining camp not a town, but a sort of a halting place in the wilderness. There is a rough, wild, uneasy appearance to the whole company, a something that says, "We are here for a purpose, but we will get out of the diggings at no distant date."

Still the life of the mining camps on the Yukon is not as rude or as bad as might be supposed, partly from the fact that the remoteness of the diggings for a long time kept away dangerous and undesirable characters, and partly from the presence of mounted police, who did their best to preserve law and order.

Dawson City, Circle City, Forty-Mile, Sixty-Mile and all the older camps in the region for years after the mining of gold was begun, maintained an enviable reputation, and after the discovery of gold in the Klondike robbed the older camps of interest and brought about a general exodus of the miners to the new diggings the same characteristics were preserved. Hence, a word descriptive of one of the older camps may be taken as fairly true of all the camps in the region. Says a miner writing from Dawson City :

Is a Moral Town.

" It may be said with absolute truth that Dawson City is one of the most moral towns of its kind in the world. There is little or no quarreling, and no brawls of any kind, though there is considerable drinking and gambling. Every man carries a pistol if he wishes to, yet few do, and it is a rare occurrence when one is displayed.

" The principal sport with the mining men is found around the gambling table. There they gather after nightfall and play until late hours in the morning. They have some big games, too, it sometimes costing as much as $50 to draw a card. A game of $2000 as the stakes is an ordinary event. But with all that there has not been any decided trouble. If a man is fussy and quarrelsome he is quietly told to get out of the game, and that is the end of it.

" Many people have an idea that Dawson City is completely isolated, and can communicate with the outside world only once every twelve months. That is a mistake. Circle City, only a few miles away, has a mail once each month, and there we have our mail addressed. It is true the cost is pretty high—a dollar a letter and two dollars for paper—yet by that expenditure of money we are able to keep in direct communication with our friends on the outside.

"In the way of public institutions our camp is at present without any, but by the next season we will have a church, a music hall, schoolhouse and hospital. The last institution will be under the direct control of the Sisters of Mercy, who have already been stationed for a long time at Circle City and Forty-Mile Camp."

Have Their Social Amenities.

It will be seen from this description that, remote from civilization and virtually under the Arctic Circle as they are, the camps are not without their social ameneties. Many an interesting romance might be written from the experiences of those who went to the Territory to seek their fortunes.

Amusing details are given of the way in which the men spend the long nights of the Arctic winter. It must be remembered that this means the greatest part of the year. Each claim extends only 500 feet up and down the streams—the 500 feet was limited by the Dominion government early in August, 1897, to 100 feet—and the tents or cabins of the miners are thus huddled closely together.

The miners are thus neighbors in propinquity, and the good fellowship which usually obtains in such communities make them neighbors in every sense of the term. Along the Klondike and in all the older camps the men resort to all sorts of games to kill time, as they express it, and checkers and cards thus become favorite pastime with the masses.

Then, too, remote as they are from current news and recent publications, the men, in a sense, keep up their interest in the world from which they are severed, and every odd book or old newspaper about the diggings goes the rounds and is eagerly perused by everybody. It is rather interesting to note that in the mining circles the Bible and Shakespeare are the two books most frequently to be seen.

Nearly every Klondiker on leaving Tacoma or Seattle is said to provide himself with a copy of the Bible or a single-volume copy of Shakespeare.   About the middle of August it was reported by the booksellers of Tacoma that there had been such a demand for these two books that their supply had been entirely exhausted and that they had been obliged to send east by wire for a fresh supply to meet the wants of those who started late in the season for the diggings.

A single instance will serve to show the trend of taste in liter-ary matters.   One party of twelve prospectors and miners from Missouri left Tacoma on August 14th and took as part of their baggage eight copies of the Bible and twelve copies of Shakespeare.

Newspapers in Camp.

Robert Krook, an old miner in the Yukon valley, gives some interesting information relative to the popularity of newspapers and the general run of the camp life.   Said he :

" No paper is too old to read.   We read all the advertisements and all the can labels.   There was a supply of canned lobsters at the camp and some man used to put up with the cans wrap-pings of sheets from the Bible.   We used to commit the chapters to memory and see who could repeat them first without a mistake.

" The food is neither extra choice nor plentiful.   But it is ex-pensive.   Bacon, ham and beans are the general rule—no French wines or champagnes.   The supplies are short at best and a man must often take bacon that he would not throw to a dog or go without.   There is usually more whiskey and hard-ware on hand than anything else.   A man only needs a certain amount of hardware, and the less whiskey he can get on with the better he is off.

"Sometimes a man has to watch his supplies pretty close, and they usually build a 'cache'—that is, a little platform set high up on light poles. He can then haul up his bacon and 'grub' and cover it with a tarpaulin. The risk of leaving the 'grub' in the cabin is that the bears get at it. They will even tear the roof off to get in, and there are plenty of the animals. They won't climb the thin posts, particularly when the bark has been peeled off.

"In regard to clothing, a man does not need much in summer, and in winter he studies comfort, not looks. In winter we wear moccasins, and in summer, while sluicing, gum boots. I have not had leather on my feet since I left. Overalls cost $2.50 in Klondike, and everything else in proportion, but it is a great country to make money in."

Strict Discipline Among the Miners.

Mr. Krook rather insinuated on returning from a protracted residence in the valley, that the discovery of gold on the Klondike had rather tended to demoralize the people and give rise to more or less unlawful proceedings. He said, though, that the miners were quite competent to adjust all matters of difference, and that, as a rule, it was woe betide the man who transgressed the laws of the camp. Continuing he said :

"Until this spring the men never put locks on the doors of the cabins, and nothing was stolen. You might go into any cabin and see a glass or a tin or two on the shelf full of gold, and no one would think of touching it. Anyone could steal if he wanted to do so, but there were good reasons why they did not. It was only after the mounted police arrived that locks and bolts became a necessity. Before that there were what we called 'miners' laws.

"Forty or fifty of the miners would call a meeting, select a chairman, and then if a man could make his own 'talk,' he did so, or he would get some one to make it for him. When both sides of the case had been heard the chairman would call for a vote. The decision was final.

" If a man gave trouble, he had to go. Now, they do not have miners' laws any more. We had no trouble during three years, because all questions were settled at these meeting of miners. All disputes about claims were argued and adjudicated in the same way."

Sharpers at Their Work.

As in all mining districts, where great fortunes are apt to be made in a few days by a lucky hit, there was early on the Klondike an element among the people who were unwilling to obey either the statutes of the government or the unwritten laws of the miners, men who apparently worried their brains to devise schemes to get hold of claims, to evade rules and to gain possession of as large a part of the miners' earnings as they could. The miners, however, soon rose up against this element at the diggings, as they had previously at Dawson City and the older camps, and determined that, come what would, order should be preserved at all hazards.

They pointed out with pride that there had been a vast difference between the camp life on the Yukon and the camp life of the days of '49, the difference being in favor of the days of '97. They made a crusade, as strenuously as possible, against gambling and the sale of liquor. Of course, it could not be expected that drinking and gaming could be entirely prevented. But the miners, realizing their own best interests, did good work in limiting the evil.

The United States statutes distinctly prohibit the importation of liquor into Alaska for purposes other than for medicine, but the law was ignored by those who recognized that there was a glorious opportunity for money making in pushing the liquor traffic. Thousands of gallons of alcohol, whiskey and brandy were landed almost every week at Dyea and other towns, from which the stock was transported into the interior. A large share of these goods found way directly to the Klondike.

Whiskey Came High.

The worst kind of whiskey found ready sale to the Indians at three dollars per bottle, and in almost every bay or nook of land where Indians lived, were sloops from which whiskey was sold in abundance, alike to natives and white men. At Dyea and Skaguay, as well as at Juneau, Wrangel, Sitka, and other towns, many saloons were run wide open. By a curious contradiction the government issued internal revenue licenses, and at the same time prohibited the importation and sale of liquors.

A word may be said of the ordinary life of one or two of the older towns as being characteristic of the country outside of the mining camps proper. Sitka, the capital of the Territory, is a quaint old place that has never yet worn off the glamor of romance and mystery which has hung over it ever since the days of Russian occupation. During the whole of 1897, however, the pathos and tragedy of romance were entirely subordinated to the wild and feverish frenzy after wealth which marked the year after the find had been made on the Klondike, and the old town took on a briskness and life that it had never known before.

Of course, Sitka is only an apology for a city, but it does have many of the conveniences and comforts to be found in the older States. Hence, the prospector or miner going to the Klondike in a measure gets used gradually to the marked change from civilization to the wilds. Henry Ellsworth Haydon has a word to say about Sitka which is worth quoting in this connection. Says he:

" Let me tell of the town as it appeared to me the winter of my visit there, with the white Chilkat blanket of the snow spread over its shoulders and trailing its fringes in the sea.

" Fancy a bracket fastened to the front of the mountains with its outer edges washed by the estuary of the Pacific Ocean, and on the bracket a number of frame buildings of all sorts and sizes—perched like birds above high-water mark.   On its eastern side vast, towering, snow-crowned mountains rise mass on mass, precipice above precipice, until their summits seem like the white, tapering finger of a giant god, reaching upward to pluck diamond stars from the ether of the winter skies.

Exposed to Wind and Storm.

" Northward, low lying hills stretching in endless companionship toward the frozen ocean, and across their desolate solitudes the wild winds of storms born in the Arctic blow their cold breath out over the little city, as if they would fain freeze the inhabitants and carry their congealed bodies into the sea.   Westward, across Gastineau Channel, Douglas Island, with its famous Treadwell mine, and Douglas City, and southward lengthwise of the bay one sees the trembling waters undulate along an ocean horizon.

" Dwellers in cities beyond the eastern slope of the Rocky Mountains, who read much and travel little, have formed queer and mistaken ideas of the condition of society in places known as the mining camps of isolated districts."

Juneau, from which so many thousands took their way to the interior, is younger, sturdier and more enterprising than Sitka, and may serve as a sort of transition from the life of Southern Alaska to the bona fide camp life of the north.   It is one of the most cosmopolitan little places, or it was in 1897, under the sun.

Men winter at Juneau who have wandered through Australian

forests, prospected Montana, Idaho, Nevada, New Mexico and
California ; been tossed about in whaling and sealing vessels on
the billowy waters of the Arctic seas ; trailed through Asiatic
deserts, hunted for diamonds in Africa, and among all sorts and
conditions of people have learned the creed of the wise and
the brave, to accept the present as the only living time, and await
with unspoken faith and hope whatever the future may bring
them.

They are pleasant to talk with, affable, courteous, intelligent,
being full of strange stories of camp and field, of quartz mines
near lonely cabins far up the mountains, and placer " diggins "
in populous places near to the sea, and all the wonderful romances
which are part of the adventurer's lot in whatsoever land his tent
may be pitched.

Many of these transient pioneers of the primeval solitudes of
sea and forest stay at Juneau until the April or May days come,
when they set out to cross the divide and launch themselves in
frail canoes or on crazy rafts, and go floating down the mountain
streams to the Yukon River. For the most part, one and all
have the same purpose. Those of the prospectors, or fortune
seekers, who have spirit and energy enough to bear up under
the trials they have to meet, make comparatively jolly parties,
and as a result life goes on noisily along the trails and in the
camps as in the older and better known towns that serve as a
threshold to the country.

Have a Rude Awakening.

When once the camps are reached, the real business of the
pilgrimage to the north begins, and many thousands realize
shortly that life at the camps is an entirely different matter than
they had anticipated. The common experience soon settles
down to a round of duties and efforts ; and the absence of all

that the fortune-seekers have been accustomed to, emphasize the unpleasant features of the new life.

There were not lacking, however, early in the days of the gold-working enterprising people, who sought to make a good thing out of the gold craze, not by mining, but by catering to the pleasures of those who delved and washed for the precious metal, primitive theatres were started at many of the camps. Omer Maris speaks of one of these playhouses at Circle City and says that it met a positive want among the people. Says he:

"The present conception of the popular taste in Alaska seems to be that the public wants a strong show, and in the attempt to meet the demand the managers cannot find anything up to the standard in books and are driven to the point of inviting new features. 'The Man from Douglass Island' was an original drama that was offered to the people of Juneau.

Barkeeper Charley.

"The title had local significance, as Douglass Island is just across the channel from the town. It was a very successful play. The hero was a barkeeeeper named Charley, and the heroine, to use the hero's own words, was a 'perfect lady,' who had a desire to see something of the town with a fancy, rather unusual in a person of that description, for incidentally 'hitting the pipe.'

"There was a bootblack, a Chinaman, an Irish policeman, a dude and a number of sports and 'ladies' in the piece. After the requisite amount of adversity and bad luck had been ground out, the hero, with the help of the bootblack, triumphed over the dude, got a 'pull' with the policeman, married the heroine and otherwise attained brilliant success as the proprietor of the 'finest joint in the town,' to quote his own language again."

Those familiar with the scenes of revelry and riot in the days of the Californian gold fever would look in vain, however, along the Yukon and the Klondike for anything similar to the playhouses of '49.

In all the diggings there was, as might be expected, more or less lawlessness that could not be suppressed either by the government officials or by the better class of the miners themselves. It early became necessary to take positive steps for the protection of the miners and their claims. The Dominion Cabinet did much to preserve order and prevent anything of an especially flagrant kind.

A detachment of Canadian mounted police, twenty-five in number, was stationed at Fort Cudahy, opposite Forty-Mile post, and the owners of the mine there applied to Captain Constantine, in command, for assistance in protecting their property.

A detachment of twelve men was called out at once and they made the trip of seventy miles to the seat of the trouble in the shortest time on record. They placed their arms and rations in a canoe, put in two or three Indians with poles to guard against rocks, and then the twelve men took a line and towed the canoe the whole seventy miles.

It was expected that there would be trouble in dispossessing the claimants who caused the trouble, but the Yukon miners are a law-abiding lot generally, and at the display of authority they submitted and the owners of the mine were given possession. As to the original question involved it was soon settled, as the owners probably got their legal rights.

Dawson City sprang up like a mushroom and was one of the most thriving of the mining towns until the discovery of gold on the Klondike directed attention thither and caused a general stampede to the new diggings. Edgar A. Mizner gives us a little peep into the life of this town.

When he visited it Dawson had a population of about 4000. This was just before the Klondike fever broke out and the men hurried away as rapidly as their legs, or the river steamers, or horses or dogs and sledges could carry them. Says he of this camp:

"And such a town! It has some of the characteristics of mining camps that Bret Harte has made into story, but it has qualities that California camps never had and never could have. The game of life is played fast, and the boisterous side of mining camps is developing as the population increases. Now Dawson would match Tombstone when Tombstone was young. There are gamblers by the score, and there are dance halls by the score.

" The principal source of fighting in frontier mining camps, disputes over the possession of claims, has been missing up to this time from the Klondike region. The Canadian mining laws seem fair, and they are regarded and are enforced as well as possible by the small official force representing the Dominion government. A section in the law prohibits a miner from 'taking up' more than one claim in a neighborhood. This provision of law leads to caution in the selection of claims, and estops land grabbers from controlling all the claims in sight."

CHAPTER XVIII.

Domestic Life in the Wilds.

Miners' Experiences not those of a mere Romantic Sojourn in the Wilderness—Absence of Conveniences and Comforts—The Older Towns Antiquated and, during the Gold Craze, Overcrowded—Graphic Pictures of Skaguay, Dawson City, Circle City, and Camp Lake Linderman—Hotel Project for the Territory that Promises to be the Means of Furnishing a Larger Quota of Comforts—Women's Influence on the Domestic Life—Some of Those Who Grace the Camps with their Presence, and the Particular Line of Work to which they Devote Themselves—Sisters of Mercy for the Sick and Dying, and Sisters of Cookery for the Well.

THE domestic life of Alaska is not the domestic life of the old, settled communities of the United States, and the thousands who flocked to the North, when the Klondike fever broke out, had a rude awakening from their dream of a merely romantic sojourn in the wilderness. Nor did it require an actual residence in the mining camps to force upon the fortune seekers the fact that they were entering, not merely a new and unknown country, but a new and unknown series of domestic experiences.

Even the oldest of the Alaskan cities—Sitka—is but the veriest excuse for a town, despite the fact that its history and its fame date from the early days of the Russian occupation. Consequently, the moment the prospectors and miners set foot on Alaskan soil, they found a lack of the conveniences and comforts to which they had been accustomed. These did not exist in the city, and their absence was accentuated by the feverish rush and turmoil that characterized the place.

It was literally a new era in the history of Sitka, as well as of Juneau and the rest of the older towns of the territory. The gold craze came in a moment, and there was no opportunity to provide for the horde of people who wended their way toward the diggings as soon as the news was received in the cities of the South. Every available place in the old towns was filled with newcomers, and on the outskirts of the cities there were little suburbs of tents, which were pitched for the temporary accommodation of the people.

In the established mining camps like Dawson City, Circle City, Sixty-Mile, and Forty-Mile, the state of affairs was not essentially different for a time. Soon, however, there was a general exodus from these towns, and then there were accommodations, and to spare. In Klondike itself, as might be supposed, it was for a long time a mere matter of the rudest huts, supplemented by tents.

No Place for Style.

In this world of antiquated or temporary structures, or of no structures at all, the domestic arrangements were cast upon just such lines as one might look for in an unsettled country. As the reader may have gleaned from the preceding pages, it was no place for dress suits or train dresses, and those who went to the gold fields soon learned that it was no place likewise for the conventionalities of ordinary life. Here and there, to be sure, was found some one who essayed to put on style. But these "fops and frumps" were early taught that they had better cast conventionality to the winds, and adopt the rude life, with its hearty, whole-souled ways, which obtains in all mining localities.

A mere word about some of the towns will enable the reader to form some idea of the "home" life that necessarily prevailed in them. Hal Hoffman, who went to Alaska on a special mission, early in August, 1897, wrote as follows of Skaguay:

"Skaguay is, at this date, a city of eleven frame or log houses, a saw-mill, five stores, four saloons a crap game, a faro layout, blacksmith shop, five restaurants, which are feeding people all the time, a tailor shop, on which is hung the sign 'bloomers fitted for shotguns;' a real estate office, two practicing physicians, another professional pathfinder whose specialty is shown by the sign painted on a board nailed to a tree, 'teeth extracted;' some 300 tents, and a population of about 2000 men and seventeen women. Four of the women are accompanying their husbands into the Klondike. The others are unchaperoned.

"A dance hall will be erected next week. Skaguay is already a typical mining camp. Its population is proud of it. They go further, and say it will be a 'hot town' next winter. Streets have been laid out. Broadway runs from high tide four miles back to the mountain base, and is walled with tents, piles of supplies, and felled trees. The gold-seekers never overlook an opportunity to make fun drown their impatience.

"The event of to-day was a foot race for a purse of twenty-five dollars, in which fifty men entered. Lanterns are flickering like fireflies among the tents to-night. One turns his glance with a shiver from the snow-topped mountains which, half a mile from camp, point 4000 feet into the pale night overhead.

Unique Miners' Meeting.

"A miners' meeting stands without a parallel among things unique. It was recently decided at such a meeting at Circle City that a man cannot lick his own dog. What a miners' meeting says goes. A teamster named Cleveland was run out of town two days ago for refusing to haul a corpse free of charge. It was the body of young Dwight B. Fowler, who fell into the river and was drowned in the clear water in sight of his companions, owing to the weight of the pack strapped to his back."

Another writer has the following to say of Dawson City about the same time:

"There are several public resorts in Dawson—each with a bar

in front, gambling tables in the rear and a dancing floor in the middle.   Yukon has struck the typical early mining camp pace. Faro and poker are the favorite means for parting with gold dust. One hears of games with $20 ante and $50 to call blind.   They don't have money in circulation.

"There is no such thing as money.   When you go in just leave your sack at the bar and say, 'Give me five hundred,' or 'Give me a thousand,' and get your chips," explained a Yukoner. "Then if you lose you can call for what you want, and it's just put down, and when you get through they just weigh out what you owe.   I have seen fellows go in with $50,000 they had cleaned up and go out with an empty sack and go to work again."

A Wretched Place.

Miss Anna Fulcomer, who lived for a year at Circle City before seeking the Klondike fields with the rest of the fortune-hunters, gives a rather graphic account of the town.   Said she, in a letter written to her sister in Chicago :

"This is a wretched place to be side-tracked in.   A poor little town with few houses, and those for the most part of bad construction!   Not the possibility of going anywhere and getting out of sight of the little aggregation of buildings without going out into the wilderness away from everybody and everything! To do this requires not a little courage and energy.   People here are not primarily pleasure seekers.   Those who have come here have come for business, and this becomes manifest in everything, from the way in which they put in their time to the way in which they dress and deport themselves.

"There is no such thing as style.   There is little visiting, except to kill time when it is no longer possible to work.   You must not forget that this is the land of the midnight sun, and that it is also the land of the midday moon.   Consequently one gets up, works, goes to bed, does everything either by sunlight or by moonlight, according to the season of the year, without the natural phenomena that in southern latitudes accompany and lend a certain character to the duties and pleasures of life. Every thing seems turned about, and one scarcely has the inclination even though he might have the opportunity, to do as they do in the old States.

"What is more, there seems little prospect of any change in domestic conditions for the better for many a long day. Even though the mining interest keeps up, the influx of people to the camps will probably be so largely in excess of the accommodations for them, and they will bring with them such a meager supply of conveniences and comforts, that the prospect is that Dawson City and Circle City will continue to be Dawson City and Circle City until capitalists, realizing the necessities of the towns, will take steps to provide ampler and better accommodations than now exist."

At Camp Lake Linderman.

Of the camps proper William J. Jones gives a fair idea. Says he of Camp Lake Linderman :

"From fifty to one hundred white tents, as many camp fires and nearly 200 people constitutes the little colony of gold-seekers who are camped here, building boats and awaiting an opportunity to sail down the river. It is remarkable to note the difference in the personnel of the men. Only the better and more substantial element is able to cope with the hardships and reach this far. It would seem that the less persevering, or what might more properly be termed the lazier classes, are to be found scattered along the trail between Dyea and Sheep Camp, bemoaning and bewailing the hardships they are undergoing. They are having a picnic if they only realize that much, as compared with what they will experience after passing Sheep Camp.

"There is one saloon at Lake Linderman, and it is doing a thriving business in a tent. Without a license or other lawful restriction and with the poorest quality of liquor, so diluted as to be unrecognizable to the fastidious taste of experienced epicures, the proprietor is coining money by selling drinks at 50 cents each. A bottle of whiskey is worth $15. As the Indians arrive in from the coast with their packs and receive there stipends, averaging about $30, they are inveigled into the saloon and made drunk. A few drinks and a bottle of vile concoction called whiskey, and they are "broke." After sobering up they are ready to "hit the trail" and get another pack. Some of the nights are made wild and hideous with the orgies of these natives."

Hotel for Alaska.

Early in August, 1897, the North American Trading and Transportation Company took the very step that Miss Fulcomer advocated and perfected a plan for the transportation to Alaska of a hotel which would accommodate about 500 people. It was the plan to have the frame work sawed, finished, and put in such shape as to be ready for erection as soon as the material could be transported to the Territory. The decision once made, active steps were taken toward carrying out the project.

The new hotel was designed for Fort Get There, on St. Michael's Island. This was nothing but a trading post of the company situated about a mile from the town of St. Michael's, and only a hundred yards from the canal. At the time the enterprise was planned, and steps were taken to carry it out, there were only about twenty white men and probably twice as many Indians there. The plan of the hotel resembled a fort, being square, with a court in the center and a tower on either corner. Speaking of the enterprise, Mr. P. B. Weare said at the time:

"A special train of ten cars will convey the hotel furnishings and the steamer J. C. Barr, recently purchased at Toledo, to Seattle, in time to catch the City of Cleveland, which sails September 10th.   The J. C. Barr, which is now being taken apart preparatory to shipment, is intended for use on the Yukon River, and will make the fourth boat the company expects to have in operation on the river at the opening of the spring season.

"I do not know how soon we can carry out all our intentions," continued Mr. Weare, "but we realize the fact that domestic life in Alaska is in a large measure a matter of hardship and privation, and we know that there ought to be ampler accommodations provided for the people.   It is not, in that cold climate, as it was in California in the days of '49, for there, if the miners had not houses, it did not entail suffering or danger to camp out with nothing but the sky overhead and a blanket wrapped around them.

"In Alaska one cannot put up with camp fires and such conveniences as can easily be carried about on a pack saddle.   It is often terribly cold and the miners, in order to survive and keep themselves in fit condition to do their work, have to have good protection from the inclemencies of the weather.

Accommodations Not Good.

"These, it need not be said, do not now exist.   Of course, in the old towns in Southeastern Alaska, there are a number of places where strangers can get fairly good accommodations, but these accommodations vanish as soon as one heads his way toward the interior.   The tramp over the mountains and through the valleys, of course, must necessarily be one fraught with all the dangers and inconveniences and hardships of a journey in the wilderness.   But at the present time even in the old mining towns—you will understand I mean by old such places as Dawson City and Circle City—the existing condition of things is such that one can scarcely speak of domestic life at all. It is simply life without the domestic.''

Yet it was into this wilderness, devoid as it was of most of the amenities of civilized life, that scores of women of education and refinement took their way, actuated by various motives. Man was not to have the Klondike country to himself. If there were no sidewalks and boulevards, no boudoirs or parlors, the women meant to go there and share with their husbands and brothers and fathers the strange experiences of the mining camps. That the news of this exodus of women to the diggings was cheering news to the miners, needs scarcely to be said.

Women Off for the Diggings.

In the middle of August, 1897, an announcement of the intention of women to go to the North was made in the following words:

" Woman's refining hand is to be laid on the camps at Dawson City and other Arctic settlements. The home comforts of civilization are to be introduced in a country in which they have been hitherto sadly lacking. This winter will bring a radical change in domestic and social conditions in that far-off part of the world and enforced seclusion will be relieved of its greatest terrors.

" Eight Illinois women have thus far announced their purpose to make the pilgrimage to the gold fields of Alaska, and this number is likely to be doubled before the last steamer of the season sails from Seattle. Similar reports come from other States, so there is a strong certainty that the Klondike district is to have an agreeable and useful addition to its present population.

" Some of these women are the wives of men now in Alaska wresting wealth from the frozen earth—these go to make lighter for their husbands the hardships of an Arctic winter. Others will make the long and dangerous journey to dig gold for themselves, to make money by keeping boarders, by ministering with needle and thread to the wants of helpless masculinity—and even by running newspapers, in which the lucky strikes, the sad failures, social doings, and all the breezy gossip of the camps will be duly chronicled.

" Then there are others—women of mercy—whose sole object in braving Arctic perils is to care for the sick and afflicted, to nurse back to life and strength the victims of accident or disease, and soothe the last moments of those who receive the final summons to the great beyond.

Promise is Fulfilled.

The promise held forth to the miners of having woman's influence in their rude life, was carried out with a fulness they little anticipated. Mrs. Caroline Wescott Romncy, a Chicago woman, early expressed her determination to go to the Klondike and pass the winter. It was not her intention to go on a pleasure jaunt, but strictly on a business venture, and on one well calculated to make the camp life brighter and better. Her main purpose was to start a newspaper at Dawson City, and she decided to take with her a complete printing outfit, so that she could issue a little sheet and supply the mining community not merely with news about local doings, but with reprinted matter, which would serve to instruct and amuse the people.

Mrs. Romney had had a good deal of experience in a similar line in Leadville and Durango in the boom days of Colorado. She was a strong believer in mines and mining, and, having worked with success in this line in Colorado, and also in Mexico, she thought she could enter the Yukon valley and by her enterprise meet a decided want in the domestic life of the community. Speaking of her project before she started, she said :

"Of one thing I am confident, there is gold in plenty in Alaska. I believe there is a fortune for me, and I am going to get it or know the cause of failure.   What is more, I am not going to work in the mines, but in the camps and for the benefit of the people.   I do not think there is any occasion for the lawlessness that has characterized almost every mining community on record. That sort of thing springs up primarily from the absence of those conveniences and comforts that in these days legitimate enterprise could easily supply."

Mrs. Gage's Enterprise.

Mention has been made elsewhere in this volume of the enterprise of Mrs. Eli S. Gage, who left her cozy home in Chicago and went to the mining region to be with her husband and lend what influence she could for the good of the camp life.   She left Chicago early in the fall of 1897 and took her way to the diggings by way of the Chilkoot Pass.   According to the plans of Mrs. Gage, as expressed before starting, she intended to keep house in Northern Alaska, doing the cooking, washing and other forms of housework herself.

There are no trained servants or domestic help in the Territory, and consequently it is a practice of the miners to shift for themselves the best way they can.   It was Mrs. Gage's opinion, which was also shared by the officers of the transportation company with which her husband is connected, that the presence of herself and other women of good character would have a great influence in brightening and making more agreeable the long winters of the northern region.

Mrs. A. W. Little also left her Chicago home and followed the example of Mrs. Gage. She went to Alaska well equipped for a winter in which the cold often gets as low as 60 degrees below zero. Before starting from her home she had an outfit of dogs and sledges prepared and in waiting for her at Dyea, to transport her over the snow-clad country to Dawson City.

Willing to Meet Danger.

Miss Pauline Kellogg, of Chicago, daughter of Judge Kellogg, a pioneer miner of Colorado, and a woman well trained in mining life, also went to the diggings in the fall of 1897. Early in her life Miss Kellogg had lived in a Rocky Mountain cabin and had become proficient in miners' work. She knew exactly what people in a district like the Yukon valley had to experience, and had a lively recollection of the hardships imposed by such domestic life as one has to encounter in camp life.

"Danger!" said Miss Kellogg, before taking the train from Chicago. "Of course there will be danger, but I have been all through Colorado when that country was new, and I think I can take care of myself in Alaska. I am not sure that I shall be much of a success in the mining role, but I do think I can be of a whole of service to the miners, and if I fail in one line I shall hope to make it up in another."

Mrs. William Chase was one of the hundreds to brave the perils of the new life to carry something of life and cheerfulness into the miners' experiences. She left her Chicago home to join her husband on the Yukon and help him and his associates. She expressed a determination to keep house, to attend to the cooking and other domestic duties herself, and so far as she could, to teach and assist the miners and prospectors who had no woman's hand to help them to do likewise.

"In this way," she said, "I can be of more use to them than by digging in a pit like a man. What makes life in the Arctic

Circle so hard to bear, I am told, is the absence of home comforts. These I propose to furnish to as great an extent as possible, and it will be much better, even if I am not very successful, than to have my husband up there alone. The miners I know will welcome me."

Their Mission of Mercy.

Mention was briefly made elsewhere of the two Sisters of Mercy who, in the early days of the gold craze started for the North to minister to those who might need their assistance. They started from San Francisco for St. Michael's Island, meaning to push on to the interior by as rapid stages as possible. Their avowed intention was to nurse the sick and solace the dying in Northern Alaska. They were Sister Mary of the Cross and Sister Mary Magdalene of the Sacred Heart.

In striking contrast to the heavy clothing and big outfits of provisions and tools of gold-seekers were the simple black habits of the sisters. They had no stores of groceries, no supply of furlined garments, no equipment of tools. Two hand satchels and a couple of trunks in the steamer's hold contained all their worldly goods. When asked if they were not afraid to venture into so cold and desolate a country with such a scanty outfit, Sister Mary Magdalene said: "The Lord will provide. We go to do his work and he will take care of us." This simple statement had an impressive effect upon the passengers and crew, and every man on the boat became a helpful ally of the sisters.

Mrs. Bessie Thomas, of San Francisco, also early left for the Klondike fields, but her mission was an entirely different one. She did not go to care for the sick and solace the dying, but to give the miners and prospectors good, wholesome dinners and suppers and keep them well. In other words, Mrs. Thomas intended to start a restaurant, and while primarily it was a business venture on her part, it was one that met a crying want of the mining camps.

It can readily be understood that with a meager supply of cooking utensils, and no skill in the art of cooking, the majority of the miners and prospectors were in rather a bad way in the matter of providing their meals. Mrs. Thomas was shrewd enough to recognize this and take advantage of the opportunity offered her. Further, there was a touch of real philanthropy in her project. Before leaving San Francisco Mrs. Thomas said:

" Miners have got to eat and I think there is more money to be made in feeding them than in slaving my life away here. I have got to earn my own living, and I do not see why there shouldn't be just as good a chance for me in a mining camp as there is for a man. There is another side to this matter, too. Here I just do my work for the pittance accorded me, and don't know I am doing anybody any especial good or myself either.

" I do know that one of the most important things in a mining community is for the men to have good, wholesome meals, properly cooked and served. In the diggings, I am told, the diet is almost exclusively one of fish and canned goods. A diet of this sort becomes very monotonous, and if a few good, whole-souled women would go up north and look after the culinary end of the camp life, there would be a great sight more happiness as well as a great deal less disease."

CHAPTER XIX.

Ethnography.

ONE of the most engrossing and perplexing problems of the
ethnologist is presented by the aboriginal native inhabitants
of the islands and mainland of Alaska. Many of them
present characteristics at variance with any prediction of ultimate
American origin. White people going into the country are apt
to regard the aborigines as a branch of the great race of North
American Indians, and that they are called Indians in common
parlance greatly favors this misconception ; but to the student,
most of them are absolute and distinct, with not a drop of American Indian blood in their veins, unless it has come from cross-
breeding with the red Indians further south.

The population of Alaska is classified as white, mixed-Indian,
Indian, Mongolian, and all others. Some figures as to its extent are interesting, as serving to correct many commonly held
mis-opinions on the subject.

Census of Alaska.

The United States Census of 1890 was the first organized
effort to get at the facts of the population of this great territory,
one-sixth the size of the nation of which it is a part. It showed
the total of inhabitants, living in 309 settlements, was 32,052,
of whom 4298 were white ; 1823 mixed-Indian ; 23,531 Indian;
and 2288 Mongolian. Of these the Greek Church claimed as

478

converts 10,335, of whom 8414 were natives ; the Presbyterian, 1334, of whom 1260 were natives ; and the Roman Catholic, 498, the natives numbering 131. This topic is more elaborately treated in the chapter on " The Spread of Christianity."

The efforts of the Czar's officers to obtain a census were crude and the results altogether valueless as statistics. Delarof's estimate, made in 1792, gave 6510 natives to Kadiak Island and the near mainland region. Baranof, in 1796, made the total in the same area 6200, but he also reported a probable total of 5000 Thlinkets, unsubdued and not enumerated. Baron Wrangel, in 1825, estimated the total population at 8481. Veniaminof made three censuses : in 1831, of the Aleuts, whom he numbered at 1515 ; in 1835, of the Thlinkets, whom he estimated at 5850 ; and in 1839, of the entire population, which he placed at 39,813 ; a remarkably close result when it is understood that nearly all the statistics of natives were the result of what might be called scientific guess-work. In 1860 the Holy Synod made a census of the Christian population of both sexes and fixed the total at 9845, exclusive of the Russian employes of the company.

Classification of Indians.

General Halleck, U. S. A., made an estimate of the inhabitants in 1868, which was extravagantly wild, even for guesswork, the total being put at 82,400, or fully 50,000 too many, as shown by the careful enumeration based on actual count in the census of 1890.

Along linguistic lines the Indians of Alaska are divided in the elements of stock and strength as follows :

Esquimeaux, inhabiting the coast from Copper River to the northern extremity of the international boundary line.

Thlinkets, occupying the coast southeast of Copper River, and known variously as Chilkats, Auks, Takus, Hootzanoos (on Admiralty Island), Sitkans and Tongass.

Aleuts, on the Aleutian Islands.

Athapascans (Tinnehs), living in the interior and known as Kutchins and Ingaliks.

Tsimpseans (of foreign extraction) on Annette Island, principal type.

Skittagans, the Haydas of Prince of Wales Island, principal type.

It will be best to examine these rather in the order of their importance than of their strength.

The Thlinkets.

Thlinket, the name given to the people by themselves, means "the people" and indicates the esteem in which this once powerful family was held by its savage tribesmen. These aborigines are lighter colored than the North American Indian, and in many more important particulars are radically different from their red neighbors.

There are many separate tribes of Thlinkets and, as many unreliable traditions of supernatural origin, a deluge and a sole surviving couple. Their propitiation of evil spirits, their Shamanism, their belief in the transmigration of souls, their worshipful regard for the spirits and ashes of their ancestors, would suggest an Asiatic origin. Their methods, tools and postures are Japanese. Their totem poles are like those of the Maoris and South Sea Islanders. Their sun and nature worship and their legends of the Thunder Bird are Aztec. Totemism is the base of their social organization, but the totem pole has no religious significance, and is not an object of worship. Its purpose seems to be purely heraldic.

A theory which would go far to explain the Asiatic characteristics of the Thlinkets and other similar Alaskan peoples, and which has found many advocates among scholars is based upon the action of the Kuro Siwo, or Japan current, which sweeps around through the ocean from the shores of the Chrysanthemum Empire and passing to the south of the Aleutian Islands washes the northwestern coast of the American continent. It has been conjectured that in some remote age Japanese junks with their crews, which in ancient times were often composed of men and women, were caught in terrific storms and partly wrecked, so that return to the home port was impossible ; that the disabled hulks, caught in the ever-flowing current, drifted helplessly around the circuit of the North Pacific and were finally, with the remnants of their ill-starred crews, cast upon the shores of the Alaskan Archipelago. Granted that all this came to pass, environment would easily account for the differentiation from the parent Asiatic stock which marks the Alaskan Indian of the days of history.

Famished Japanese Sailor.

This hypothesis of an Asiatic origin, fanciful as it may seem in some ways, is not altogether without the support of facts. Within the memory of living men a Japanese junk was cast ashore near the mouth of the Columbia River, and from the wreck was rescued the sole survivor of its crew, a famished and sea-crazed Japanese sailor, who was able to relate before he died the story of the awful storm, which drove himself and his companions into the wilderness of the ocean on which he drifted for eight months, his comrades dying one by one along that awful unmarked trail through the billows. Perchance, the hardier men of another age might have endured such a terrible voyage with death and still survived with vigor enough to found a new race in a new land.

31

In many ways the Thlinkets strongly resemble the Japanese. They have the same small hands and feet and their features are much like those of the Mikado's people. Their babies are fat and chubby, and were a Thlinket and a Japanese infant to be dressed exactly alike and placed side by side it is likely none but the mothers could tell certainly which was which merely by looking. They resemble the Japanese, too, in not being robust and in their extreme veneration for old age—wherein they differ much from some whites.

Physically the Thlinkets are magnificent specimens from the waist upwards. But they are pigeontoed and bowlegged and as awkward as aquatic birds upon the land. This is their heritage from generations of canoeing ancestors, whose warped postures in their frail, rude boats have thus stamped a trait upon their descendants.

Singular Customs of the Natives.

Though the Thlinkets are pretty well civilized, they still retain traces of their ancient savage customs. Some of the oldest hags still wear the laviette, a metal or wooden plug piercing the under lip and supposed to enhance the beauty of the wearer. Tatooing, once almost universal, has nearly disappeared, but they all paint for great dances and "potlatches," and in summer men and women daub and blacken their faces as protection against the insect pests. Polygamy and polyandry are now practically extinct, though both were formerly common. They are superstitious to a degree, and until Captain Merriman, U. S. N., whom they called a great "tyee" or king, because of his impartial and successful administration of the government, broke the power of the shamans, or medicine men, witchcraft and its attendant horrors were common. Now a witch is never heard of.

Though strong, the Thlinkets are not a hardy people nor as a rule long lived. Consumption is common and generally makes a speedy end of its victims. They are being fast thinned out by disease and dissipation. The whites have proved a curse to them in both directions. They are great gamblers and a true Thlinket will bet everything he owns, from his wives up. They drink white man's rum when they can get it, which is not seldom, and otherwise their own home-made "hoochinoo." And they go on fearful sprees.

Slavery is another of the ancient customs which has been outgrown. Prisoners of war were always made slaves, unless they were butchered to make a Thlinket holiday in the days of the nation's savagery, and their lot was of the hardest. One of the least enjoyable portions of these slaves was to be killed at the grave of the master, especially if the latter happened to have been a chief. Cannibalism, which was not uncommon among the Indians at an early date, is also now happily a thing of the past. Akin to this barbarity was the exposure of female infants, but this abominable practice has also been abandoned.

How Great Events Were Celebrated.

The "patlatch" is an ancient and honorable custom which has passed into innocuous desuetude with most of the Thlinkets. Formerly every great event was celebrated with a "patlatch," and as the festivity was an expensive one, requiring the utmost lavishness in entertaining, not only in the distribution of meat and drink, but of blankets and other presents, it sometimes made a man poor to be rich. Now the ambition of these Indians seems to be to live and dress as much like the whites as possible. They retain the barbarian's love for gaudy things, however, feathers being their especial pride for decoration, and a Thlinket in full dress is a gay sight indeed.

As a people they are brave in a relative sense—that is, they can fight like demons when cornered, or when opposed to a weak enemy ; but are not overprone to pick quarrels with those stronger than themselves. They are venturesome to reckless- ness in their sea voyages, making trips in their small boats which would daunt a white man in his larger craft. They have given up war, but the old spirit still makes them among the hardiest sailors of the Pacific. In manner they are dignified, but cour- teous, and they are extremely hospitable. Withal they are great sticklers in matters of ceremony, and a fancied slight has been known to end in bloodshed. In their habits they are the oppo- site of lazy, and nearly all the able-bodied men among the coast residents now work in the salmon canneries or salteries, or pur- sue hunting and fishing for gain. They have a decided taste to get money, and some of them are exceedingly thrifty. Princess Thom, one of their great characters, was a sort of Thlinket Hetty Green, and literally had more wealth than she knew what to do with, but still was insatiable for more.

Fondness for Display.

Though the native religion of the Thlinkets was a kind of nature worship, or feeble polytheism, these Indians proved plastic material in the hands of the missionaries, and most of the older ones are now members of the Russian Greek Church. Their great fondness for display is well gratified by the rich robes and vestments, the candles and the pictures which enter so largely into the service. Most of them speak Russian, and they are all familiar with the trader's jargon known as " Chinook." One of the results of their religious training by the Russian Fathers has been the abandonment of their ancient and almost universal burial rite of cremation, the only exceptions to which were the Shamans, or medicine men.

All the Thlinkets are divided into two clans, the Wolf and the Raven. A man never marries into his own clan, and the children are always designated as of the mothers clan. Besides the distinction of clans there are numerous tribes of Thlinkets.

The Chilkats and Chilkoots, who are really one tribe, are the great people of the Thlinkets. They have always been great traders and have possessed more wealth than any other tribe. They were opposed to white trade with the Tinnehs, and for fifty years stood as a barrier across the passage to the Yukon Basin, playing the middleman with the Tinnehs in the fur trade. The white men cheated the Chilkats, the Chilkats cheated the Tinnehs. Whom the Tinnehs cheated, unless it was the animals whose furs they took, is not of record. The Chilkats were good warriors as well as thrifty traders until in 1892 the saloon invaded their country and rum wrecked the once powerful tribe.

They are a more than commonly intellectual people. Their chief "klohkutz" drew for Professor Davidson the first known map of the famous Chilkat and Chilkoot passes. They long knew the art of forging copper, and they possess in a high degree the art of dyeing. Their elaborate dance robes, made from antelope wool and gayly colored, have a considerable commercial value as "Chilkat blankets." In their weaving they display a skill little inferior to that which has made the Navajo blanket famous. As wood carvers, also, they exhibit no mean skill, as is evidenced by the decorations of their totem poles and canoes. Their folk lore, myths and traditions exhibit a wonderful poetic sense for so primitive a people and, indeed, this is true in no less degree of the Haydas and Tsimpseans.

Dietary of the Chilkats.

One of the Chilkats' greatest delicacies is what is known as the salmon berry, a fruit salmon-red in color and shaped like blackberries. This fruit has a musky and at the same time an unpleasant flavor for white people, but the Chilkats call them their greatest relish. They eat large quantities of them in an oil, the preparation of which, to say the least, is peculiar.

In making this oil the women gather up all the salmon heads and bury them underneath the ground, where they leave them for several days, until they become very odoriferous and " ripe." Then they dig the fish heads up, place them in an old boat and throw red-hot stones among them to try out the oil. After the stones cool the Chilkat women get into the boat and squeeze out the oil from the fish heads by tramping and stamping upon them with their bare feet. The oil is then dipped up, and, being poured over the salmon berries, makes—to the Chilkats—an appetizing dish, which they partake of with great and evident relish. It is not likely that any of the tenderfeet journeying up into the gold diggings of the Klondike will stop at any Chilkat public houses on the way for a dish of salmon berries dressed in oil.

The Chilkats reckon their wealth in blankets, and a wealthy man will often accumulate as many as 1000 blankets. To add to their stock of blankets through life they would undergo any hardship, in many cases actually starving themselves to add to their accumulations.

Hootzanoos Make Hoochinoo.

The ordinary food eaten by the Chilkats is fresh or dried salmon, but when hungry they will often consume large quantities of lard and other fat. A storekeeper of Juneau tells of one able-bodied Chilkat who came into his store and purchased a four quart tin of hog lard and cotton seed oil combined and ate every drop of it before leaving the store.

The Hootzanoos at Killisnoo make an outright claim to having come from over the sea. They first distilled " hoochinoo," or native rum, making it in old coal oil cans from a mash composed of molasses and yeast. They learned the trick from the whites. They are the giants of the race.

The Hoonas, on the icy strait, a warlike tribe, have been longest preserved by environment from contact with the whites. Not for that reason but because they deserved it they have always had a bad name. In this respect, their brethren, the Auks, are like them, though they are not a quarrelsome tribe. They live along Douglas Channel.

The Sitkans as at present constituted contain many members of decidedly mixed breed, descended from outcasts, renegades, malcontents and wanderers. They are the farthest from the pure blood of any of the tribes. Once the greatest term of contempt in the Thlinket nation was: "As great a blockhead as a Sitkan." Not until 1821 were they permitted by the whites to settle on the shore, and several times after that act of clemency they repaid it by attacking the station. However, they were generally quickly overcome. Rum and contact with lawless whites have done much to destroy them. They are the best dressed and most intellectual of the tribes.

Traits of the Stickines.

The Stickines who inhabit the valley of the Stickine River, near its mouth, are a peaceable tribe at present, though they have made trouble for the whites in the past within the latter half of the century, having captured a trading vessel and murdered the crew. They possess many of the traits of the other Columbian coast tribes, believe in the Thunder Bird as if to suggest a southern origin, and are shrewd traders, and hard drinkers and gamesters when they get a chance.

Kenaians is a name applied by the Indians to the natives inhabiting the country north of Copper River and west of the mountains, except the Esquimos and Aleuts. They are generally peaceful and well disposed, though ready to avenge affront or wrong. They are good hunters and traders.

The Haydas (Skittagetans) were and are the flower of the native races. They are taller, fairer, and with more regular features than any of the other Columbian coast tribes, and nearer to the Thlinkets in characteristics than to any other people, but they are aliens to the Thlinkets, nevertheless, physically and mentally, in speech and customs. The Thlinkets call them "Di-Kinyo," the people of the sea. They are the northmen of the Pacific. Once, their forays extended as far south as Puget Sound, and they seized a schooner in Seattle Harbor and murdered the crew.

Old Traditions and Legends.

Their origin is the puzzle of ethnologists. They have a tradition of a deluge and a sole surviving raven from which their people sprung. Some identify them as the descendants of the Aztecs whom Cortez drove out of Mexico. Their legend of the Thunder Bird is the same as that of the Aztecs and the Zunis. They have images and relics similar to those found in Gautemalan ruins. But they have modern Apache words in their speech and dances and picture writing like the Zunis. Their resemblance to the Japanese is also very marked, and as the Japanese current touches directly on Queen Charlotte's shores, junks may have been stranded there in the days when the Japanese built sea-going junks and traveled afar. They have Japanese words in their speech, they sit at their work and pull their tools towards them like the Japanese. They are imitative, too, like the Japanese. In many of their customs, their bark weaving and their carving they resemble the Maoris of New Zealand and the South Sea Islanders. They have carried the totem pole to its highest development. Their folk lore is highly poetical.

The Aleuts, or inhabitants of the Aleutian Islands, have been so mixed with Russians, Indian and Kamschadale stock that it is difficult to find pure blooded men or women in the settlements, The predominant features among them to-day are small, wide-set dark eyes, broad and high cheek-bones, causing the jaw, which is full and square to often appear peaked; coarse, straight, black hair; small neatly-shaped feet and hands and brownish yellow complexion.   In many particulars they closely resemble the Esquimo.   Some few of the half-breeds are handsome physical specimens of the human race.   The average stature of the men is five feet four or five inches, though some are over six feet. They resemble the Konos of northern Japan.

The Aleuts, as a people, have been Christians for over a hundred years and many of them read and write.   They adopted the Christian faith with very little opposition, willingly exchanging their barbarous customs and wild superstitions for the agreeable rites of the Greek Church and its refined myths and legends.

Old Dwellings and New.

When first known to the whites they lived in large yourts or "oolagha-moo," dirt houses, partly underground, going in and out with the smoke through a hole in the top.   One of these ancient yourts, whose foundations were lately standing on Unalaska Island, was eighty-seven yards long and forty wide.   In these dirt houses the primitive Aleuts dwelt by fifties and hundreds for the double purpose of protection and warmth.   To-day nearly every Aleutian family has a hut or "barabkie," or a neat frame cottage, the latter owing to the Alaska Commercial Company in most instances.   The "barabkie," though built partly underground, is a vast improvement over the yourt, has a window at one end and a door at the other and is embellished within with pictures of the church and patron saints.   Here the Aleut spends most of his time, when not engaged in hunting, either drinking cup after cup of boiling tea or stupefying himself with " quass," a native beer or with home-distilled rum.

The Aleuts are remarkably polite, not only to the whites but to one another. The women are great gossips, despite the few topics of conversation which they can have, and they visit freely and pleasantly among themselves. It is only when under the influence of liquor that they lose their amiability and show something of the old savage nature. They used to be great drunkards, but the church is gradually weaning them from the disastrous habit.

Heavy Burdens and Short Lives.

As parents they are extremely indulgent while their children are under ten years of age, but after this time they become strict disciplinarians and hard taskmasters, putting burdens upon young shoulders that are heavy enough for adults and always exacting implicit obedience. The infant mortality is excessive as a result of the bad habits and sanitation of the people. The race is short-lived, owing to utter disregard of the laws of health. They are all more or less tainted with scrofula. They marry young and without the least evidence of sentimentality. And yet some of the women are decidedly pretty.

The men are sea-otter hunters, first, last and all the time, except as necessity may force them temporarily to some other occupation. In the chase they are bold and skillful and they venture far out to sea in their skin " bidarkas " and kayaks with an indifference which forever secures them against competition by the whites. The sufferings they undergo from cold and scanty food while in the chase can be better imagined than described. They haul their boats out of the water every night and bivouac along the coast in biting gales, in rain, sleet and fog, without covering and almost invariably without a fire.

CHAPTER XX.

Native Religion and Traits.

The Alaskan Indians a People of Curious Customs and Habits—Are Intelligent, Inventive, and Imitative—Are Adepts in the Vices of the White Men Who Visit Them—Are Natural-born Drunkards and Gamblers—Totem Poles Their Pride in the Olden Times—The Significance of these Barbaric Symbols of the People—Are Rich in Oral Traditions—The Theological and Cosmological Belief of the Indians—Odd Notions of the Aboriginal Thinkers—Samples of the Rites Practiced—Cannibalism and Shamanism—Law and Home Life—Description of the Innuits of the North.

THE Alaskan Indians are a unique people in a strange setting. The visitor to the Territory will be surprised at their manners, their speech, their looks and their customs, and above all, at their intelligence. The Hon. Vincent Colyer, once Special Indian Commissioner to Alaska, said in his report: "I do not hesitate to say that if three-fourths of the Alaskan Indians were landed in New York, as coming from Europe, they would be selected as among the most intelligent of the many worthy immigrants who daily arrive at that point."

This may seem a rather unusual tribute to a people whom we are accustomed to regard as mere savages. The words of Colyer, however, are not unduly eulogistic. There is a wide disparity among the natives, of course; but, from the extreme southern point of Alaska to the Arctic Ocean, these children of the wilderness are characterized by a shrewdness and a cleverness that, despite the traces of barbarism to be seen, differentiate them in a marked degree from the other aboriginal inhabitants of America.

As was said in the chapter on ethnology, it is a grave question among scientists whence the natives came, opinion differing in a very marked degree. Some contend that they came from the

491

central portion of the continent, and others maintain that they are of Mongolian origin. Be this as it may, the natives are there, and they will of necessity be a curious study to all the people from the Southern States who may visit the Territory. The strangers in the country will find in the natives characteristics of many races, and will see unmistakable indications of the shreds of culture and education which they derived from the Russians.

Natives First Teachers.

The Russians, being the first occupants of the land, naturally became the first teachers of the natives. These Indians are an inventive and emphatically an imitative people. In this regard they show a close resemblance to the Chinese and Japanese. The natural aptitude of the people for following examples is well illustrated by the exceptional skill they manifest in the matter of weaving delicate fabrics, making graceful canoes and carving their totem poles, those symbols of savage life which may be found wherever a group of Indians have settled.

This aptitude for imitation is also shown by the way in which the natives pick up the vices of the white settlers in the country. As might be expected, the examples set them are often not of the best, inasmuch as the class of people who go to a wild and unsettled country like Alaska are apt not to be of the highest stamp.

The natives have thus thrust before them very often deplorable practices and vices, which they pick up and follow as assiduously as do their instructors. The road to wrong is thus made smooth for them, and it is not strange, therefore, if those who now flock to the gold diggings find the savages adepts in many of the reprehensible practices commonly followed in more civilized communities.

The Indians, for example, are ardent lovers of intoxicants. The Russians, shortly after Bering crossed the Pacific with his band of hardy adventurers, learned to make a cooling and comparatively harmless drink from rye meal mixed with water, which they put in a cask and allowed to ferment.   From this time this drink was their luxury.   But it was not a great while before native ingenuity led them to mix in their beverage a little sugar, flour, dried apples and hops, and the result was that they had an intoxicating drink that would put the worst form of fire water to the blush, so far as its effects were concerned.

Receive a New Tutor.

Then a discharged American soldier taught them how to distil liquor, and native ingenuity again led them to manufacture their own stills, which they made from kerosene cans, with the addition of the hollow stem of the seaweed.   The art of making intoxicants they have never forgotten, and the prospector and miner to-day will find the natives filling themselves up with these drinks and running amuck, in which condition the crazy natives are well fitted for any deeds of violence or viciousness.

Again, the Indians are inveterate gamblers, but whether they learned this from their white instructors is a question.   The natives are as simple in the games of chance by which they gamble away everything, from their wives to their dinners, as they are in their domestic arrangements and their habits.

The favorite game is played with a number of small sticks, which are cut of different sizes and colored different tints.   These are named crab, whale, duck, otter, fox and the like.   They are shuffled up and then placed under bunches of moss, and the game consists in guessing under what pile of moss the whale, or duck or what not may be.   This, it will be seen, is literally a children's game, yet it is for the natives a serious matter, for very often on a guess a savage will lose home, possessions, everything.

The natives of Alaska fall into various families, but for the purpose of setting forth their most striking customs and characteristics, they may be divided into two great divisions, the Thlinkets, or people of Southern Alaska, and the Innuits, or people who live in the extreme northern regions. The Innuits, by the way, are not infrequently called Esquimaux. The minor divisions of each of these great classes present few differences. There is, however, a very sharp contrast between the two great classes themselves.

Forests of Totem Poles.

Wherever one finds a Thlinket settlement, he will find a forest of totem poles. The significance of these poles has often been made a matter of question, but it is commonly believed now that the poles have no religious significance, and are not objects of idolatrous worship. They are rather to be considered as a sort of heraldic designs, distinguishing families, very much in the same way that the herandic devices of the nobility of Europe distinguish families.

Totemism becomes thus, the base of the natives social organization, and the totem pole becomes nothing more or less than a tribal mark distinguishing the dwellings and belongings of separate families or clans.

It is interesting to note that only animal totems occur. The natives thus practically live under the guardianship of some one or other of the wild beasts or the birds or the fishes that abound in the Territory. The crow or raven represents woman, the creative principal. The wolf represents the aggressive or fighting creature. These two forms of totem are the most prevalent along the coast.

That these totem poles are simply a family designation, as was said above, is borne out by the fact that men do not marry women of their own totem. The Thlinkets were not slow in making totem poles representative of the two great nations with which they had most to do, Great Britain and the United States. They fashioned one totem with a unicorn, and it stood for " King George men ; " and they made another with a spread eagle, and had that designate the " Boston men," an ingenuous tribute, perhaps, to Boston as the hub of the universe.

Some Indian families thus live under the special protection of the bear, the whale, the frog, the wolf; and it is an easy matter to recognize the family by the rude conventionalized carvings to be found before their doors. Some of these poles are very elaborately carved from top to bottom, often reaching fifty or sixty feet in height and being three or four feet in diameter.

Rich Oral Mythology.

Centreing largely about these poles, the natives have an oral mythology, which is often of the most fabulous character. These legends are religiously handed down from father to son and are rehearsed to the visitors with all the semblance of conviction on the part of the narrators. Like many other things characteristic of the Indian's life and belief, these totem poles are largely becoming relics of the past and symbols merely of what used to be. This is due partly to the work of the missionaries and partly to the natives' intercourse in a commercial way with the white man.

In the early days the Indians were devout believers in witchcraft, evil spirits, and all that sort of superstitious invention, and many were the horrors that they committed in obedience to this form of religious belief. Out of this grew various kinds of torture, and not infrequently, the poor savages would die under the efforts of their friends to remove them from the influence of imaginary demons.

Dr. Dall, one of the closest students of the Alaskan Indians, gives a very good account of the religious beliefs of the Thlinkets. Says he:

" Their religion is a feeble polytheism. Yehl is the maker of wood and waters, he put the sun, moon and stars in their places. He lives in the East, near the head-waters of the Maas River. He makes himself known in the east wind, Ssankheth, and his abode in Nasshak-Yehl.

Men Groped in Darkness.

" There was a time when men groped in the dark in search of the world. At that time a Thlinket lived who had a wife and sister. He loved the former so much that he did not permit her to work. Eight little red birds, called kun, were always around her. One day she spoke to a stranger. The little birds flew and told the jealous husband, who prepared to make a box to shut his wife up. He killed all his sister's children because they looked at his wife.

" Weeping, the mother went to the seashore. A whale saw her and asked the cause of her grief, and when informed, told her to swallow a small stone from the beach and drink some sea water. In eight months she had a child, whom she hid from her brother. This son was Yehl.

"At that time the sun, moon and stars, were kept by a rich chief in separate boxes, which he allowed no one to touch. Yehl, by strategy, secured and opened these boxes, so that the moon and stars shone in the sky. When the sun box was opened, the people, astonished at the unwonted glare, ran off into the mountains, woods and even into the water, becoming animals or fish. He also provided fire and water. Having arranged everything for the comfort of the Thlinkets, he disappeared where neither man or spirit can penetrate.

" There are an immense number of minor spirits called Yekh.

Each Shaman has his own familiar spirits to do his bidding, and others on whom he may call in certain emergencies. These spirits are divided into three classes—Khiyekh, the upper ones ; Takhi-Yekh, land spirits ; and Tekih-Yekh, sea spirits. The first are the spirits of the brave killed in war, and dwell in the North. Hence a great display of Northern Lights is looked upon as an omen of war.

Responsibility of Mourners.

" The second and third are the spirits of those who died in the common way, and who dwell in Takhan-Khov. The ease with which these latter reach their appointed place is dependent on the conduct of their relations in mourning for them. In addition to these spirits, every one has his Yekh, who is always with him, except in cases when the man becomes exceedingly bad, when the Yekh leaves him.

" These spirits only permit themselves to be conjured by the sound of a drum or rattle. The last is usually made in the shape of a bird, hollow, and filled with small stones. These are used at all festivities and whenever the spirits are wanted."

As might be expected from this form of religious belief, a large share of the attention of the worshippers is given to propitiating evil spirits, and the religion of the natives of southern Alaska thus practically resolves itself into a form of devil worship. This, doubtless, is the origin of Shamanism, which really consists in making offerings to evil spirits in order to prevent them from doing mischief to the people.

The religion of the Indians, therefore, has a certain similarity to that of the old Tartar race before the gospel of Buddha was introduced. Indeed, forms of belief, very similar to those just given above, may still be found among some of the peoples in Siberia.

The one whose duty it is particularly to propitiate the evil

32

spirits is the great medicine man, or sorcerer, or Shaman of the tribe.   He, it is supposed, has control not only of the spirits, but, through the spirits, of diseases, and of the elements.   Dr. Dall points out the fact that the honor and respect in which a Shaman is held depends upon the number of spirits supposed to be under his control.   It is curious to note that whale's blubber, one of the greatest delicacies among the Indians of the North, was put under ban by a Shaman.   To this day it is regarded with abhorence by the Thlinkets in the South.

It can readily be seen that the Shaman is virtually a ruler among his people and that by prostitution of his power he can make himself a terror.   Bancroft, in his " Native Races on the Pacific Coast," thus speaks of Shamanism :

" Thick, black clouds, portents of evil, hang threateningly over the savage during his entire life.   Genii murmur in the flowing river.   In the rustling branches of the trees are heard the breathing of the gods.   Goblins dance in the vaporing twilight, and demons howl in the darkness.   All these beings are hostile to man and must be propitiated by gifts and prayers and sacrifices, and the religious worship of some of the tribes includes practices which are frightful in their atrocity.   Here, for example, is a right of sorcery as practised among the Haidahs, one of the northern nations.

Sample Religious Rite.

" When the salmon season is over and the provisions of winter have been stored away, feasting and conjuring begin.   The chief, who seems to be the principal sorcerer, and indeed to possess little authority save for his connection with the preter-human powers, goes off to the loneliest and wildest retreat he knows of or can discover in the mountains or forest, and half starves himself there for some weeks, till he is worked up to a frenzy of religious insanity. . . . . At last the inspired demoniac returns to his village naked, save a bearskin or a ragged blanket, with a chaplet on his head and a red band of alder bark about his neck.

"He springs on the first person he meets, bites out and swallows one or more mouthfuls of the man's living flesh, wherever he can fix his teeth, then rushes to another and another, repeating his revolting meal till he falls into a torpor from his sudden and half masticated surfeit of flesh. For some days after this he lies in a kind of coma, like an 'overgorged beast of prey,' as Dunn says; the same observing that 'his breath during that time is like an exhalation from the grave.' The victims of this ferocity dare not resist the bite of the Taamish; on the contrary they are sometimes willing to offer themselves for the ordeal, and are always proud of their scars."

The Indians are thus held in abject fear of the Shamans, and it is possibly due to this fact that the missionaries of the Christian church were so cordially welcomed and their ministrations and teachings so gratefully received. In a large measure these old beliefs of the natives are passing away.

Witchcraft Still Exists.

Still, Miner W. Bruce assures us that despite the efforts of missionaries and teachers, and the influence of civilization, witchcraft is believed in still to a greater or less extent. Evil spirits, he says, are still believed to take possession of the old, the decrepit and the deformed, and sometimes also of the young. These supposed unfortunates then have to be exorcised, and it becomes a matter of duty on the part of the Shamans to dispossess them of their tormentors.

One of the curious things that will be noticed by the traveler in Alaska, is the natives' method of disposing of the dead

Many years ago cremation was generally practiced along the whole coast. This, however, has fallen into abeyance, except among those tribes who have not yet been visited by missionary influences. Wherever the influence of the Christian church has been felt the natives have adopted a modified form of disposal of the dead, based on our common custom.

The dead are usually placed in boxes, but as these boxes are not long enough to permit the whole body to recline at full length, the joints are severed so that the corpse may be placed in a sitting posture. Then the box is put away in some more or less remote place and usually kept above ground. There is a little bit of sentiment attached to the practice of the savages of placing their dead on some high point so that the departed spirit can look out upon the plains and valleys which were his former haunts.

Often, also, some of the personal effects of the deceased are placed beside him in the box. The Shamans, or medicine men, it must be remembered, are never cremated. Their bodies lie in state for four days, one day in each corner of the building. Then the corpse is conveyed to the dead house, placed in an upright position, and surrounded with all the blankets and paraphernalia that the Indian's idea of comfort suggests as necessary for the spirit land. It is a common practice of the people to dispose of the bodies of witches and slaves with the greatest secrecy.

Cannibalism was Prevalent.

It should be mentioned here that directly connected with and growing out of Shamanism is one of the most horrible of customs or practices, namely, cannibalism. This was commonly practiced by the whole people on the death of the chief, and the members of the tribe would enter with zest upon their horrid repast. Frequently, too, on the death of a chief a number of slaves were sacrificed that they might accompany their lord to the hereafter.   The bodies of these slaves, it is supposed, were cooked and eaten.

Within the days of the American occupation of the land, medicine men have been known to devour portions of corpses under the belief that they would thus acquire control of the spirit and gain influence over demons.   Happily, however, these enormities are growing fewer and fewer, and it is not improbable that at an early day, under the influence of Christian teaching, the superstitious rites and abominable practices of the savages will entirely disappear.

War dances and religious dances are also features of the Indian's life.   Dr. Sheldon Jackson describes one he witnessed at Fort Wrangel in 1879.   Says he :

"One afternoon we were invited to the house of Toy-a-att, a leading chief and Christian, to witness a representation of some of their national customs.   When everything was prepared, dressed in a hunting shirt, with face blackened and spear in hand, Toy-a-att appeared in the war dance.   Retiring amid much applause, he reappeared in the form of a wolf and with mask, rolling eyes and snapping teeth, gave the dance of the invocation of the spirits for successful hunting.

"Then he put on a horrible mask to represent the devil, and with hideous rattles, gave the devil or Tamanamus dance.   Then with dress and mask and large hat, with tinkling bells on the rim, and eider-down in the crown, which down he showered around the room as blessings upon his guests, and rattles in his hands, he gave us the religious dance of the Shamans, or medicine men. After the series of national dances, he came out and made a speech, apologizing for the feebleness of his representations."

A word more specifically about the Shamans.   When they are ill their relatives are expected to fast in order to promote his recovery. Their commands are absolute law. Every Shaman has any amount of paraphernalia, which includes a large assortment of masks—one for every spirit or demon over which he is supposed to have any power. The Shaman's hair is never cut. As was said above, on death his body is never burned or buried, but is put in a wooden box on four high posts.

Attending the funeral are certain performances, which begin at sunset and last till sunrise. Those who participate assemble in the Shaman's lodge and unite in a song, to which time is beaten on a drum. Then follows a form of religious dance, which in a measure includes or suggests all the ceremonies known to the art of Shamanism.

By these ceremonies, it is believed, the different spirits represented by the Shaman's various masks are all for the moment inspired.

Turning from these weird rites and superstitious beliefs, it is a pleasure to note that very many of the natives are clever artisans, if not artists. Their totem poles, as has been said, are often very skilfully carved. Arrow heads, spear heads, and silver and copper ornament likewise go to show that the natives are not destitute of artistic taste. The baskets of the Indians are also of ingenious design and coloring. These are made from grasses and roots.

The women do the weaving, and often the blankets they make are very beautiful in design and workmanship. The women sit day after day at their rude hand looms, and not infrequently it takes six months for an industrious workwoman to make a single blanket. The visitor to Alaska, however, is very apt to be imposed upon, as a large percentage of the blankets that are offered for sale, and said to be of genuine Indian make, are spurious. The real article, Mr. Bruce says, is now becoming very scarce.

CHAPTER XXI.

Spread of the Christian Faith.

Empress Catherine Takes the Initiative in Bringing a Purer Religion to the Savages—Work of the Early Russian Missionaries and the Progress of Their Work—Schools Early Established—Introduction of the Lutheran Church Due to the Efforts of Commercial Bodies to Provide for Their Employes—Sad Result of the Transfer of the Territory to the United States—Deed Interest Shown By the Natives—Some Striking Literature from the Wilds—Methodists Follow the Presbyterians in Their Missions —Great Hope for the Future.

THE cross has been planted in the wilds of Alaska for over a century ; and, strange to say, the Empress Catherine of Russia personally took the steps necessary to carry a purer religion into the barbaric rites and superstitious practices of the savages.

It was on June 30, 1793, that Catherine issued an Imperial order that missionaries should be sent to her American colonies. That order was obeyed immediately, as autocratic mandates are, and eleven monks set sail as soon as their equipment could be provided from Ochotsk for Kadiak Island.

This little band of Christian workers was in charge of Archimandrite Joasaph, elder in the order of Augustin friars. In 1796 Joasaph was made bishop and returned to Russia to receive consecration. That year was signalized by the erection of the first church in Alaska.

The newly-consecrated bishop and the missionaries coming with him were shipwrecked and lost on the return trip in 1799. All save one. This solitary monk remained alone in the Russian colonies for eleven years before another soul was sent to assist him in his work. Then, in 1822, three more priests were sent, who reached the colonies safely.

The one man, however, of all others, who did most to spread Christianity in Alaska during the days of the Russian occupation was Innocentius Veniaminoff. He began his labors at Unalaska in 1823. For seventeen years he worked as an ordinary priest, and then he was made bishop. Step by step he advanced from one position to another until he became Metropolite of Moscow, which is the highest position in the Greek Church. He died in the spring of 1879, and, it is safe to say, was sincerely mourned, not merely by his countrymen, but by the savages, among whom he had worked in Alaska, and to whom he had brought the blessings of civilized life.

What is more, he was the one Russian priest sent to Alaska who left an untarnished name in that country, and who evinced anything like the true missionary spirit. As a result of his exertions, the Russian Church at one time had seven missionary districts in Alaska, with eleven priests and sixteen deacons. In the year 1869 the Russian Church in Alaska claimed a membership of 12,140.

Helped by Fur Company.

It is one of the bright spots on the records of the Russian Fur Company that it contributed annually $6600 to the support of the missions. The sum of $2313 was annually received from the Mission Fund of the Holy Synod, and $1100 for the support of the work was received from the sale of candles in the church. The balance came from private individuals.

There was no opportunity for ostentation and display, and consequently the church work was conducted as economically as efficiently. The result was that in 1860 the church had a balance or surplus of $37,000, which was loaned out at five per cent. interest.

In evidence of the practical side of this early missionary work one may point out the fact that a school system was soon developed in the wilderness. The first school was established by Shelikoff on the Island of Kadiak. Three things alone were taught—language, arithmetic and religion. This was about the year 1792, and it was not a great many years thereafter that a similar school was established at Sitka. In 1841 an ecclesiastical school was opened in Sitka, and in 1845 this was made a regular seminary.

Object of the Schools.

Established as they were, under religious auspices, these schools were all of a parochial nature and their main object was to further the spread of the Greek Church. In 1860 we find a colonial school opened, with twelve students, which two years later had gained twenty-seven students.

Even in those far off districts and virtually among savages it is pleasing to find the first steps taken in a movement which has only of recent years become popular in civilized communities, namely, the education of women. In 1839 a girls' school was established in the wild regions of Alaska, which, in a certain sense, was also an orphans' home. It was patronized largely by children of the employes of the Fur Company.

Separate schools for the natives were also established, one being opened in 1825 on Unalaska Island. A similar school at Amlia Island had thirty in 1860. As far north as the lower Yukon, school-houses were also built.

The suspension of all these schools followed almost immediately upon the occupation of the country by the United States Government.

During the Russian domination the Russian-American Fur Company employed many Swedes, Finlanders and Germans, and to this fact is due the introduction of the Lutheran faith in Alaska. A church was built in Sitka in 1845, which was still running in 1852 under the charge of the first Lutheran minister sent to Sitka to provide for the population indicated. He was succeeded by the Rev. Mr. Wintec, who preached in the Swedish and German languages. Mr. Wintec remained until 1867, when the Russian Government withdrew his support, and he returned to Europe.

During the life of this early Lutheran Church, however, the work was done as carefully and as economically as by the Greek Church, and the denomination soon accumulated many thousand dollars in church property. It should be observed that the Protestant Churches of Russia, while allowed no self-governing and self-sustaining organizations, are still recognized under the Ministerium of Public Instruction Provision is made for their support, which comes direct from the public treasury.

Decline of Church Work.

It seems that when, in 1867, the great Territory of Alaska became part of the dominion of the United States it was to fall away from God's providence. At least, for many years nothing was done either to preserve or extend the work that had already been done. This in spite of the fact that when the purchase was made by Secretary Seward the matter of evangelizing the savages was discussed by almost every church organization throughout the country. Says the Rev. Sheldon Jackson:

" It was expected that the churches of the United States, with their purer religion and greater consecration, would send in more efficient agencies than Russia had done. But ten years rolled around and the churches did nothing. Ten years passed and hundreds of immortal souls, who had never so much as heard that there was a Savior, were hurried to judgment from a Christian land. Ten years came and went and thousands were left to grow up in ignorance and superstition, and form habits that will keep them away from the Gospel, if it is ever offered them."

The Indians themselves, however, had experienced something of the blessings which the Greek Church had brought them and noticed with regret that their brethren in the districts where formerly the Russian priests ministered were retrograding.

So, in the spring of 1876, Clah, Su-gah-na-te, Ta-lik, John Ryan, Lewis Ween, Andrew Moss, Peter Pollard, George Pemberton and James Ross, all Tsimpsean Indians, went from Fort Simpson to Fort Wrangel to obtain work. Here they secured a contract to cut wood for the government, and here on the Sabbath it was their practice to meet together for worship, as in the old days before Alaska became a portion of the United States.

This little band of devoted Indians is responsible for the re-birth of Christianity in the Territory. Its members found a warm friend and protector in Captain S. P. Jocelyn, of the United States Infantry, who was then in command at that station. He took a hand in the movement, secured a room for worship on the Sabbath, and helped the Indians in every possible way.

All this in face of the futile efforts being made in the United States. It may be interesting to note some of the projects in the old settled States that came to naught.

Some Apathetic Projects.

The Rev. Dr. Saunders, of the Board of Domestic Missions of the Presbyterian Church, offered a resolution soon after the purchase of the Territory that a band of missionaries be sent by the church to Alaska. A similar proposition was made to the Committee on Home Missions of the same church. From 1869 to 1877 the Rev. George H. Atkinson repeatedly agitated the question of sending missionaries to the Territory.

These efforts in the Presbyterian Church were backed up by Major-General O. O. Howard, of the United States Army, and the Hon. Vincent Colyer, Secretary of the Board of Indian

Commissioners.   This last friend of the Indians even succeeded in getting Congress to appropriate $50,000 for educational purposes in the Territory, but no one was found willing to go to the wilds of the North and administer the fund, and so it was not used.

In 1875 and 1876, however, the Rev. Sheldon Jackson, accompanied by Mrs. A. R. McFarland, went to the Territory and

MISSIONARY AMONG THE ALASKA INDIANS.

renewed the work for the Presbyterian denomination.   The missionaries met at various houses, in vacant stores, and even in the huts of the natives, and held religious services, and especially lent their aid in support of the little band of Indians mentioned above, and in 1879 there was such interest in Christian work in the districts they visited that services of a revival nature were frequently held and were largely attended by the Indians.

It is curious to notice how quickly and sincerely the savages took to the new life and its literature.	Dr. Jackson gives a list of some inscriptions he copied from an Indian cemetery, where once were found, as indications of religious belief, nothing but the totem poles of the savages.	Among these inscriptions were the following :

" His end was peace." " There is hope in his death." "Jesus pity me." " Take my hand and lead me to the Father." "I have been poor in the world and wicked, but all is over now." "Take me home to God." " Said to his father, trust in God." " He departed trusting in Jesus." " Of such is the kingdom of Heaven." " His last act was to sing a hymn and offer a prayer to God."

Still more interesting and significant is the following creed or statement of belief, or religious compact, which the Indians drew up and signed :

1. " We concur in the action of Mr. I. C. Dennis, Deputy Collector of the United States Custom House, appointing Toy-a-att, Moses, Matthew and Sam to search all canoes and stop the traffic of liquor among the Indians.

2. " We, who profess to be Christians, promise with God's help to strive as much as possible to live at peace with each other, to have no fighting, no quarreling, no tale-bearing among us. These things are all sinful and should not exist among Christians.

3. " Any troubles that may arise among the brethren, between husbands and wives, or if any man leaves his wife, these brethren, Toy-a-att, Moses, Matthew, Aaron and Lot, have authority to settle the troubles and decide what the punishment shall be, and if fines are imposed, how much the fincs shall be.

4. " The authority of these brethren is binding upon all, and no person is to resist or interfere with them, as they are appointed by Mr. Dennis and Mrs. McFarland.

5. " To all the above we subscribe our names."

These little incidents show that the natives were ripe for good Christian work, and those who had the courage to brave the dangers and hardships of the North in the interests of the church sent home the most favorable reports as to their reception and the most heartfelt regrets that the great Christian church of the United States should be so dilatory and apathetic in its mission work in the Territory.

And it must not be supposed that these children of Nature were slow of understanding or lacking in natural gifts. We quote, as an example of Indian eloquence and Indian earnestness, the following, which was reported in the Port Townsend *Weekly Argus*. The speaker was Chief Yoy-a-att, whose name occurs in the religious compact given above :

" The white man's God we knew not of Nature evinced to us that there was a great first cause. Beyond that all was blank. Our god was created by us, that is, we selected animals and birds, the images of which we revered as gods.

" Natural instincts taught us to supply our wants from that which we beheld around us. If we wanted food, the waters gave us fish ; and if we wanted raiment, the wild animals of the woods gave us skins, which we converted to our use. Implements of warfare and tools to work with we constructed rudely from stone and wood. Fire we discovered by friction.

Change in the Dream.

" In the course of time a change came over the spirit of our dreams. We became aware of the fact that we were not the only beings in the shape of man that inhabited this earth. White men appeared before us on the surface of the great waters in large ships, which we called canoes.

" Each day the white man becomes more perfect in the arts and sciences, while the Indian is at a standstill. Why is this ?

Is it because the God you have told us of is a white God, and that you, being of his color, have been favored by him? My brothers, look at our skin. We are dark. We are not your color; hence you call us Indians. Is this the reason that we are ignorant? Is this the cause of our not knowing our Creator?

We ask of our father at Washington that we be recognized as a people, inasmuch as he recognizes all other Indians in other portions of the United States. We ask that we be civilized, Christianized and educated. Give us a chance, and we will soon show to the world that we can become peaceable citizens and good Christians."

In view of this direct appeal from the Indians themselves it is rather lamentable that the Christian Church of the United States for more than a decade not merely allowed all the work done by the Russians to lapse, but even brooked the introduction of evil practices and evil ways among the Indians. It must not be forgotten that these savages were apt scholars not less in the vices of civilization than in its virtues.

Took Naturally to Whisky.

In illustration of this it may be said that early in the days of the American occupation the savages learned to distil whisky, calling their rudely made stills hoo-chi-noo. The natives made the whisky by distillation from molasses and their stills were very simple affairs. They consisted of two discarded kerosene oil cans and the long, hollow root of the sea weed for a pipe. The still took its name from the tribe that first manufactured it. The tutor of the savages in the art of making whisky was a discharged soldier.

From 1877, when Dr. Jackson and Mrs. McFarland began the work of the Presbyterian missions of Alaska at Fort Wrangel, interest never died out. Steps were taken in the United States to render assistance and the little band of Indians who joined together in Christian work before the missionaries' arrival were their constant helpers. Communication was had as often as possible with interested people in the South, and soon these fearless workers for Christ had the satisfaction of knowing that, in a large measure, wherever their efforts were directed, they had put an end to witchcraft, and to many of the grosser practices of the Indians, and had thus brought better hopes, better manners and better morals among the natives.

Methodists Begin Work.

About the same time that this movement was inaugurated by the Presbyterian denomination, a similar movement was started by the Methodist Church. Dr. Jackson pays a tribute of appreciation to three men, whom he deems remarkable workers in the cause of religion in Alaska. These are the Rev. Innocentius Veniamimoff, of the Greek Church, who, commencing as a humble priest in Alaska, was made Bishop and then Primate of the Greek Church of all Russia ; Mr. William Duncan, of the Church Missionary Society of London, who built up the model Indian village of Metlahkatlah ; and the Rev. Thomas Crosby, missionary of the Methodist Church of Canada at Fort Simpson, on the edge of Alaska.

It was in February of 1862 that Mr. Crosby left his old parish for work among the Indians in the Territory. He began by teaching an Indian school at Nanaimo in 1863, and in 1867 he took a circuit extending up and down the coast among the Indians for 180 miles, and up the Fraser River to Yale. Two years later he inaugurated a regular system of typical revival meetings among the natives, and hundreds of the Flathead Indians became interested and professed conversion.

Mr. Crosby had several efficient allies. Among these was a

Mrs. Dix, who was a full-blooded Indian woman, the daughter of a great chief, and a chieftaness in her own right. When a child she was frequently taken up a great river in a canoe and taught to worship a large mountain peak. Her mother's god, Dr. Crosby says, was a fish. Desiring to learn something of the white man's God, the Indian girl began to attend religious services in Victoria, following it up systematically for seven years. But, as she afterwards stated, she found no light or comfort.

A New Recruit.

In 1868 a great medicine man named Amos, who, in his incantations, had torn in pieces with his teeth and eaten dead bodies, commenced attending the Methodist Church. Amos became one of the first converts and soon a class leader. Through him Mrs. Dix became a disciple of Christ, and later on an ardent worker for the betterment of her people.

Another instance of Indian conversion may be given as a sample of the interest the natives took in the efforts made to instruct them in Christian life. An old, grey-haired, blind Indian, hundred of miles away, heard of the work being done by the Methodist missionaries, and took his grandson and started for the coast. They paddled many a lonely mile in their canoe, and many were the suns that set upon their bleak evening camp.

When near the coast, it is related, they were met by a Christian. The blind man was ever repeating to himself as he groped along: "Jesus Christ came into the world to save sinners." The attention of the Christian was arrested and his interests awakened. He stopped the little party and got from the old man the story of his wanderings. Then the Indian was directed to a mission station and went on his way rejoicing. He, too, during his life, and his grandson after him, were energetic and enthusiastic assistants of the missionaries.

33

Under Methodist auspices schools of various kinds have been successfully established.   A day school in winter was soon running, which had 120 pupils, and it is not too much to say that the little band of energetic spirits who gathered about Mr. Crosby soon reached whole tribes and led them steadily, even though slowly, to a higher form of civilization.

Under the influence of Christianity the Indians began to abandon their large houses, which were the common abode of several families, and build separate houses for each family.   Within two years from the time the work began sixty such dwellings had been erected by Indian mechanics, and the old houses, that had been scenes of so much depravity and corruption, were fast disappearing, with other remnants of the Indian's old life.

No apology is offered for the insertion of the following simple but touching native address, which tells much of the spirit of the Indians and the earnestness with which they welcomed the new life that was brought to them:

" We, the chiefs and people of the Naas, welcome you from our hearts on your safe arrival here, to begin in earnest the mission work you promised us last spring.

Hope for the Young.

" Our past life has been bad, very bad.   We have been so long left in darkness that we fear you will not be able to do much for our old people, but for our young ones we have great hopes.   We wish from our hearts to have our young men, women and children read and write, so that they may understand the duties they owe to their Creator and to each other.

" You will find great difficulties in the way of such work, but great changes cannot be expected in one day.   You must not get discouraged by a little trouble, and we tell you again that we will all help you as much as we can.

" We believe this work to be of God. We have prayed, as you told us, and now we think that God has heard our prayers, and sent you to us ; and it seems to us like the day breaking in on our darkness, and we think that before long the great Sun will shine upon us and give us more light.

" We hope to see the white men that settle among us set us good example, as they have had the light so long, they know what is right and what is wrong. We hope they will assist us to do good that we may become better and better every day by following their example.

" We again welcome you from our hearts, and hope that the mission here will be like a great rock never to be moved or washed away. And in order to do this, we will pray to the Great Spirit that His blessing may rest upon this mission and upon us all.

" (Signed)   CHIEF OF THE MOUNTAINS
and six other Chiefs."

CHAPTER XXII.

British Columbia and Northwest Territory.

Region is One of Vast Extent and Diversified Features—Has a Magnificent
Ocean Frontage—A Land of Great Rivers which Afford Internal High-
ways—Greatest of All is the Columbia—Has a Large Ocean Trade Even
Now—Experiments in Fruit Growing Successful—Construction of Rail-
ways Has Given an Impetus to Development—Many Districts Famous
for Their Grain and Others for Their Mineral Deposits—Gold Mines in
Abundance—Klondike Within the Canadian Territory—Some of the
Mines Now Worked—Silver Not Wanting.

THE vast stretch of British territory lying immediately adjacent to Alaska, British Columbia and Northwest Territory,
properly calls for a description in the present work, since it
contains many of the most valuable gold fields about which there
was such excitement in the year 1897. The Klondike district, it
will be remembered, is at least thirty-five miles within the real or
alleged boundary between Canada and the United States.

British Columbia is the most westerly province of Canada, extending from the 49th parallel on the south to the 60th degree
of north latitude, and from the summit of the Rocky Mountains
westward to the Pacific Ocean, Vancouver Island and Queen's
Charlotte's Islands being included within its bounds. The Province contains the immense area of 383,000 square miles. It is
a diversified country of immense mountain ranges, fertile valleys,
splendid forests and magnificent waterways.

The position of British Columbia on the north Pacific Ocean,
bearing a somewhat similar relation to the larger portions of the
American continent that Great Britain does to Europe for the
trade of the world, makes it one of the most important and valu-

able provinces of the Dominion, both commercially and politically.

The Province has a magnificent ocean frontage of 1000 miles. This coast line abounds in harbors, sounds, islands and navigable inlets. Principal among these harbors are English Bay and Coal Harbor, at the entrance to Burrard Inlet, a few miles north of the Fraser River. Vancouver is the terminus of the Canadian Pacific Railway, and is situated between these harbors. Victoria, on Vancouver Island, also has a magnificent outer harbor at which all the ocean liners dock, and an inner harbor for vessels drawing up to eighteen feet. It has also another harbor at Esquinalt, three miles to the southeast.

This latter harbor is about two miles long and nearly two miles broad in the widest part. It has an average depth of six to eight fathoms and thus affords an excellent anchor for vessels. The Canadian government has built here a dry dock with a length of 450 feet and a width of ninety feet, which will accommodate vessels of the largest size.

Magnificent Rivers.

Like Alaska, British Columbia and Northwest Territory have some magnificent rivers, principal among which are the Fraser, the Columbia, the Thompson, the Kootaney, the Skeena, the Stickine, the Laird, and the Peace. The Fraser River is the greater water course of the province, rising in the northern part of the Rocky Mountains, and running about 200 miles in two branches in a westerly direction, and thence in one stream due south for nearly 400 miles before turning to rush through the gorges of the coast range to the Straits of Georgia.

The total length of the river is therefore about 740 miles. On its way the Fraser receives the tributary waters of the Thompson, the Chilicoten, the Lillooet, the Nicola, the Harrison, the Pitt, and a number of smaller streams. For the last eighty miles of its course it flows through a wide alluvial plain, which has largely been deposited from its own silt.

The Columbia River rises in the southeastern part of the province, in the neighborhood of the Rocky Mountains, near the Kootanay Lake. On this lake has already been established a regular steamboat service. The Columbia runs north to just beyond the 52d degree of latitude, and then turns suddenly and runs due south into the State of Washington. The loop thus made is commonly known as "The Big Bend of the Columbia." No less an area than 195,000 square miles is drained by the Columbia River.

Network of Lakes and Creeks.

The Peace River rises some distance north cf the north bend of the Fraser and flows eastwardly to the Rocky Mountains, draining the plains on the other side. In the far north are the Skeena and Stickine Rivers, both flowing into the Pacific, the latter, of course, being in a country valuable for its gold deposits. The Thompson River has two branches, which are known as North Thompson and South Thompson. The former rises in small lakes in the Cariboo district, and the latter in the Shuswap Lakes in the Yale district.

British Columbia, undeveloped and little known, as it is, is already an important Province of the Dominion. Its trade, which is ever rapidly increasing in volume, has assumed immense proportions, and reaches to China, Japan, Australia, Europe, Africa and South America. The principal seaport—Vancouver, the western terminus of the Canadian Pacific Railway—is the gateway of the ncw and shortest highways to the Orient, the Far North, the Tropics and the Antipodes. The voyage from Yokohama, Japan, to London has already been made in twenty-onc days by this route, beating all previous records; and the

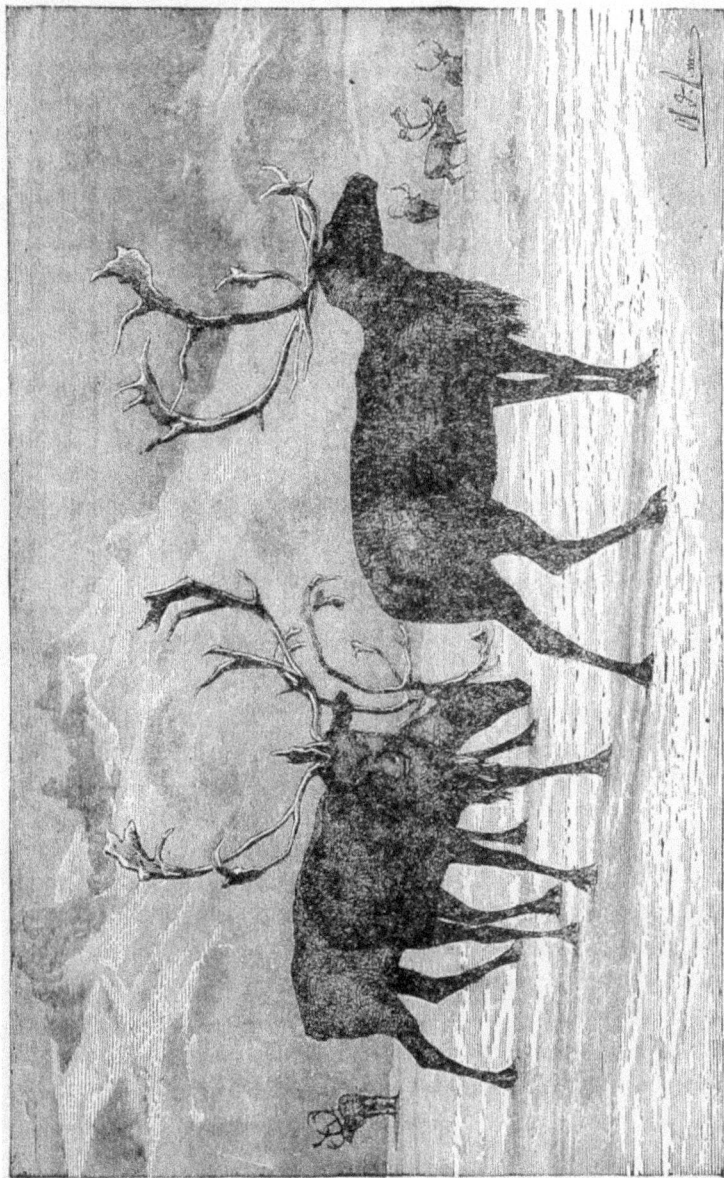

REINDEER OF NORTHERN ALASKA.

journey to and from Australia, via Vancouver, is speedier and more pleasant than by any other route.

British Columbia attracts not only a large portion of the Japan, China and Australian rapid transit trade, but must necessarily secure much of the commerce of the Pacific Ocean, the steamers of the Canadian-Australian Line touching at the Hawaiian and Fijian Islands. Its timber is unequalled in quantity, quality or variety ; its numerous mines already discovered, and its great extent of unexplored country, speak of vast areas of rich mineral wealth ; its large fertile valleys indicate great agricultural resources, and its waters, containing marvelous quantities of the most valuable fish, combine to give British Columbia a value that has been little understood.

Boundaries of British Columbia.

The vast Territory of British Columbia is divided into six districts, the New Westminster, the Cassiar, the Cariboo, the Lillooet, the Yale, and the East and the West Kootenay.

The New Westminster district extends from the international boundary line on the South to 50° 15' on the North. Its eastern boundary is the 122° longitude, and its western the 124° where it strikes the head of Jarvis Inlet and the Straits of Georgia. In the southern portion of this district there is a good deal of excellent farming land, particularly in the delta of the Fraser River. The soil there is rich and strong, the climate mild, resembling that of England, with more marked seasons of rain and dry weather, and heavy yields are obtained without much labor. Very large returns of wheat have been got from land in this locality—as much as sixty-two bushels from a measured acre, ninety bushels of oats per acre, and hay that yielded three and one-half to five tons to the acre, and frequently two crops, totaling six tons.

Experiments have of late years been made in fruit growing, with the most satisfactory results—apples, plums, pears, cherries and all the smaller fruits being grown in profusion, and at the Experimental Farm at Agassiz, figs in small quantities have been successfully produced. This part is fairly well settled, but there is still ample room for new comers. Those having a little money to use, and desirous of obtaining a ready-made farm, may find many to choose from. These settlements are not all on the Fraser ; some are at a distance from it on other streams. There is considerable good timber in the western and south-western portions.

The chief towns of this district are Vancouver and New Westminster. Vancouver is situated on a peninsula, having Coal Harbor, in Burrard Inlet, on the East, and English Bay on the West. It is surrounded by a rare country, both in beauty and climate. In the far distance it is backed by the Olympian range. On the north it is sheltered by the mountains of the coast, and it is also sheltered from the ocean by the highlands of Vancouver Island. While it is thus protected on every side, it enjoys the sea breeze from the Straits of Georgia.

The inlet affords unlimited space for sea-going ships, the land falls gradually to the sea, rendering drainage easy, and the situation permits of indefinite expansion of the city in two directions. It has a splendid and inexhaustible water supply brought across the inlet from a river in a ravine of one of the neighboring heights.

The Canadian Pacific Railway was completed to Vancouver in May, 1887, when the first through train arrived in that city from Montreal, Port Moody having been the western terminus from July of the preceding year. In 1887, also the Canadian Pacific Railway Company put a line of steamships on the route between Vancouver and Japan and China, and in 1893 an excel-

lent service was established between Vancouver and Victoria
and Australia, via Honolulu and Suva, Fiji.

These three important projects are giving an impetus to the
growth of the city by placing its advantages entirely beyond the
realm of speculation, and the advancement made is truly
marvelous.

New Westminster was founded by Colonel Moody during the
Fraser River gold excitement in 1858. It is situated on the
north bank of the Fraser River, fifteen miles from its mouth.
It is accessible for deep water shipping and lies in the centre of a
tract of country of rich and varied resources. It is connected
with the main line of the Canadian Pacific Railway by a branch
line from Westminster Junction and with Vancouver by an electric
railway.

This town is chiefly known for its great salmon trade and its
lumber business. The agricultural interests, however, of the
district are now coming to the front and the city has the promise
of stability and importance.

Wide Stretches of Fertile Lands.

The Cassiar district occupies the whole western portion of the
province from the 26th degree of longitude. While its argicul-
tural capabilities have not yet been fully determined, it is known
to possess a number of tracts of very fertile land, notably that
occupied by the Bella Coola Colony, which has the promise of
great prosperity.

The district contains some of the richest gold mines yet discovered in the province, and indications are numerous of further
mineral wealth to be developed. There are some prosperous
fish canning establishments on the coast, and parts of the district
are thickly timbered. Communication with the Cassiar District
is principally by water. Steamers start at regular dates from

Victoria for the Skeena River, Port Simpson and other points on the coast within the district.

The Cariboo district lies between Cassiar on the west and the Canadian Northwest on the east, its southern boundary being the 52d parallel. This district contains the famous Cariboo mines, from which $50,000,000 in gold have already been taken.

It is said that there is still in this district a promising field for the miner. The immense output of the placer diggings being the result of explorations and operations necessarily confined to the surface, the enormous cost and almost insuperable difficulties of transporting heavy machinery necessitate the employment of the most primitive appliances in mining.

Obstacles a Hindrance.

These obstacles to the full development of the marvelously rich gold fields of Cariboo have been largely overcome by the construction of the Canadian Pacific, and the improvement of the great highway from that railway to northern British Columbia, with the result that the work of development has recently been vigorously and extensively prosecuted. During the past few years several costly hydraulic plants have been introduced by different wealthy mining companies which are now operating well-known claims with the most gratifying results, and there is every prospect of a second golden harvest, which in its immensity and value will completely overshadow that which made Cariboo famous thirty years ago.

The development work for the season of 1896 served to materially advance the interests of this district. Many hundreds of men found employment in 1897, and it is said that no one wishing to do honest work for fair pay need there be idle.

The quartz mines have not as yet been exploited only in a very superficial way, but the rich surface showing on Burns,

Island and Bald mountains, all tend to prove that further research and a fair use of capital will make the quartz mines of the Cariboo district among the great producers and dividend payers of the world. Gold abounds in every valley, and in every stream that empties into it, and there is no estimating the unusual activity in the Cariboo mining circles, some of the richest places merely awaiting the advent of capital for that development which the new condition of affairs has rendered easily possible.

Cariboo is not without agricultural resources, and there is a limited area in scattered localities in which farming and ranching are carried on ; but this region will always prove more attractive to the miner than to the settler. The early construction of a railway from a point on the main line of the Canadian Pacific, through the district, when completed will open up many desirable locations and largely assist in developing the immense mineral wealth already known to exist.

The Yale district is on the east of Lillooet and New Westminster. It extends southward to the international boundary and eastward to the range of high lands that separates the Okanagan Valley from the Arrow Lakes. This district, it is said, affords fine openings for miners, lumbermen, farmers and ranchmen.

Is Famous for Grain.

Okanagan is famous as a grain growing country. For many years this industry was not prosecuted vigorously, but of late there has been unusual activity in this respect, and samples of wheat raised in the district were sent to the Vienna Exposition, where they were awarded the highest premiums and bronze medals. One of the best flouring mills in the Dominion is now in operation at Enderby. It is said that the flour manufactured at this point is equal to the product of any other section of North America.

Considerable attention is now being given to the various kinds of fruit culture, and an important movement is on foot looking to the conversion of the grain fields into orchards and hop fields. Attention has been more particularly turned to the production of Kentish hops, and during the past four years hops from this section have brought the highest prices in the English market, competing successfully with the English, the Continental, and those grown in other parts of America.

The Earl of Aberdeen, Governor-General of Canada, has a large fruit farm near Kelowna, on the east side of the lake. His Excellency has also over 13,000 acres near Vernon, in the Coldstream Valley, where general farming, hop growing and fruit raising are carried on. His orchard of about 125 acres is the point of attraction for visitors to Vernon. An excellent quality of cigar wrapper and leaf tobacco is grown about Kelowna, shipments of which are yearly increasing, but the production has not yet become general.

Has a Vast Acreage.

The West Kootenay district is the next east of Yale, extending north and south from the Big Bend of the Columbia to the international boundary, embracing, with East Kootenay, an area of 16,500,000 acres. West Kootenay is noted chiefly for its great mineral wealth. Rich deposits of various metals have been discovered in different sections and new finds have been made almost weekly for years. It is described by those who have visited it as a country of illimitable possibilities. It is as yet, however, only in the earliest stages of development. Its vast hidden wealth is thus largely a matter of conjecture.

Great progress has been made, though, and many camps have been established throughout the entire district, and equipped with all the necessary machinery for mining operations. In the

Lardeau, Big Bend and other parts of the district the promise is that the output will be very large in the near future.

The output of ore in 1896 in West Kootenay approximated $6,000,000, and with the additional transportation and smelting facilities now being afforded this amount will doubtless be largely increased during 1897. Capitalists and practical miners have shown their unbounded confidence in West Kootenay by investing millions of dollars in developing claims, equipping mines, erecting smelters, building tramways, etc., and an eminent American authority speaks of it as "the coming mining empire of the Northwest."

In 1896 the population of West Kootenay was trebled, and the year witnessed the creation of a number of new mining camps which astonished the world with their phenomenal growth and prosperity. There are valuable timber limits in different parts of the country, and saw-mills are in operation.

Mines Easily Reached.

The mining districts are easily reached from Revelstoke, on the main line of the Canadian Pacific Railway, about midway between the eastern slope of the Rockies and the Pacific coast. From this point a branch line south is completed to Arrowhead, at the head of Upper Arrow Lake, from which the fine new steamers of the Columbia & Kootenay Steam Navigation Co. are taken to Nakusp, near the foot of the lake, where rail communication with the towns of the Slocan, the principal of which are New Denver, Three Forks and Sandon, the centre of a rich mining region, has been established, and there is an excellent steamboat service on Slocan Lake.

Steamers can also be taken from Arrowhead past Nakusp to Robson, at the mouth of the Lower Kootenay River, along the bank of which unnavigable river the C. P. R. runs by its Colum-

bia & Kootenay branch to Nelson, the metropolis of the Kootenay mining district, in the vicinity of which are the celebrated
Silver King and other mines.

From Nelson steamers ply to all the mining towns on the
Kootenay Lake—Pilot Bay, Ainsworth, Kaslo, etc. From Robson the steamers continue down the Columbia to Trail, from
which point Rossland, the centre of the new gold fields of the
Trail Creek district, is reached by railway, and to Northport in
the State of Washington.

The East Kootenay district comprises the larger part of the
famous Kootenay region of British Columbia, which is entered
from the East at Golden, on the Canadian Pacific Railway.
Here, too, mines are worked successfully, and prospectors are
constantly seeking for new fields. The district contains a valley
nearly 300 miles long from the internationally boundary line to
the apex of the Kootenay triangle of the Big Bend of the
Columbia, with an average width of from eight to ten miles.

An Attractive Valley.

In the centre of this valley are enclosed the mother lakes of the
Columbia River, which lie 2850 feet above sea level. The soil
is reported to be rich. Judge Sproat describes the country as
one of the prettiest and most favored valleys in the province,
having good grass, a fine climate, established and promising
mines, excellent waterways, and an easy surface for road making.

There are numerous mines at work in different sections of the
district, chiefly in the Lower Kootenay country, in the north of
which are the Kaslo-Slocan mines; in the centre, those around
Nelson and Ainsworth, and in the south those of the Goat
River and Trail Creek districts. There are no richer gold fields
than those of the latter mentioned district, of which Rossland is
the centre. Several mines are already operated extensively and

are paying large monthly dividends, while new discoveries indicate that the full richness of this region cannot yet be even
approximately estimated.

Large shipments of ore are being made from Le Roi, War
Eagle, Josie, Nickel Plate, Crown Point, Evening Star, Columbia
and Kootenay, O. K., Jumbo, Cliff, Iron Mask, Monte Christo,
St. Elmo, Lily May, Poorman and other leading mines, while
the Centre Star and other properties have large quantities on the
dump ready for shipment. With increased home smelting facilities, the output of the camp will be immensely increased.

The most notable silver mines are in the famed Slocan district,
from which large shipments of ore have been and are being
made—the general character of its ore being high grade galena,
often carrying 400 ounces of silver to the ton, and averaging 100
ounces and over. The principal mines are the Slocan Star,
which paid $300,000 in dividends in 1896, Enterprise, Reco,
Good Enough, Whitewater, Alamo, Ruth, Two Friends, Dar-
danelles, Noble Five, Washington, Payne, Idaho, Mountain
Chief and Grady groups.

During the summer of 1896, some of the richest discoveries
in the Kootenay were found in the Salmon River country, between the Lower Kootenay River and the international boundary.
In the North, in the Illecillewaet, Fish Creek and Trout Lake
districts are rich properties which are being worked, and around
Lardeau, some valuable placer gold mines and extensive deposits
of galena are being developed. Between the Gold Range and
the Selkirks is the west side of the Big Bend of the Columbia
River, that extends north of the 52d parallel.

CHAPTER XXIII.

Advent of Winter.

Confirmation of Stories About the Wealth of Klondike and Alaska—Perils
of the Passes—Dark and Bright Sides of the Picture, as Seen by Argo-
nauts—New Diggings Opened—Copper River and Cook's Inlet—New
Strikes in the Yukon Basin—Two Experiences in Crossing Chilkoot
Pass—Over the White Pass—Belated Gold Seekers Camping on the
Trail—Woes of the Horses—New Routes—Tramway at Dyea—Via the
Snow Train—At St. Michael's—In Dawson and Skagway—Glacier Slide
and Flood—Mt. St. Elias Scaled.

THE advent of winter in Alaska in the boom year of 1897
found several things definitely settled for the argonauts,
which before had been in some senses matters of debate,
if not of doubt. For one, there was no longer any question
that the Klondike was the richest gold field in the world. For
another, it was settled that to get to the diggings was no holiday
jaunt. But it had also been demonstrated that the trip was
practicable, and, for men who chose to use common sense in
outfitting and traveling, even easy, in comparison with some
frontier experiences of other pioneers.

It had cost much money and misery to gain this knowledge—
the world was the richer by the measure of the bitter experiences of individuals.

It was estimated more millions had been spent between the
middle of July and the first of October in procuring outfits
and transportation to the Klondike than had been dug and
washed out of the golden placers in the entire year. And
much of this treasure had gone to waste, too—the trails from
the ocean over the mountains were strewed with wreckage, till
they looked not unlike the path of a routed and panic-stricken
army. "Tenderfeet" had played their historic part.

34   529

The physical waste had also been something appalling. Not so many lives had been sacrificed as in some other famous gold stampedes, for the way was not so long nor the perils so many as in the case of California or Australia or the Rand; nevertheless the total was a startling array of casualties. Lives had gone out in icy torrents or under avalanches, murder and the swift vengeance of the vigilantes had been done, and the tragic element had been further sustained by the uncounted scores of those who had broken health and spirit in the mad rush through frontier privations and perils only to fall by the wayside.

On the Bright Side.

That was the dark side of the picture. On the other hand, thousands of men and not a few women had got through to Dawson and its neighborhood, and many more, in good health, with ample supplies and unflagging energies, were already well along on the journey to the mines when October set in. Reports from the Klondike indicated that the fears of wholesale starvation among the mining camps during the winter were unfounded. The commercial and trading companies had succeeded in getting in large stores of staple supplies, and the prospect was for abundant and profitable employment for those who, by preference or fate, might be forced to work for others. Preparations for ample policeing of the Yukon basin had been made, and law and order, unusual in primitive mining camps, were promised. Engineering, science and capital had come to the solution of the transportation problem, and the days of relatively rapid and easy traffic over the passes and through the wilderness seemed just at the dawn. If the picture had a dark side for the '97rs who had tried to get through and failed, it had a compensatory bright side for those who were looking forward to trying their fortunes in 1898.

This later history of Alaska is being written daily in the experiences of thousands. Much that is new one day will be old the next, so rapidly does the Klondike kaleidoscope revolve. Some of the more remarkable incidents of the Alaskan autumn of '97 follow. They are all part of the wonderful chronicle; though the relative importance of each to the prospective gold seeker may be varied by after events, their place as facts in the marvelous development of the new El Dorado is fixed.

In New Diggings.

The close of the season brought the news of many new diggings. Peace, Stewart and McMillan rivers attracted especial attention of prospectors during the fall, and many parties went in to explore the new fields. The most interest probably centered, however, in the Cook's Inlet and Copper River countries. The former field seemed to be exceptionally rich. Early in October over one hundred miners reached Sitka from the Inlet and every one had his " pile." Most of the metal came from Mill Creek, Lınk Creek, Bear Creek, Cañon Creek or smaller streams in that vicinity.

The clean-up represented the work of only one season on the claims. The men who brought out the most were those who had worked their claims the longest. George T. Hall, who represented the Alaska and Klondike Mining Company as expert and chief engineer, said the gravel in the Inlet region would average $1.50 a yard and there was no end to it in sight.

The comparatively temperate winter climate of the southeastern coast region attracted early attention to the Copper River as a handy make-shift for those who had sought to go into Dawson via the passes and had been stranded at Skagway or Dyea by lack of transportation over the crowded trails. Several parties were reported organizing for winter prospecting tours in that region early in the fall and the chances were thought to be that another year might see a formidable rival to the Klondike in a more accessible basin. The chief drawback to these ventures lay in the stories of the savage native tribes, related to be fiercer warriors than any others on the Alaskan coast, but the most appalling of these tales were freely discounted by veteran frontiersmen and, at the worst, it was argued, a well-equipped body of determined men could probably find a way to keep their gold and get out with it, if they made a strike.

In the Klondike.

In the Klondike new discoveries were reported on Victoria and Bear Creeks which were as rich as those on the original stream, but both fields were small and every claim was quickly located. Miller Creek and Minook Creek also had "booms" and in fact every gulch was the scene of more or less excitement as the rush for gold swept over the country from one bonanza to another. Hunker Creek and Gold Bottom (suggestive) Creek were among the most highly esteemed of the later fields. J. F. Maloney, of Juneau, estimated some of the Hunker claims at $2000 to the box.

Dominion Surveyor William Ogilvie, who is an acknowledged authority, was one of the latest to come out from Dawson en route to Ottawa on official business. In a report on the gold-bearing quartz prospects of the Yukon valley, he said:

"It is a most difficult country to do quartz prospecting in. Only at a few points along the creek is any rock exposed. The tops of the higher hills and ridges are void of vegetation, except arctic mosses and lichens, but all the rest of the country is covered with a thick layer of moss which, again, supports scrub spruce, some scrub white birch, and a thick growth of northern shrubbery. This completely conceals the surface of the rocks, and to remove to a sufficient extent to search for quartz prospects would entail a vast amount of labor—much more than the ordinary every day prospector can afford.

Quartz Mining.

" The cheapest and most expeditious methods of quartz prospecting here would be by diamond drill. A light, portable machine of that description, a compact light engine and boiler sufficient to work it, could be easily made and set up at various points along the various creeks. From the cones thus obtained experts could readily determine what the probabilities and prospects were. This requires capital, but I have no doubt a company formed with this object in view, prospecting in this way, would find it a profitable investment.

"All the gold I have seen taken out of El Dorado and Bonanza, for that matter of other creeks, too, bears no evidence of having traveled any distance. Many, it might be said the majority, of the nuggets found are just as regular and irregular in shape as if they had been hammered out of the mother lode, instead of being washed out of the gravel.

"I have seen no evidence of glaciation anywhere in that district, so I cannot help coming to the conclusion that much of the mother lode from which this gold came will yet be found along the valleys. Whether it is concentrated enough to pay for the expense of quartz mining can only be determined by proper search. I cannot help thinking that much of it will.

" Now let us take a glimpse of the country south of the Stewart River, some sixty-five or seventy miles further up and about 400 miles in length. Its tributary will easily double this. This gives us in the neighborhood of 1000 miles of stream. On a great deal of surface prospecting has been done and fine gold found everywhere.

" Now, where fine gold is found coarse gold has generally been found, too. Assuming this to hold good in the Stewart valley, we will have here one of the largest, if not the largest, mining areas in the world, upwards of one hundred miles farther up the Pelly Joinso. On this fine gold has been found, too. Above is the Hootalinqua, upon which fine gold has been found. Still farther south the Cassiar district, in British Columbia, was a famous gold field. Farther on yet the Cariboo district was famous.

Where the Gold Is.

" Now, draw a line through these several points and produce it northwestward, you will find that the Forty Mile gold bearing area, Mission Creek and Seventy Mile Creek, below Forty Mile, Birch Creek, Minook Creek, and still farther down the Klondike is either in this line or close to it. The general trend of these points lies in the direction of an arc of a great circle of the earth and it is probable that gold will be found along its production as far as Bering Sea. It is likely the gold found in Siberia is a part of the same system.

" This shows a most extensive area of vast possibilities. What it wants for its proper development is increased transportation facilities, with the certainty of sufficient food supply to sustain the number of people required. At present and during the past, a visit to the country entailed a long period of time and considerable expense and much uncertainty as to whether or not one can remain there more than a few weeks. Give us increased, quicker, and cheaper ingress and egress, with a certainty of food in this part of Canada, and Alaska will furnish employment to untold thousands."

All the discoveries were not confined to gold. William Miller, a veteran from the diamond mines of South Africa and Brazil, wrote late in the summer that he had found a blue clay near Dawson which was practically identical with that of South Africa. From this he argued the probability of finding diamonds. One paragraph in his letter said :

"You have undoubtedly heard much of the great wealth of this land, but the best has never yet been told. It is my honest opinion that diamonds will yet be found in this country, for I have found a blue clay that is practically identical with that of South Africa, with other characteristics that in Africa would be taken as a certain indication that shiners were in the neighborhood. I have not made a systematic search for stones, but I propose doing so later. Just now I am too busy panning gold to spend any time prospecting for a bird in the bush."

Situation at Dawson.

Joaquin Miller wrote from Dawson on the "anniversary" day, as follows :

"An agent of the Rothschilds told me that he offered £1,250,000 for ten claims together, but did not get them. I think he is going out without making any purchases. The most of the ten claims have not even had a pick in them yet, far as I can see. They look like a marsh with mud and moss. You sink at least six inches in the soft and sloppy brown mud as you walk over it. This marsh is a muck as you can see by claims that are partly open up and down the gulch, and below this muck of three or four feet is the frozen ground of five or ten feet thickness, in which the gold is found.

"The prices asked for claims are absolutely steep. A lawyer from Juneau offered $100,000 for a claim yesterday, but was laughed at by the owner, who simply camps with his claim and does not work enough to hold it. He is waiting to get $250,000 for it, he says.

"Captain Healy told me that neither Montano nor Idaho ever

showed anything like the gold in sight in the Klondike mines. He said there would be more gold taken out of this Yukon country than ever has been taken out of all the States together. Of course, they all say that they are the richest in the world, and that they are practically exhaustless, but they advise men to keep away if they are not miners. It is to our interest to have a great rush this way, but I don't want weak men of any sort here. This is no place for a man who knows nothing about mining. Only miners, and sound good miners at that, should come to the Klondike."

Tales of the Passes.

All sorts of stories come in about the passes and the principal towns at their coast ends. All of them, perhaps, were somewhat exaggerated, according to the temperament and good or bad luck of the relator, but all probably had a fair foundation of truth. A "tenderfoot" would naturally view a foot-and-hand journey through a mountain pass, whose principal points were precipitous paths, mud, snow, rain, sleet, ice and tempests, as something terrifying and terrible ; an old frontiersman might as naturally see nothing unusual or inappropriate in the same conditions. The varied reports, however, emphasized the truth that it is hard work to get to the Klondike, and if a man does not want to rough it to the fullest extent he had better stay at home in civilization, though, if he is willing to take risks and endure hardships, he can get into Dawson with reasonable speed and safety.

After Joaquin Miller was fairly afloat on the Yukon and nearing Dawson he wrote back his impressions of the Chilkoot Pass in these words :

"As for the hardships, I find they have been mightily multiplied. As for the perils there are really none to speak of now.

Of course, if disposed to fret or find fault, you can make the journey down the Yukon dreary and hard. On the other hand, if you have any heart for nature, strange scenes, vast lands and indescribable skies you will find delight in every day from the time you touch land where the steamer sets you down at Dyea till here in sight of the Klondike as we are now.

The Hardest Climb.

" I must frankly admit that the Chilkoot Pass is a fearful climb for a man to make with a load on his back. But it is not nearly so bad as the climbing of Mount Hood, Mount Shasta or any other one of the ten or a dozen peaks that I have climbed, and hundreds of others have climbed and are still climbing, and all just for fun. You see, all these things depend a deal on the light in which you are willing to view them. For my part, while I, as a truthful chronicler, confess that the so-called twenty-four miles of the Chilkoot seemed to me to be about forty, with my pack on my back, and also confess that my feet were lame and legs weary, and my back felt as if the weight of a century lay upon me, yet I enjoyed every spot of it as entirely as ever I enjoyed the ascent of any steep I ever made, aye, and more entirely, for here I had a purpose and was bearing a man's, and a strong man's, pack in the battle of life; not climbing for the view or honors of it.

"And one notable difference between the perils and hardships of to-day and the days of old is the safety from savages. We used to be in constant danger, and no man went about by day or lay down at night in the Sierras without a gun or two at his side, and, trained to the old life, I am constantly finding myself choosing my bed when we camp on the river bank for the night with cautious guard against a possible arrow by light of our camp fire. But the men with us who have been years on the Yukon select resting places with regard only to comfort. The few Indians in this vast region are not only harmless, but very honest and inactive. There are no snakes, and I, so far, have found no insects of any sort that bother anybody, excepting the mosquitoes and flies."

S. C. Dunham's Hard Luck.

Samuel C. Dunham, the statistical expert of the United States Department of Labor, who had been assigned by Commissioner Wright to investigate the chances for the remunerative employment of American labor and capital in the Yukon country, had a different experience in getting over the pass from Dyea. His official report, sent in from camp on Lake Linderman, contained the following :

" I left Dyea Monday morning at 11 o'clock and arrived here Tuesday evening at 7. My four Indians started ahead of me, but I have not seen anything of them since the start and am waiting for them here. When I reached the foot of the summit a terrible storm was raging on the pass, and I presume the Indians went into camp somewhere on the other side to await better weather. It has been storming—rain, sleet, and snow alternating—constantly on the summit since Tuesday morning and the situation is aggravated by a piercing wind of thirty miles velocity. I had an awful experience coming across the summit. I started out with my handbag strapped on my back, thinking that as it weighed only forty pounds I could carry it. I managed to struggle along to the head of navigation for canoes, six miles from Dyea, and was there forced to employ an Indian packer, paying him $10 to carry my grip to Sheep Camp, twelve miles from Dyea.

" I spent the night at Sheep Camp, which is merely a collection of tents, and started for the summit at 8 o'clock Tuesday morning in a drizzling, cold rain. I employed a packer to carry my grip from there to Lake Linderman, paying him $16. At the foot of the summit we met perhaps a hundred Indian and white packers who had cached their packs on the trail above and were returning to Sheep Camp to await an abatement of the storm. We were warned that it was dangerous to attempt to get over, but as the wind was blowing the way we were going, we decided to go ahead, as I felt sure my packers had gone on, and I wished to be here when they arrived. The distance from the foot of the summit to the top is said to be three-quarters of a mile, but it seems like five miles.

On the Trail.

" The trail ascends at an angle of forty-five degrees, skirting precipices, where a misstep would hurl one a thousand feet below, crossing the face of glaciers as smooth as glass, and in many places traversing the polished surface of great granite bowlders hundreds of feet in extent. Every hundred yards or so mountain torrents, fed by the glaciers, and on the present occasion augmented by the rainfall, rush across the trail and have to be waded, the water often coming to the knees. Add to this a gale blowing fifty miles an hour, with sleet and snow rushing horizontally through the air and the temperature at thirty degrees, and you will have a faint idea of the horrors of my passage across the summit. After struggling up a steep ascent of twenty-five or thirty feet, I would be forced from sheer exhaustion to rest for a moment, but would scarcely stop before the chilling wind would cut me to the marrow, and I would have to continue my course to keep from chilling to death. Before I reached the summit I was wet to the skin and my boots were full of water, and the added weight of the water made it almost impossible to proceed.

" I finally reached this camp, at the head of **Lake** Linderman, about 7 o'clock in the evening, having been eleven hours in covering twelve miles, so exhausted that I could scarcely drag one foot after the other. I had a letter from a friend to a gentleman who is in camp here, and I was kindly received by him and made as comfortable as possible in his tent. As my Indians had not arrived I had no change of underclothing, and was forced to accept his offer of a suit of warm, dry underclothing, and these, supplemented by half a teacupful of rum, brought some warmth back to my body. I remained in bed all day yesterday, too thoroughly worn out to move. I had some fear of pneumonia, but, with the exception of some soreness, am feeling fairly well this morning."

Via the White Pass.

T. A. Davies, writing of the White Pass route just before the trail was closed, drew a none too inviting picture of that famous gateway to the Klondike. He said that the foot of the first hill, four miles out of Skagway, 3000 gold seekers were in camp at one time trying to "get in." Some succeeded, more turned back disheartened, and many were still on the ground, unable to move, when he passed there in the middle of September. The camp had come then to be known locally as "Liarsville:"

" At the foot of this hill tons of abundant provisions can be seen—wagon loads of oranges, apples and onions—which speculators had intended taking to the Klondike, hoping to realize handsomely thereon. Among piles of goods are seen numerous boats, originally intended for immediate use on arrival at the lakes, but now they are left to rot with the other useless supplies. A great many improvised signs on trees tell of persons having goods for sale all along the trail. From the foot of the first hill to the summit of Porcupine hill is a gradual rise of four miles, and then a descent to what is known as the First Bridge over the Skagway river. To the third crossing of the river the passage is simply a repetition of the first three miles —mud and dead horses on every side. At the third bridge the first camp of any size is reached. A cut-off around one of the larger hills has been blasted out of the solid rock, and this is followed until the ford is reached. This ford is the last crossing of the river.

"A climb of an hour and the summit of White Pass is reached, half way from Skagway to Lake Bennett. About three inches of snow have fallen. The wind blows a gale and dashes snow, sleet and rain in the face of the prospector. The snow and sleet are so blinding, even at this season of the year, that it often is necessary for the prospector to double on his track for the purpose of finding the trail.

On to Lake Bennett.

"Leaving the lower end of Shallow Lake, the beginning of the last tramp toward Lake Bennett begins. The trail runs through timber, meadows and marshes, affording a pleasant diversity of scene. This is by far the best portion of the Skagway trail. Within a distance of ten miles twenty marshes are crossed. On every hand evidence of the final rush to reach water before the freeze may be observed. Immense pack trains are hurrying along. Blockades of horses and goods are of hourly occurrence, and the oaths of the men turn the air blue. The prize is almost lost in sight, and the men feel that it must not be lost by delay at the final point. Prices suddenly become very high. Oats sell for $40 per 100 pounds. Two miles from Lake Bennett and the sound of hammers, axes and saws is heard. Crowds of men, felling trees and cutting timber, are eloquent of the struggle to get material for boats.

" Reaching Lake Bennett the beach is covered with tents, their occupants impatiently waiting to get away. A strong breeze disturbs the surface of the lake, and the boats put out as they are completed, with all manner of rigging. One that I noticed had a bed blanket for a sail. The wind takes the boat in a direct line towards the mighty Yukon, and it soon passes out of view. The proverbial honesty of mining camps does not prevail at Lake Bennett. Instances of stealing are so common that every one leaves a guard on duty with his goods all the time.

"A few days ago three men started down the Yukon together. After going thirty-five miles, two of them landed to see a friend on shore. The third stole the entire outfit and went on down, compelling the two who had landed to tramp back through a wilderness of woods to Camp Bennett, which they had reached during my stay there.

Universal Demand for Boats.

" Boats are one great commodity at Lake Bennett.. Everybody wants one. A small, wheezy sawmill attempts to supply lumber and boats. All the lumber it can cut—1000 feet a day —is readily gobbled up at 75 cents a lineal foot. An ordinary river boat sells for $300, larger ones for $400 and $500. A passenger for Dawson City without goods can buy a passage in one of these boats, or rather a place big enough to sit down in, for $100. Most of the boats carry four or five passengers in addition to the regular supply of goods. When a party finds that it has room left in a boat a sign is placed on a convenient tree offering passage for men, and possibly for goods, at stated price.

" Leaving Lake Bennett and walking half a mile to the southwest, the worst section of the rapids in the portage between Lake

Linderman and Lake Bennett is reached. Here many prospectors, after a hard struggle in the mountain passes, have lost all in attempting to shoot the rapids without unloading their boats. It is here, also, that one comes to a little rude inclosure, bearing a sign telling that all that is earthly of J. W. Mathes is there buried. A year ago Mathes and a party of his fellows got so far on their way to the Yukon gold fields. Mathes fell and broke one of the small bones in his leg. Being already discouraged and disheartened, and believing that he never would reach the gold fields, anyway, and that if his companions were obliged to bring him back to the coast they would blame him for their lost fortunes, he placed a revolver to his head and killed himself. The site of his grave is now one of the best-known landmarks on the route to the Klondike gold fields. At this place the boats are usually unloaded and the goods carried around the dangerous rapids, and the boat then floated down empty. The afternoon I reached the portage one party had unloaded all its goods and was letting the boat through the rapids after the usual methods of lining it down. One man remained in the boat to steer, and three men on shore held the rope to keep the rapids from carrying it out of reach. When fairly started the rope broke and the boat went down the rapids like a shot. By rare good fortune a friendly current carried the boat to a sand-spit and it was saved: but instances are numerous where men have not been so fortunate."

Woes of the Horses.

The demand for transportatiou over the passes was the cause of bringing in hundreds of horses to be used as pack animals and supplement the Indians in the arduous work of getting supplies and outfits from the coast to the head of river navigation. The experiment was in the main profitable to the owners, for the prices for packing made a horse pay for himself in a comparatively few trips, but the mortality among the poor beasts was something unparalleled.

On the Skagway trail, or White Pass, as many as 1200 horses were in use at one time after the trail was fairly opened in the middle of September, but of this number it was estimated not one hundred would be alive in a month's time. Even then (September 15th) 600 dead horses could be counted along the trail. Many of these were the victims of accidents, but by far the greater number had succumbed to exhaustion and disease. Poor food, and not too much of it, made them weaker day by day, and pneumonia, the result of getting chilled at night, swept them off by scores.

On the Dyea trail, or Chilkoot Pass, not so many norses were employed, and the visible mortality was consequently less, but at that, at least 150 dead animals lay beside the trail when T. A. Davies passed over it in September. The unfortunate beasts had been left to perish where they fell from fatigue.

Enormous Prices for Transportation.

The loss of horses had a material effect on the packing tariff. A contract for the entire White Pass trail was almost an impossibility to make, and the aggregate price sometimes reached as high as one dollar a pound. The largest long contract reported during the fall was for $30,000, with the Canadian Government, for moving the supplies for twenty-five of the Canadian mounted police.

An official survey ordered by the Dominion government to locate if possible a new and more practicable trail to the upper Yukon, reported an easy and comparatively short cut to Selkirk or Dawson from the seaboard and one suitable for cattle, wagon or railroad. J. M. McArthur, who was in charge of the party, made the following preliminary report of its work when he passed through Juneau in September.   It will be seen the old Dalton trail was made use of for some distance :

"From the extreme left of the Chilkoot Pass the party headed northwest for a point about 100 miles inland, where Dalton and others have a trading post.   Thence they took a course north to a chain of small lakes called Hootchie Eye.   So far the course was over what is known as the Dalton trail, which, from the Hootchie Eye, continues down the river sixty miles to the Lewis River, but from the Hootchie Eye, Dalton struck out due north for Fort Selkirk, into a country 120 miles across, never before explored by a white man and totally unknown.   Such is the wonderful instinct of this man that the entire party came out of the wilderness at a point directly beyond the buildings at Fort Selkirk, in the Yukon, at the mouth of Pelly River, where the Yukon proper begins.

" Plenty of grazing for the cattle was found.   The country is characterized by comparatively low and rolling mountains, over which the party went.

Advice from Wrangel.

United States Commissioner Kenneth M. Jackson, writing of the various routes into the Klondike, had this to say which may not come amiss as a pointer for those who choose to take time to pick their way to the diggings :

" Of all the routes into the Yukon country I would advise the one via Wrangel, the Stikine River, and Lake Teslin, as presenting less difficulties and hardships.   By next spring the only portion of this route that cannot be made by steamboat or rail will be over an easy pack trail from the Stikine to Lake Teslin, a distance of about 135 miles, and upon which the British Columbia government is now spending money, and over which a wagon or railroad will be constructed very soon.   From Lake

35

Teslin down the Hootalingua to the mines one or two steamboats will be running next year. I advisedly caution persons from attempting the trip till next spring, and when they do start, if possible, arrange to buy a year's supply of grub per capita when they get to the coast. One can get better information as to what is needed here than at home."

W. A. Pratt, sent in by the Yukon Mining and Trading Company, of Wilmington, Delaware, reconnoitered what he declared was a practicable route for a railroad from the head of Taku Inlet to Lake Teslin. The Canadian Pacific had a party out during the fall running a line for a railroad from Lake Teslin to Telegraph Creek.

Tramway at Chilkoot Pass.

Out of the many schemes for rail transportation over the mountains, the first to take definite shape in action was that for a tramway over Chilkoot Pass. The engineer's plans were in working order early in October, and the first of the material had then begun to arrive on the ground at Dyea. Seventeen miles of inch wire cable will be used in constructing the eight and a half miles of aerial tramway by which freight will be transported seven and a half miles, lifted to the summit of the pass and let down again to Crater Lake. The road will be a broad guage, with a daily capacity of 120 tons of freight, or the outfits for 120 men. The contract calls for the completion of the road by January 15, 1898, and then it is expected the journey from tidewater to Dawson can be made in less than forty days and with an immense economy in men and money.

Among the novel schemes—which at the same time had an air of practicableness—for getting into the Yukon basin during the winter season, was that of the snow locomotive, invented by George T. Glover, of Chicago. The snow train had been in successful operation in the pineries of Michigan for two years, hauling on runners great loads of logs and making fairly good time over considerable grades.

When the reports of probable starvation in the Klondike region made it a matter of instant importance for the General Government to prepare for the exigency by ascertaining the best and speediest means by which supplies could be transported from the coast to Dawson, General Alger, the Secretary, to whom the Glover log locomotive was familiar, bethought himself at once of the snow train, and at the same time Mr. Glover bethought himself of the Secretary of War. The result was a series of conferences in Washington, the matter was laid before the Cabinet, and it was practically agreed that, if it became necessary to succor starving argonauts, the Glover snow locomotive should be used. It was estimated that a train carrying 100 tons of freight and passengers could be pulled by this locomotive over the passes, across the plains and down the river on the snow and ice, from Fort Wrangel to Dawson, in less than ten days, and could keep lowering the record as the road became worn, until not more than six days each way would be consumed."

Caught on the Trail.

Of all the thousands who started for Dawson by the various routes before the winter had laid an embargo on the mountain passes or blocked the Yukon with ice, it was variously estimated from the civilized end of the line that from 6000 to 7000 succeeded in reaching their destination. How many others were forced to winter at intermediate points was beyond accurate computation—the region to be covered was too vast and there were too many vicissitudes of climate and trail to be figured on.

On the White Pass trail, late in September, there were at least 1200 gold seekers, of whom probably not more than 300 succeeded in reaching the lakes, the rest being caught by snow and ice. On the Chilkoot Pass trail there were probably as many more in all stages of progress and predicament. Perhaps half this number made out to get to Dawson, or at least well down the rivers. All those who remained behind had only the alternative to build log cabins on the trail and camp for the season, or leave their goods and make a perilous struggle back to civilization. Camps approaching the dignity of small towns were established at Lake Bennett and Lake Linderman when the first snow came, and many went into permanent winter quarters at once, reasoning that it would not cost more in supplies to winter there than in Klondike, and that by saving their health and remaining at the advanced post they would have a good start in strength and distance and could be the first " in " in the spring. Among the 300 in camp at Lake Linderman were a number of women and children.

Snow and Low Temperature.

One of the proprietors of the saw mill at Lake Bennett reached Juneau on October 7th, and reported a heavy snowfall on the headwaters of the Yukon when he left. On the morning of October 3d the thermometer showed eight degrees below zero, and the boats in the river had to be cut out of the ice.

The McKay party, which contained a number of women, had reached Lake Bennett and was about to start down the river, insufficiently clad and provisioned. The Canadian police were debating stopping the party, considering the attempt to make Dawson would be little less than suicide.

Captain Tuttle, of the United States revenue cutter Bear, sent in an official report from St. Michael's, dated September 16th, in which he said :

" There are in port seven seagoing vessels and six river steamers, with one steamer and one barge in process of construction on the beach. About 300 people are encamped on the beach awaiting the completion of these vessels. At least seven vessels are expected to arrive, many of them with passengers. There is no possible chance of these people reaching the Upper Yukon this season, and they must winter here or at some point inside the mouth of the Yukon. While there will be an abundance of provisions, the trading companies having their main depots here, trouble is likely to arise from those who have no provisions and no means to purchase them. This, however, is a small matter when taken into consideration with matters above Fort Yukon on the Yukon River.

" On September 13th the river steamer Hamilton returned from its up-river trip, having been unable to reach Circle City. Captain Hill reported the river so low as to prevent his reaching his destination."

Danger of Starvation.

Probably enough more argonauts reached St. Michael's after Captain Tuttle's letter left to raise the total number prepared to winter there to 600 or even twice that number.

Captain Tuttle closed his official communication with this suggestion :

" Laws in regard to the inspection of steam vessels are entirely disregarded, as no inspector of hulls or boilers has visited this place. At least sixteen such vessels are now running in this part of Alaska. If I should seize them starvation would ensue to those who are depending upon these vessels to bring them provisions. At the same time hundreds of people are traveling on these vessels, which are without the safeguards to life that the law provides they shall have.

"A deputy collector of customs is stationed at St. Michael's, who is required to attend to all customs business. Frequently

there are several vessels in port discharging bonded goods at the
same time. It is impossible for one man to attend to all this
business. After leaving St. Michael's there is no customs officer
in charge of these goods. Vessels frequently get aground, and
it is necessary to discharge their cargoes before they can be
floated. Great opportunities are afforded to defraud the customs. There should be a customs officer on every vessel carrying bonded goods, and provision should be made to have the
vessels inspected as the law requires."

At Dawson City.

The prospect for Dawson City at the beginning of winter indicated a population in the town of about 7000 and in the tributary country of half as many more. Considerable building enterprise had been displayed and log houses were multiplying for
residences, while commodious business houses were rising along
Water Street. The new Mission house was expected to be in
full readiness for its works of charity by the time the ice season
was fairly settled. The new opera house or music hall was in
full blast and in general the promise was for a bustling, thriving
town. Lots in Water Street the first of October sold for $10,000
and lots for cabins at proportionate rates.

St. Michael's, old Yukoness thought, stood a chance to be the
winter haven of the easily disheartened overflow from Dawson.
When the low water in the river delayed the arrival of the boats
with provisions many took fright and started down stream to
meet the supplies or force their way through to the sea. Others
formed parties to go out en route to civilization and the combination made quite an exodus. The Klondikers who stayed behind,
however, were not troubled by the departures—they meant fewer
mouths to feed and more claims to " go around."

Returning steamers from Sitka, Juneau and other ports late in the fall brought full complements of gold seekers who had been beaten by delays or the climate in getting over the passes, and preferred to spend their money in steamer fares to reach the homes and flesh pots of civilization for the winter, rather than in paying boom prices for bare subsistence in such already over-crowded towns as Skagway, Juneau and Dyea. It is estimated that over a thousand argonauts had returned to Washington, Oregon and California in this way by the middle of October.

Among the last to go in by way of St. Michael's was Lieu-tenant Colonel Randall, U. S. A., who took with him twenty-five soldiers from Fort Russell and an outfit of 150 tons of stores and provisions. Part of the detachment was to be stationed at St. Michael's and the rest were to go up the river near the inter-national boundary.

Growth of Skagway.

Skagway is one Alaskan town which owes its existence to gold, though there is none of the precious metal there, except what has been brought in and "dropped" by argonauts rushing to the Klondike placers. It owes its standing as a town to the lucky fact that it is the natural landing place for the White Pass, and to the additional fact that owing to the lateness of opening this trail several thousands of men had to linger in Skagway for many weeks waiting and struggling for the coveted chance to get out.

Two pictures of the place are of interest. In August, 1897, Hal Hoffman wrote of this "half-way to Klondike and stuck" town, as it was then familiarly called:

"This is a city of eleven frame or log houses, a saw mill, five stores, four saloons, a crap game, a faro layout, blacksmith shop, five restaurants which are feeding people all the time, a tailor shop, on which is hung the sign, 'Bloomers fitted for shotguns,' a real estate office, two practicing physicians, another professional pathfinder whose specialty is shown by the sign painted on a board nailed to a tree, ' Teeth extracted,' some 300 tents, and a population of about 2000 men and seventeen women. Four of the women are accompanying their husbands into the Klondike, the others are unchaperoned. A dance hall will be erected next week. Skagway is already a typical mining camp. Its population is proud of it. They go further and say it will be a ' hot town ' next winter. Streets have been laid out. Broadway runs from high tide four miles back to the mountain base and is walled with tents, piles of supplies and felled trees. The gold seekers never overlook an opportunity to make fun drown their impatience."

As it Developed.

In the latter part of September, when all but the hardiest or those who had determined to winter there or to take up a more permanent residence had left the town, another wrote of the same place in different terms thus :

" Skagway is a conglomoration of all nationalities. All kinds of buildings—or, rather, lack of buildings—are in evidence. Sidewalks are unknown. One just wades and wades. The first requisite is a pair of rubber boots—good, long ones.

" Along each side of Broadway, the main street, are ranged the business houses. There are about twenty saloons, eleven blacksmith shops, thirty restaurants and bakeries and fifty miscellaneous lines, dance halls, hotels, custom houses, etc., while the Territorial Surveyor and his deputies find room to do a good business. This is a mecca for speculators. On one corner the dismayed prospector, outfitted completely for the Yukon, has decided to abandon his trip and is selling his flour for, perhaps, 50 cents a sack. Within fifteen minutes after he has sold the flour the speculative purchaser is offering it from $2 to $3 a sack.

"Skagway now has about 100 frame buildings, a population approaching 1200, and committees for almost every conceivable purpose—from a committee on removing dead horses to committees to look after the numerous correspondents of Eastern dailies—the latter committee having even more to do than the dead-horse committee. Skagway really is an orderly town from a frontier standpoint. Comparatively few robberies are reported, when the great number of miscellaneous specimens of humanity who have rushed in there is considered. To be sure, probably not half the population know when Sunday comes—in fact, there is no Sunday in Skagway, Saloons and all kinds of gambling games—keno, faro, black jack, poker, roulette—flourish by day and by night, seven days and seven nights each week, without interruption. The only cloud that appears on the gambler's horizon is the appearance of the deputy United States marshals, which usually threatens a seizure of liquors, providing there is an overabundance. The liquor traffic presents a peculiar complication at this place. It is not an offense, as the law is administered, to sell whisky except to Indians ; but if the liquor is found in a person's possession the liquor is liable to seizure. Most saloon men, therefore, carry very small stocks 'in sight,' tha balance being conveniently ' cached' in nearby places."

Glacier Slide and Flood.

On the morning of September 18th a terrible glacier-slide and deluge swept down the Chilkoot Pass and three men lost their lives, one, Morris Choynski, a cousin of Joe Choynski, the pugilist. His body alone was recovered. About twenty-five campers had pitched their tents on the dry ground in the bed of the river when suddenly the cry went forth, about seven o'clock in the morning, that the glacier was falling. Every one made for the hills, and the coming torrent, two miles away, sounded like thunder or the roar of heavy artillery. On came the waters in a wall twenty feet high, moving massive rocks like pebbles and sweeping everything before them. All the tents and goods along the river were lost and only twenty-two of the campers succeeded in saving their lives. A deposit of sand from one to two feet thick marked the path of the awful flood.

Reports of disasters on the lakes and at the fords on the trails were numerous as the severe weather drew on, but happily most of them proved to be unfounded rumors. Many upsets occurred, and a quantity of supplies was lost from the boats, but the number of serious casualties was remarkably small considering the number of men exposed and the excellent opportunities for accidents.

Comparative Absence of Crime.

Crime and its corollary, Lynch law, happily made few public appearances along the trails when the great rush was stopped by the winter. The known cases had a semblance of ample cause and effect, according to frontier ethics, except in the one instance of the Buchanan-Kossuth murder and suicide at Skagway, which was entirely a cold-blooded affair and one in no way chargeable upon the argonauts. The comparative peacefulness and honesty which reigned along the trails, considering the great temptations to greed and high temper which marked the conditions, were a marked tribute to the character of the gold hunters of '97.

One of the most important contributions which Alaska made to the sum of human knowledge in 1897 was something which could not be weighed in the gold scales or discounted at a bank, and yet in another sense was of more permanent value to the world than the diggings themselves. This was the successful scaling of Mount St. Elias, the corner post of Alaska and hitherto regarded as the one inaccessible spot remaining on the North American continent.

The successful ascent was accomplished by the party headed by Prince Luigi Amadeo, of Savoy, a nephew of King Humbert, of Italy, and a mountain climber of world-wide fame and experience. The party included Clevalier M. Cagni, Francesco Gonella, President of the Turin Section of Alpine climbers; Vittorio Selle and Dr. Fillippo de Fillippi, all noted Alpine experts. The party measured the height of the peak, up to that time estimated only, and that within a range of several hundred feet, ascertaining the exact elevation to be 18,120 feet, an important geographical and engineering fact.

The first expedition to attempt to scale Mt. St. Elias was led by Lieutenant Schwatka in 1886. Two years later William Williams and the Messrs. Forham, of London, England, made the attempt and failed. T. C. Russell, of the United States Geological Survey, made two attempts—one in 1890, and the second a year later. Both were unsuccessful, though the explorer reached a greater height than any of his predecessors, turning back only at an altitude of 14,500 feet, or 3620 feet below the summit.

Klondike has a Permanent Interest.

The interest in Alaska and its gold deposits, widespread and universal as it is, will very likely increase with the advance of time. A marvelous region is this northwest Territory, great in natural wonders and great in wealth. Where gold is, there men will go, whether to the tropics or the Arctic regions, the heats of the equator or the realms of endless frost.

That hundreds, perhaps thousands, will lose their lives is only to be expected, yet thousands of others will rush forward as men do in battle to take the places of their comrades who have fallen.

Some will survive the dangers, outlive the trials, and by overcoming almost miraculous obstacles, will gain the coveted treasures. Stories that never have been surpassed in tragic interest are yet to be told concerning Klondike, and very likely all that has been written and said in the past will be overshadowed by events that are yet to come.

It is not likely that Mt. St. Elias will be ascended again for many years, perhaps not in the present generation, but these mountains, valleys and gulches are sure to be explored. The enterprising Yankee will be found in every nook and corner of Alaska, and if there is any money there to be found he will pick it up. Every man will think it possible for every man to fail except himself. Many will trust to luck and later will be sorry for it. Others will go about mining intelligently, understanding exactly what they are doing, and they are the ones who will succeed and bring home the yellow nuggets.